STATISTICS
IN POLITICAL AND
BEHAVIORAL SCIENCE

DENNIS J. PALUMBO
BROOKLYN COLLEGE OF THE
CITY UNIVERSITY OF NEW YORK

STATISTICS
IN POLITICAL AND
BEHAVIORAL SCIENCE

APPLETON-CENTURY-CROFTS
EDUCATIONAL DIVISION
New York MEREDITH CORPORATION

to

Sachiko

Preface

Recent years have witnessed an increasing use of statistics in political and the behavioral sciences. This book attempts to fill the needs of the political scientist by introducing the student of politics to the principal statistical concepts using examples taken largely from contemporary research in political science. At the same time it is comprehensive in coverage and therefore may be of value to other behavioral sciences as well.

This is not a book in mathematical statistics. Its mathematical content is relatively elementary. The reader with some knowledge of algebra and geometry will have an easier time with the book than someone who does not have such knowledge. But the book does not assume a great deal of mathematical background on the part of the reader. In an attempt to reach the student who is not mathematically oriented, the objective throughout has been to explain as carefully as possible on the verbal level all of the mathematical theorems and derivations that are used. The mathematical proofs for some of the theorems are given but not for all of them. What is lost in rigor, precision, and generalizability by this procedure should be more than adequately compensated by a gain in understanding for the student without mathematical background.

Statistics cannot fully be understood apart from the general methodological problems contained in scientific method. It is inappropriate to think of it simply as a "tool" for research. In some of its newer forms, such as the growing convergence between Bayesian decision theory, simulation, and game theory, statistics becomes more than a way of describing data or of making inferences to specific populations, but also a way of interpreting phenomena. At these points statistics converges with substantive theory and holds the most promise for behavioral science as a whole and for political science in particular. To a great extent, the development of the computer has stimulated this, since it has relieved the student and the researcher of the necessity of doing long and tedious computations. These developments have also made it possible, if not actually necessary, to write a book of this kind, which emphasizes fundamental principles rather than "how to do it."

There are many ways that a book on statistics might be organized. It might be divided into the two main divisions conventionally used: descriptive and inferential statistics; into parts in accord with the level of measurement assumed; in accord with whether one, two, or n (multi) variable analysis is involved; or in accord with whether experimental, cross-sectional, or longitudinal research design is being used. This book tries to incorporate several of these dimensions into a single design. The first three chapters deal with basic concepts—concepts that are used throughout the various aspects of statistics. Chapters 4 and 5 introduce the foundations of probability theory needed for statistical inference. Chapters 6 and 7 deal with the two major subdivisions of inferential statistics: making estimates and testing hypotheses. Chapters 8 through 14 deal primarily with methods of analyzing relationships among variables. The final chapter deals with some special problems of sampling and survey research. In general an introductory course would be able to cover Chapters 1 through 10. Chapters 11 through 15 could be covered in a more advanced course.

I have given attention in the writing of the book to those concepts and problems that I have found in my own research and teaching to be most significant for political science. The book draws on several years' experience in teaching the subject both at the graduate and undergraduate levels and upon experience in applying the techniques in research. During the course of these years I have incurred a large number of debts: to students, whose reactions to particular explanations helped put my own knowledge on a sounder footing; and to colleagues, research collaborators, and research assistants, whose stimulation and encouragement made me probe further into the field. I am happy that I can now publicly acknowledge these debts, while absolving the persons named of any culpability for the errors still contained in the book. A number of people read the entire manuscript and offered very helpful comments. Benjamin Walter of Vanderbilt provided a scintillating and careful review of the manuscript. Many of his suggestions have been incorporated in the book. Kenneth Janda of Northwestern read an earlier version of the manuscript and made several very valuable suggestions. My former colleagues, Thomas Davy and Oliver Williams of the University of Pennsylvania also read the entire manuscript as did several of my colleagues at Brooklyn. Both the written comments and discussions of these people proved to be very helpful. In particular I would like to express a debt of gratitude to Marty Landau, Bob Abrams, and David Abbott, all of Brooklyn. I also am most grateful to the Research Center in Comparative Politics and Administration of the Department of Political Science at Brooklyn, and to the Chairman of the Department, Benjamin Rivlin, for financing both the typing of the manuscript and duplication of enough copies for me to give the book a trial run in two of my classes. Many students in these classes made useful suggestions and I regret that I cannot name them individually here. But I would like to thank two students in particular: Milton Heumann and Rita Immerman, both of whom read much of the manuscript and provided written comments. Finally, I am indebted to my research assistant, Mary Ann MacDonald, who not only typed the entire manuscript but also helped check the arithmetic accuracy of all the examples and thereby saved me from several errors.

I am indebted to the Literary Executor of the late Sir Ronald A. Fisher, F.R.S., to Dr. Frank Yates, F.R.S., and to Oliver & Boyd, Ltd., Edinburgh, for permission to reprint Tables III, IV, V, and XXXIII from their book *Statistical Tables for Biological, Agricultural and Medical Research*.

D.J.P.

Contents

1
Introduction

The word "statistics" derives from the Latin word *status*, or state. The Latin phrase *statisticum collegium* means lecture course on state affairs. The association of the words statistics and state came about because governments were originally responsible for the collection of data concerning the number of people they governed. To collect taxes, it was necessary to know something about the number and wealth of people living in the country. The word statistics still refers to a collection of data, although the discipline of statistics has moved much beyond this original and simple connotation. In contemporary literature, statistics is defined as that body of methods that enables decisions to be made in the face of risk and uncertainty. It is a giant step to go from the notion of statistics as simply a collection of figures to the decision-theory definition. Let us consider how this step can be made.

It is true that statistics is based on data that have been measured in quantitative terms. Numbers may be used to represent the phenomena being studied, such as voting behavior, social phenomena, the behavior of consumers, and the opinions of people. The data might themselves be called statistics. The word then has a plural sense. But statistics as a discipline refers to the procedures used in extracting information from the data. When used to refer to the discipline, the word is used in the singular.

There are many ways to extract information from data. We may want to *describe* them in short and precise terms. Our interest usually does not stop there. We may also want to *analyze relationships* between two or more variables. This sometimes cannot be done directly because the data available are incomplete or comprise only a sample of the total. Hence, we may want to generalize, or *infer* something about a larger *population*,[1] based on the incomplete data. Such inferences are probability, or uncertain statements. Nevertheless, we may want to *make decisions* in the face of these uncertain states of affairs. For political scientists in particular, the behavior in question often involves opposing interests in which participants are trying to gain some end. Statistics can help predict the *probable strategies* they will follow. Finally, an adequate theory must account for *changes* in strategies or variables or both, and we are therefore often interested in saying something about what happens over the course of *time*.

[1] The words "variable" and "population" are given precise definitions below.

Each of these objectives about data is the concern of statistics. A more complete definition of statistics should try to encompass them all. In terms of its objectives statistics can be defined as a body of methods that *describes data* in an efficient manner, *analyzes relationships* among variables, *infers* things about a wider universe, *makes decisions* in the face of uncertainty, *analyzes strategies* of action, and makes statements concerning *changes* in behavior over the course of *time*. This definition is a bit long-winded, but it also is more accurate than a single decision-theory definition. It is a useful definition because it mentions all the things that are covered in the course of this book.

Statistics has proved to be of value in a number of sciences from quantum physics to history. In political science also there are numerous questions that statistics can clarify. In this book, the use of statistics is illustrated by questions as diverse as the following: How much inequality in voting power exists in the various counties of the United States? Which candidate are voters likely to vote for in a presidential election? What are the degrees of conflict or consensus that exist in various legislative bodies in the United States? Is the role perception of legislators related to the degree of competition for legislative office? Is the structure of communications related to the performance of groups? How much influence do elections have on the voting behavior of congressmen? What is the relative importance of various demographic factors in explaining variations in public policy? Do voters tend to follow their narrow self-interest in voting on referendums? Does position on the ballot influence the outcome of elections? Is centralization of power in organizations related to the performance of these organizations?

These do not represent all the questions for which statistics can be used. This book is not intended to be a comprehensive survey of how statistics is now being used nor of its potential use in political science. Nor does it take the space to try to provide precise and actual answers to all the questions mentioned above. The principal purpose of this book is to convey knowledge about statistics.

STATISTICS AND SCIENTIFIC METHOD

Although the definition of statistics already given is accurate as far as its various objectives are concerned, it may be better to conceive of statistics through the wider perspective of its relationship to the scientific method. The heart of the contemporary model of scientific method has been called the "hypothetico-deductive" process (Kaplan, pp. 9f.),[2] which we interpret

---

[2] Generally, reference footnotes are not used in this book. References used are given at the end of each chapter under the heading "Readings and Selected References." Where a specific page or pages are referred to, they appear in the text along with the name of the author. In the present case, the work of Abraham Kaplan given at the end of the chapter is the reference.

as the relationship between the two principal modes of reasoning — deductive and inductive. Deductive, or analytical, thinking is exemplified by such disciplines as logic and mathematics; inductive thinking by empirical sciences, such as political science, sociology, and psychology. However, both modes of reasoning are necessary for the development of a discipline as a science. In fact, the two operate more or less simultaneously in the mind of the scientist. Deductive reasoning makes assumptions and then derives consequences or conclusions from them analytically. The vehicle used is the concept, an abstraction, an idea, several steps removed from what can be observed with the senses. When a number of concepts have been interrelated in lawlike generalizations, we have a theory through which predictions about the world are made and tested empirically. Inductive reasoning is the process of moving from observations of events in the world to more general statements. Its principal vehicle is measurement, or the operational definition[3] of abstract concepts. Through the inductive method, concepts are explicated or further clarified, and theories are confirmed, modified, or refuted.

There are many complex ideas in the last paragraph that we cannot consider in detail here, since our purpose is neither epistemology nor scientific philosophy. However, the idea of the hypothetico-deductive method is useful in explaining the nature of statistical theory. Statistics uses a deductive system of thought to make inductive generalizations. It uses the deductive system of the mathematical theory of probability, which is entirely analytical. It tells us what to expect under idealized conditions. These idealized models help to infer something about a universe of events based on conclusions drawn from a sample of observable events.

The last point can be clarified by the following example taken from Cohen and Nagel. "There are at least two persons in New York City with the same number of hairs on their heads." How could we go about verifying this statement? To verify it by deduction, we begin with the premise that there are more people in New York City than there are hairs on any one person's head. If this premise is true, the conclusion follows necessarily. To demonstrate this, we simplify by assuming that there are 51 persons in New York City and that the most hairs any one person has on his head is 50 (we exclude the possibility that a person is completely bald). If the first person has 1 hair, the second 2, ... and the fiftieth has 50 hairs, each person has a different number of hairs on his head. But what about the fifty-first person? He must have the same number of hairs as one of the other 50. The conclusion, therefore, is true, provided that the premise on which it is based is true.

Conclusions reached in this manner are called analytical a priori statements; that is, made prior to observation. They are valid because of the form of the argument rather than because of conformance or nonconformance to empirical reality. To verify the argument about hairs on the heads of New York City residents by the inductive method we observe and count.

[3] This term is defined below.

Let us assume that we cannot observe all individuals in New York City, which for all practical purposes is true, and therefore must take a sample. Assume that in a sample of 100 people not one person has the same number of hairs as any other person in the sample. Could we conclude from this that no two people in New York City have the same number of hairs on their head? The most that can be concluded is that this statement is *probably* true. Conclusions reached through deductive reasoning are absolutely true or false, given the veracity or falacy of the premises. Conclusions reached inductively are probably true or false, never absolutely true or false. To determine the probability of being wrong in making an inductive generalization, statistics uses mathematical models that are purely deductive.

The reader is not expected to comprehend the nature of these mathematical models at this point; he should, however, have a fairly good understanding of them when he finishes this book. The point that should be emphasized here is that these mathematical models — just as the classical if–then syllogism — apply in a given situation only if the assumptions on which they are based correspond at least approximately to the real-world situation. There are two major classes of assumptions made in most of these mathematical models. One pertains to the level of measurement used in the research, and the other to the sampling process. Since these assumptions are crucial to the whole of statistics, we study them in detail in the rest of this chapter.

STATISTICS AND MEASUREMENT

Constants and Variables

Statistics is based on numbers; that is, before it is possible to use statistics, the phenomena under study must be measured. Concepts themselves cannot be added, multiplied, or compared; quantitative operations cannot be performed on concepts. We cannot add the concepts tables, men, and legislatures. We can measure and perform quantitative operations on properties of these concepts, such as their number, weight, beliefs, or amount of centralization. When properties of systems are measured, symbols represent the properties. These symbols might be classified into two categories: constants and variables. A constant is an unchanging value, such as the ratio of the circumference of a circle to its diameter, represented by the Greek letter π and equal to $3.1416+$.[4] There are no known constants in political science. For this reason, we generally speak of variables. A variable is a quantity

---

[4] Throughout this book we symbolize constants with letters, such as a, k, X, and Z. Variables are symbolized by capital or small letters followed by a subscript, such as X_i. The meaning of the subscript is explained below.

that can take on different values, such as the size of cities, or the percentage of voters in each census tract who are Democrats. By convention, variables are represented by letters of the alphabet followed by a subscript to indicate that the property to which the letter refers can take on different magnitudes. For example, in the variable, size of cities in the United States, to be represented by the symbol X_i, the subscript i indicates that for one observation X might be equal to 2 million, for the next observation it might be 25,000, and so on, through the entire list of cities in the United States. The *range* of a variable consists of all those numbers which might be put in the place of the symbol. Suppose that we are discussing just five cities whose respective sizes are 50,000; 128,000; 49,000; 325,000; and 2,400,000. The letter X_i represents these five cities and refers to the five cities in general. When the specific cities are referred to, we write

$$X_1 = 50,000 \quad X_2 = 128,000 \quad X_3 = 49,000 \quad X_4 = 325,000 \quad X_5 = 2,400,000$$

A letter followed by a numerical subscript, such as X_5, therefore refers to one of the particular elements in the set of observations. A letter followed by a letter subscript, such as X_i, refers to the variable in general. When two letters are used together, the operation of multiplication is implied. For example, aX_i means the multiplication of the constant a times the variable X_i. If the variable X_i is multiplied — where the subscript i refers to the particular five values given above — by the constant a — where a is 2 — we mean the following by the notation aX_i: 2 × 50,000, 2 × 128,000, . . ., 2 × 2,400,000.

Measurement and Data Collection

The process of measurement involves two operations: the development of an instrument to do the measuring and the collection of the data itself. The problem of developing measurement instruments is complex. It is a process of going from an abstract concept to a physical operation. The physical operation enables us to determine to what extent the phenomenon, to which the concept refers, is a factor in the behavior studied. Some concepts have a fairly direct and clear operational meaning. For example, if it is the size of cities being studied, the problem of measurement is simple. Size of cities has a direct and clear meaning, although it could be taken to mean geographic as well as population size. If we are concerned with more sophisticated concepts, such as centralization or conservatism, the problem of developing a measurement instrument is more complex. The instruments most frequently used in political science to measure these more complex variables are attitude scales and indexes constructed from census data or other sets of recorded figures.[5] Examples are the F scale of authoritarianism;

[5] The reference by D. C. Miller cited at the end of this chapter contains a number of sociological scales and indexes.

McCloskey's conservatism scale; Hoffebert's party competition index; Dawson and Robinson's index of liberalism of states; various morale scales; and the like.

Once a measurement instrument has been constructed, the next step is the collection of data. In social science, most data are collected by interviews, questionnaires, or participant observation. The manner of collecting data is usually covered in books on standard research methods. It involves, among other things, common sense, rules of thumb, and statistics. In many cases, there are no scientific principles involved. Conducting a good interview, asking the right questions, or finding good records is as much an art as a science. Experience is frequently the best teacher, since there are a multitude of rules of thumb to be learned, and rules of thumb are not systematic. There are rules of thumb involved, for example, about whether or not and under what circumstances it might be best to use a telephone survey, mailed questionnaires, or personal interviews. The rules involve things like cost, objectives, and the level of error that can be tolerated. Statistics enters into these questions. Statistics can predict the margin of error to be expected, given different sample sizes. Rules of thumb would indicate how much nonresponse to expect under different conditions and statistics would designate a corrective method for nonresponse. The final state in the research process, analysis of data, is the stage with which this book is most concerned.

Levels of Measurement

After the data have been collected, the next step is to analyze and describe them in the context of the theory and hypotheses with which the research began. The kinds of information that can be extracted from the data depend on how well the variables under study have been measured. A greater amount of precision of statement is possible the higher the level of measurement achieved. Four levels of measurement are most often distinguished: nominal, ordinal, interval, and ratio.

Nominal Scales. If the symbols represent concepts themselves rather than properties of the concepts, the resulting level of measurement is called a nominal scale of measurement.[6] For example, separating individuals according to political parties — Democrats or Republicans. This is categorizing, the major mode of thinking in political science. The mere act of placing an individual in one or the other of these categories might be considered measurement. In order to place individuals in a category, it is necessary to define the

---

[6] Some authors would argue that nominal scales are not measurement and that measurement begins only with the higher levels of scales.

characteristics of the members of that class. There are two ways to do this. Each category can be defined in terms of some other concept; for example, a "Democrat" may be defined as a "liberal," although this would not be a very good definition. This is conceptual defining; that is, defining one concept in terms of another. Scientific method requires, however, that we define "Democrat" operationally; for example, a Democrat can be defined as a registered voter in the Democratic party, or as someone who has voted for the Democratic party candidates in at least eight of the last ten elections.[7] Regardless of what particular definition is used, the process of defining in this manner is called operational defining, because it indicates what to look for in the empirical world to determine the classes into which each observation should be placed.

When categories have been set up and the characteristics of each category defined operationally, two requirements should be met in order to assume that a nominal scale of measurement has been achieved. The categories must be mutually exclusive, and they must be exhaustive. Being mutually exclusive requires that individuals or objects be placed into one and only one category. If it is possible to place the same individual or object into two different categories, then the classification scheme is not good. Being exhaustive requires that all individuals to be classified be placed into one or the other of the categories. If some of the individuals cannot be so placed, then the categorization scheme is not good. In the example above, the two categories of Republican and Democrat would not be sufficient, since some persons might be neither.

The mathematical and statistical operations that are thought to be applicable to data measured on a nominal scale are very limited. Generally, we can obtain proportions, percentages, and ratios (Chapter 2), or we can measure and test relationships between variables (Chapter 8), but not with a great deal of precision.

Ordinal Scales. If measures can be so refined that properties of individuals or things are measurable, then we can develop what is called a quantitative as opposed to a qualitative scale. If degrees of party identification, intelligence, power, and centralization can be measured, then we have measured properties of things rather than just placed them into categories. For example, if properties of legislatures, such as the degree of party competition, size, amount of centralization, and party cohesion, are measured, we have measured properties of legislatures. In this case, the numbers assigned to the properties are meant to reflect the magnitude of the property. The requirement here is that a one-to-one, or isomorphic, relationship exists between the properties of the thing being measured and the properties of the number

---

[7] Note that one operational definition is not more correct than another except insofar as it leads to better predictions. Thus, the process of defining is somewhat an arbitrary process.

system that is used to do the measuring. The properties of the number system with which we are concerned here are:

1. They are ordered, each number stands in a definite relationship to every other number.
2. There is an equal distance between each of the natural numbers.
3. The number system has an origin or zero point.

A measurement scale can have one, two, or three of these properties. We refer to these properties as types of scales. If the scale conforms to the first property of the number system, it is an ordinal scale. If it conforms to the second in addition to the first, it is an interval scale. If it has all three properties, it is a ratio scale. Let us consider these in a little more detail.

If the objects being measured can be arrayed in serial order with respect to the property being measured and if numbers can be assigned to the values of this property so that the order of the property agrees with the order of numbers, the result is an *ordinal scale*. Assume, for example, that the property of interest is authoritarianism. Assume further that individuals have responded to a series of questions and that the responses can determine degrees (more or less) of authoritarianism. If the following conditions hold, the individuals are *ordered* in terms of authoritarianism:

1. The scale is asymmetrical (if $A > B$, then $B \not> A$).
2. The scale is transitive (if $A > B$, and $B > C$, then $A > C$).
3. The scale is connected (the properties $>$ and $<$ hold over all values of the scale).
4. The scale is irreflexive ($A \neq A$ for all A).[8]

Simply put, if authoritarianism is measured on an ordinal scale, we can say that individual A is more authoritarian than B; and if individual B is also more authoritarian than C, then A is more authoritarian than C; that all individuals measured stand in the relations of more or less to all other individuals on the scale; and finally, that there is no true value or score of authoritarianism other than that obtained by the relations of more or less. (The last condition, the irreflexive condition, is discussed in more detail below.)

An important fact about the ordinal scale is that the exact distance between members of the scale is not known. This can be illustrated in terms of points on a line:

A	B	C	D	E	F
1st	2nd	3rd	4th	5th	6th

We know that A is greater in the property being measured than B, but exactly how much greater is not indicated. It is not true, for example, that $B + C = D$. Hence, an ordinal scale shows relative position in terms of greater ($>$) or less ($<$) but not the particular value for any person.

[8] A slash mark through the equal sign $\neq$ is read "is not equal to."

Let us briefly consider the way an ordinal scale for an attribute like authoritarianism is constructed. One of the statements on the F scale of authoritarianism is "Obedience and respect for authority are the most important virtues children should learn." The respondent is asked to circle a number in accord with how strongly he agrees or disagrees with the statement, as follows:

Strongly agree	Moderately agree	Agree	Disagree	Moderately disagree	Strongly disagree
1	2	3	4	5	6

The authoritarian scale, like most attitudinal scales, has a large number of statements like this. In order to place individuals on the scale, all the numbers circled by a respondent are added and this sum is used to rank individuals in order of authoritarianism from most to least. If the measurement instrument has 50 statements of the kind given above and if each has the same response choices, the highest possible score is $6 \times 50 = 300$, and the lowest is $1 \times 50 = 50$. The individual whose score is 300 is ranked first in authoritarianism; the individual who scores the next highest, even if it is 260, is ranked second; and so on. Most often, several individuals come out with the same score. Ties result in what is known as a *partially* ordered scale. In Chapter 8 we show that a different statistical method of analysis is appropriate for completely ordered scales from that for partially ordered scales.[9]

With an ordinal scale, any transformation that preserves the order of the individuals on the scale is possible. Each score on the scale can be multiplied by 2 and still maintain the same order among the individuals on the scale. The kind of statistical operation that can be performed on ordinal-scale data is limited; for example, neither a mean nor a standard deviation of data can be measured on an ordinal scale. Measures of strength of relationship between ordered variables are generally restricted to what is called *order* statistics (see Chapter 8).

Interval Scales. If, in addition to being able to order individuals in terms of greater or lesser on some property, we also know the exact distance between each of the individuals — and this distance is the same for all values of X_i — an interval scale of measurement has been achieved. The interval scale has, in addition to the properties of the ordinal scale, properties that enable us to apply all the following axioms:[10]

[9] We discuss validity and reliability later in this book. For a discussion of the question of equal-appearing interval scales and other measurement questions, see the book by Torgerson cited at the end of the chapter. See also C. Alan Boneau, "A Note on Measurement Scales and Statistical Tests," *American Psychologist*, vol. 16, pp. 260–261, 1961.

[10] These axioms are considered in more detail in the section on the rules of summation below.

1. *Uniqueness.* If a and b denote real numbers, then $a + b$ and $a \times b$ represent one and only one real number.

2. *Symmetry.* If $a = b$, then $b = a$.

3. *Substitution.* If $a = b$ and $a + c = d$, then $b + c = d$; and if $a = b$ and $ac = d$, then $bc = d$.

4. *Commutation.* If a and b denote real numbers, then $a + b = b + a$, and $ab = ba$.

5. *Association.* If a, b, and d denote real numbers, then $(a + b) + c = a + (b + c)$, and $(ab)c = a(bc)$.

With an interval scale, any particular individual (or unit) has a specific value regardless of the values of other individuals (or units). Examples of interval-scale measurement are temperature, time, and income. Consider temperature as an illustration. The temperature of cities in the United States for a particular day can be measured. A temperature can be assigned to one city by the instrument used (thermometer) regardless of the temperature of any of the other cities. This is what is meant by the reflexive property of an interval scale. An ordinal scale is irreflexive, since the value of one observation has meaning only in relationship to the other observations. Furthermore, if the temperature in one city is 80°F and the temperature in another is 60°F, the first is exactly 20°F more than the second. In contrast, if one person has a score of 280 on a scale of authoritarianism, such as that described above, and another has a score of 260, the first person is not exactly 20 units more than the second unless the instrument meets a number of criteria, including constancy in the number of questions used and response choices allowed. There are no accepted standards for these, and it is therefore questionable whether existing attitude scales form an interval scale.[11]

In addition to the statistical operations appropriate for nominal-scale and ordinal-scale data, with interval-scale data we can apply more precise operations, such as measures of the degree to which variables might be described by a mathematical equation, and the strength of the relationship between variables. These are considered in Chapters 9 and 10.

Ratio Scales. It is not possible with interval-scale data to compare ratios of differences. To do this, it is necessary to have an origin or zero point; for example, the measurement of IQ might be said to be an interval scale, but there is no natural zero point in current IQ scales. Properties such as weight, time, area, and length have natural zero points. It is possible to say that 100 pounds is to 50 pounds as 2 pounds is to 1 pound (the ratio of difference is 2). But it is not meaningful to say

$$\frac{\text{Individual with an IQ of 200}}{\text{Individual with an IQ of 100}} = \frac{\text{individual with an IQ of 100}}{\text{individual with an IQ of } 50}$$

---

[11] See the book by Torgerson cited at the end of this chapter for a discussion of this complicated question.

It is true that the numbers on both sides of the equation reduce to 2, but it is not true in the same sense that a genius is to a normal man as a normal man is to someone with a damaged brain.

The level of measurement that can be achieved is not inherent in the thing being measured, but it is a function of our ability to devise instruments, and of the general development in our ability to conceptualize. The property we call *temperature*, for example, illustrates that scales can be improved as the basic concept is modified. Temperature is now measured on a ratio scale through the Kelvin thermometer; whereas not too long ago, it was measured on an interval scale by either the Centigrade or Fahrenheit thermometers. In the latter two the zero point was arbitrarily selected with reference to the boiling and freezing points of water. In the Kelvin thermometer, there is a natural zero point that represents the complete absence of heat in contrast to arbitrary zero points of the Fahrenheit and Centigrade thermometers. The development of the Kelvin scale came only with developments in the concept of temperature and its relationship to physical laws.

A ratio scale can be converted into an ordinal scale, and an ordinal into a nominal. We can always change from a higher to lower level of measurement, but the opposite progression is more difficult.

In this description of types of scales, we noted a relationship between level of measurement and the kinds of statistical methods thought to be legitimately applicable to each. Some writers tend to play down this question (see Games and Clare; Hays). It is an important enough question to warrant a few additional words at this point.[12]

It is true that the various statistical methods do not depend directly on the level of measurement used. Statistics deals with numbers and is not concerned with whether the numbers represent the essence of the matter being studied. Given numbers, any kind of statistical measurement can be performed. From this perspective, statistics is viewed as a body of concepts and procedures and is therefore contentless. We could discuss each of the methods contained in this book without once referring to political or other kinds of empirical behavior, and the book would then be a book on mathematical statistics. Similarly, we could treat the numbers obtained from a scale of authoritarianism, such as that described above, as if they constituted a ratio scale and use the statistical measurements appropriate to such scales. The numbers used in questionnaires for scales, such as authoritarianism, do form a ratio scale. But this does not mean that the attitude dimension does, unless there are accepted standards for such attributes. Hence, strictly speaking, we should not apply the correlation measure that assumes an interval scale (described in Chapter 9) to data that do not form an interval scale. However, there is empirical evidence to show that it does not matter a great deal in most cases whether we apply to attitude scales the correlation measure

[12] The beginning reader will do well not to try to understand everything that is said here and return to this point again after he has completed the book.

for the interval level of measurement rather than that meant for the ordinal level. Given this, we might distinguish only two levels of measurement: qualitative, or categorical, and quantitative. The latter would include ordinal, interval, and ratio scales, and a distinction would not be made among them.

All this is perfectly correct. But it is still useful to make the distinction among types of scales, especially if we adhere to the view that statistics is not just a tool of research but an integral part of the scientific method. It is thus useful to keep levels of measurement in mind so that we are alerted to the many kinds of error that may creep into our analysis. We can distinguish some of these as being measurement error, clerical error, errors of observation, and sampling error. Sampling error is what inferential statistics deals with, and it is the major concern of this book, although some of the other sources of error are considered. Furthermore, we see in Chapter 13 that statistics and measurement theory come together in the technique called *factor analysis*. Finally, from one point of view it might be said that statistics itself is nothing more than measurement. We measure aspects of the phenomena we are studying with statistics and then apply probability theory to determine the extent to which the same things are true of a wider universe of values.

Continuous and Discrete Variables

It is important to make an additional distinction between types of variables. Nominal-scale data constitute discrete variables, because they can take on only integer, or whole-number, values. For example, if the members of a community are listed in terms of their religious affiliation as Protestant, Catholic, Jewish, or other, the total in each category will be a whole number. It is not too meaningful to say that there are 3 1/2 Protestants in a community unless one is only partially a Protestant, in which case we have introduced the idea of degree and have thus moved to a continuous variable. A continuous variable can take on either integer or decimal (real-number) values, such as 3.126 or $\sqrt{2}$ (the latter is generally called an *irrational* number, because the decimal does not terminate). An example is the weight of each member of a community or his height. In the case of continuous variables, if we have extremely accurate measurement instruments, we carry out the decimal to many places if it is important to do so. Otherwise we round off to the nearest tenth or hundredth, depending on how much precision is wanted. The reason for making the distinction between discrete and continuous variables is made clearer later.

All measurement can be considered approximate, and the higher the level of measurement, the more accurate the approximation. In most cases, however, a very high degree of accuracy is not required, and rough approximations are sufficient. There are conventions in rounding for approximations that we describe now, because this is a question that occurs frequently.

Suppose that we are measuring the number of man-hours worked by members of a government department in the following kinds of activities: (*a*) contacts with the community, (*b*) meetings with other department personnel, (*c*) telephone communications, and (*d*) filling out reports. Let us also suppose (unrealistically) that we are able to follow each man around every day for a week and record how much time he spends in each activity. Finally, let us suppose that the first person observed spends his time during the week as follows: (*a*) 20 hours, 35 minutes; (*b*) 5 hours, 10 minutes; (*c*) 2 hours, 40 minutes; (*d*) 3 hours, 55 minutes. None of these is exact, because we have not recorded seconds. Let us say that, for the purposes of this study, man-hours are sufficient and that we do not need minutes, but the latter have been recorded to obtain greater accuracy. The first convention in the accuracy of measurement concerns rounding. We generally round up if the last significant digit is equal to or greater than 5 and down if it is less than 5. For example, if the unit of measurement is dollars and we have a measurement of \$3.57, the last significant digit is 7. We then round the 57 to 60 for precision to a tenth of a dollar. For the example concerning man-hours, assuming that precision to the nearest hour is all that is necessary, rounding yields (*a*) 21 hours, (*b*) 5 hours, (*c*) 3 hours, and (*d*) 4 hours.

Another convention concerns rounding when different measurements are added, subtracted, multiplied, or divided and each is measured to a different degree of precision. For example, suppose that we have two observations of 1.2 and 2.232. The sum, difference, or product of these two approximate numbers can be no more precise than the least precise measurement used. Thus, $1.2 \times 2.232 = 2.7$ and not 2.6784; $65.22 + 10.1135 = 75.33$ and so on. The quotient of two numbers is carried out to the number of significant digits equal to the smaller of the numbers. Thus, $12/6{,}500 = 0.0018$, not 0.001846. The square root of a number can contain no more significant digits than the number itself. Thus, $\sqrt{36.34} = 6.028$.

DESCRIPTIVE AND INFERENTIAL STATISTICS

Another important assumption in statistics concerns the method of sampling. Most of the inferential statistics discussed in this book are based on the assumption that the data we are dealing with have been obtained by random sampling, which is defined briefly at the beginning of Chapter 6 and in detail in Chapter 15. Here we note that, if random sampling or another form of probability sampling is not used, statistics can be used only in a descriptive manner. This means that a specific sample can be summarized with statistics, and we can measure the relationship between variables for particular data but we cannot make inferential statements about a population based on the particular data. In order to get a clearer idea of what this means, we define some key words: population, sample, parameter, and statistic.

Population in statistics does not refer strictly to people. It designates the total set of observations that can be made. For example, if we are study-ing the weight of American males, the population is the set of weights of all American males. If we are studying cities of the world, the population is all cities in the world. If we are studying legislators, the population may be all legislative behavior. The last example requires further explanation, be-cause it is obvious that it refers to something not capable of being listed. We distinguish between finite and infinite populations. *Infinite* populations are hypothetical rather than empirical.[13] Hypothetical populations refer to things that transcend time and place. For example, if we study contemporary legislative behavior in the United States, the population is all American legislative bodies. This is a finite population. We might be concerned with American state legislative behavior, in which case the population is the 50 state legislatures. This also is a finite population. Any population that, in principle, can be specified in a list is therefore a *finite* population. Any popu-lation that cannot be so listed is an infinite population.[14]

The set of whole numbers is an infinite population, because no matter what number is named, a larger one can also be named. Which of the following populations is (are) infinite? All college students in the United States; all presidents of the United States; all movies produced in 1968. None. Each could be listed. There might not be a list of all college students in the United States, but it is perfectly clear that, if such a list could be made up, it would have a beginning and an end.

Usually we are not able to study a population. We must be satisfied with a *sample*, which may be defined as a subset, or part of the complete set of observations. For example, a sample of the set of numbers 1 to 10 may be (1,5,9) or even (2,3,4,5,6,7,8,9,10). It is necessary to be satisfied with a sample if we are studying infinite (analytical) populations, because it is not possible to study the entire population. Frequently it is necessary to draw samples from finite populations because they are too large to study.

Consider the following question as an illustration. Suppose that we are interested in whether the process of socialization is related to the "role per-ception" of legislators. Suppose that we are interested in whether a "per-missive" socialization process is related to a "delegate" as opposed to a "trustee" perception of the representative's role.[15] Finally, suppose that we

[13] Sometimes the word "universe" is used to refer to analytical populations, but in this book we use the terms "universe" and "population" interchangeably.

[14] Later in the chapter we discuss the way such a population as American state legisla-tures might be conceived of as a sample from the analytical and infinite population of legislative behavior.

[15] A "delegate" perception is defined as the belief that the representative should do exactly as his constituents desire, and a "trustee" perception is defined as the belief that the representative should use his best judgment of what is in the interest of his constituents. We might define "permissive" socialization as environments that encourage individuals to make their own decisions. The example is hypothetical; no studies of this kind have been made, although studies of legislative role perception have. See, for example, Heinz Eulau et al., *The Legislative System*, John Wiley & Sons, Inc., New York, 1962.

are able to study 500 representatives (any level of government), and we find that of those exposed to a permissive socialization process, 54 percent perceive their role as a delegate; and of those not exposed to a permissive socialization process, 74 percent perceive their role as a delegate. Do the 500 representatives constitute a sample or a population? A sample. Is the population they were drawn from finite or infinite? It depends on the question we are asking. They would constitute a sample from a finite population if we wanted to generalize to the actual population of all representatives in the United States and the sample were drawn so that we knew the probability of each member's being included in it. They would constitute a sample from an infinite population if we wanted to generalize to the population of all representatives' role perception, past, present, and future.[16]

When we describe characteristics of a population rather than a sample, we refer to population *parameters.* The word "parameter" is used in different ways. It may refer to a limiting factor, an unknown quantity, or a population value. In this book, it designates a population value.

When we describe a characteristic of a sample, we refer to a sample *statistic.* For example, the mean income of all males in the United States might be the population parameter we are interested in. The mean income of a sample of 400 males selected from this population would be the sample statistic. It is conventional to designate population parameters by lower case Greek letters or capital English letters, and sample statistics by lower case English letters, with some exceptions. One of the exceptions is the mean. The sample mean is usually designated by $\bar{X}$, and the population mean by the Greek letter μ (mu). The symbols conventionally used to designate some of the

Table 1-1 Two Major Purposes of Statistics

I *To Describe Characteristics of a Particular Sample of Data, Such as:*	II *To Infer Characteristics of a Population, Such as:*

Central Tendency:
  Mean $(\bar{X})$, Median (m), Proportion (p)
	Central Tendency:
	  Mean (μ), Median (Md), Proportion (P)

Dispersion:
  Variance (s^2), Standard Deviation (s)
	Dispersion:
	  Variance (σ^2), Standard Deviation (σ)

Skewness: (SK)	Skewness: (α_3)
Kurtosis: (K)	Kurtosis: (α_4)

Relationship:
  Correlation (r)
	Relationship:
	  Correlation (ρ)

The definitions of most of the symbols in parentheses are given in Chapter 2.

[16] The conditions under which we might be able to generalize to either population are considered below.

principal population parameters and sample statistics are given in Table 1-1, which also shows the two major purposes of statistics: to describe something about a specific set of data (a sample) and to infer the same thing about a population. Note that we measure the same things in both cases.

Inferential statistics, in turn, can be broken down into two major sub-categories: (*a*) making estimates of unknown population parameters (such as μ), called *estimation statistics*, and (*b*) making tests of hypotheses, called *test statistics*. These two aspects of statistical inference are considered separately in Chapters 6 and 7.

THE USE OF SUMMATION IN STATISTICS

Before concluding this chapter, let us consider one of the elementary things that frequently hinder an understanding of statistics; that is, the extensive symbolism. When a few rules are understood, many statistical concepts that seem formidable become simple.

The first set of rules concerns the operation of summing, which is central to statistics. Summation is designated by the Greek letter sigma, $\sum_{i=1}^{N} X_i$, where the lowercase *i* under the sigma and the *N* above indicate that we should sum all values of *X* from the first to the *N*th values. X_1 represents the score of the first individual, X_i represents that of the *i*th individual, and *N* represents the number of individuals. Thus

$$X_1 + X_2 + \cdots + X_i + \cdots + X_N = \sum_{i=1}^{N} X_i$$

For example, if we wanted to sum a variable X_i that could take on the values 4, 5, 6, 10, and 15, the symbol $\sum_{i=1}^{5} X_i$ means to sum the values of the variable X_i beginning with the first, which is 4, and ending when all five values have been summed, or $4 + 5 + 6 + 10 + 15 = 40$. X_3 refers to the third score, 6, and so on.

If we are dealing with a matrix of numbers, such as those presented in Table 1-2, double summation notation is used: $\sum_{i=1}^{r} \sum_{j=1}^{c} X_{ij}$, where *i* refers to the rows, and *j* to the columns. The rows range from 1 to *r*, and the columns from 1 to *c*.[17] The double summation notation means to sum all the values

---

[17] There is no requirement that *i* designate rows and *j* columns. We could allow *j* to designate rows and *k* columns and write $\sum_{j=1}^{r} \sum_{k=1}^{c} X_{jk}$. The latter is used by some authors. In this book the notation described in the text is used.

in the matrix across columns. For the numbers in Table 1-2, we have

$$\sum_{i=1}^{2} \sum_{j=1}^{3} X_{ij} = (10 + 2 + 8) + (15 + 16 + 2) = 53$$

Table 1-2 Matrix

		Columns		
		1	2	3
	1	10	2	8
Rows				
	2	15	16	2

If we want to designate a particular value, we write X_{21}, which refers to the score in the second row, first column, or 15. Note that this is read "X sub two, one," not "X sub twenty-one." If we want to sum just the first row, we write

$$\sum_{j=1}^{3} X_{1j} = (10 + 2 + 8) = 20$$

and if we want to sum just the first column, we write

$$\sum_{i=1}^{2} X_{i1} = (10 + 15) = 25$$

Thus, $\sum_{j=1}^{3} X_{ij}$ indicates the score in some row, and $\sum_{i=1}^{2} X_{ij}$ indicates the score in some column.

Frequently, we can eliminate the symbols indicating the range of a variable and the subscript, because it is quite clear in the context, and just write $\sum X$. This practice is followed in this book, and subscripts are used only if ambiguity would result from their elimination.

The work of summation can often be greatly simplified if the following three rules are observed:

Rule 1. $\sum_{i=1}^{N} (X_i + Y_i)$ is the same as $\sum_{i=1}^{N} X_i + \sum_{i=1}^{N} Y_i$; that is, we get the same result if we add each X and Y and then sum the result or if we add the X's and Y's separately and then sum.

Rule 2. $\sum_{i=1}^{N} aY_i$ is the same as $a \sum_{i=1}^{N} Y_i$; that is, the sum of a constant times a variable is the same as a constant times the sum of a variable.

Rule 3. $\sum a$ is the same as na; that is, the sum of a constant is the same as n times the constant; that is, the sum of $(2 \times 4) + (2 \times 10) + (2 \times 5) = 2(4 + 10 + 5) = 38$. Note that

$$\sum_{i=1}^{N} (X_i \times Y_i) \neq \sum_{i=1}^{N} X_i \times \sum_{i=1}^{N} Y_i$$

To illustrate these rules of summation, we give a simple example. Suppose that our task is to perform the following calculation:

$$\sum_{i=1}^{N} (X_i - K) - \frac{\sum_{i=1}^{N} K}{\sum_{i=1}^{N} KX_i} \tag{1-1}$$

where X_i is a variable and K a constant. Rule 1 states that the term in parentheses can be broken down into two parts:

$$\sum_{i=1}^{N} X_i - \sum_{i=1}^{N} K$$

Rule 3 states that the second of these, as well as the term in the numerator in (1-1), can be written

$$NK$$

and rule 2 states that the term in the denominator in (1-1) can be written

$$K \sum_{i=1}^{N} X_i$$

Hence, the entire sentence (1-1) can be rewritten

$$\left(\sum_{i=1}^{N} X_i - NK \right) - \frac{NK}{K \sum_{i=1}^{N} X_i} \tag{1-2}$$

The reason for rewriting (1-1) as (1-2) may not be immediately apparent. An illustration shows why it helps. If we have a large number of figures to work with, we can obtain an answer much more easily using (1-2) than (1-1), for it involves fewer operations. Instead of subtracting the constant from each variable and summing, we can sum the variable and subtract the product of N times the constant from it. Table 1-3 shows that the same answer is obtained with both (1-1) and (1-2) but the computations are easier using (1-2).

Table 1-3 Illustration of Simplified Computation by the Use of Summation Rules

Variable X_i = {5, 6, 8, 10, 12, 20}

Constant K = 2

N = 6

By Formula (1-1):

$$\sum_{i=1}^{N}(X_i - K) - \frac{\sum_{i=1}^{N}K}{\sum_{i=1}^{N}KX_i}$$

$[(5-2)+(6-2)+(8-2)+(10-2)+(12-2)$
$+(20-2)] - (2+2+2+2+2+2)/$
$[(2\times5)+(2\times6)+(2\times8)+(2\times10)+(2\times12)$
$+(2\times20)]$

$= 49 - \dfrac{12}{122}$

$\doteq 48.9017$

By Formula (1-2):

$$\left(\sum_{i=1}^{N}X_i - NK\right) - \frac{NK}{K\sum_{i=1}^{N}X_i}$$

$[(5+6+8+10+12+20)-(6\times2)]$

$- \dfrac{6\times2}{2\times(5+6+8+10+12+20)}$

$= 49 - \dfrac{12}{122}$

$\doteq 48.9017$

READINGS AND SELECTED REFERENCES

Alker, Hayward R., Jr. *Mathematics and Politics*, The Macmillan Company, New York, 1965, Chaps. 1–3.

Blalock, Hubert, Jr. *Social Statistics*, McGraw-Hill Book Company, New York, 1960, Chaps. 1–2.

Cohen, Morris, and Ernest Nagel. *An Introduction to Logic and Scientific Method*, Harcourt, Brace & World, Inc., New York, 1934.

Games, Paul, and George Klare. *Elementary Statistics*, McGraw-Hill Book Company, New York, 1967, Chap. 1.

Hays, William L. *Statistics for Psychologists*, Holt, Rinehart and Winston, Inc., New York, 1963, Introduction, Chap. 1.

Kaplan, Abraham. *The Conduct of Inquiry*, Chandler Publishing Co., San Francisco, 1964, especially Chaps. 1–4.

Kemeny, John G. *A Philosopher Looks at Science*, D. Van Nostrand Company, Inc., Princeton, N.J., 1962.

Miller, Delbert C. *Handbook of Research Design and Social Measurement*, David McKay Company, Inc., New York, 1964.

Parl, Boris. *Basic Statistics*, Doubleday & Company, Inc., Garden City, N.Y., 1967, Chaps. 1–3.

Spurr, William A., and Charles P. Bonini. *Statistical Analysis for Business Decisions*, Richard D. Irwin, Inc., Homewood, Ill., 1967, Chap. 1.

Torgerson, Warren S. *Theory and Methods of Scaling*, John Wiley & Sons, Inc., New York, 1958.

Western, D. W., and V. H. Haag. *An Introduction to Mathematics*, Holt, Rinehart and Winston, Inc., New York, 1959.

2

Basic Concepts:
The Frequency Distribution
and Measures of Central
Tendency

Several of the basic concepts used in the analysis of data were listed in Table 1-1 but not defined there: measures of central tendency, such as the mean μ, median Md, and proportion P; measures of dispersion, such as variance σ^2 and standard deviation σ; measures of skewness and kurtosis; and measures of relationship, such as correlation ρ. We describe measures of central tendency in this chapter, along with some other basic concepts. The measures of dispersion and skewness and kurtosis are considered in Chapter 3. The measures of association, or relationship, are discussed in later chapters. Before we begin, however, a few things about notation should be emphasized. In this chapter and the next we use the notation appropriate to describing population parameters (Greek letters and capital English letters), and note in the appropriate places where a distinction must be made for measuring the same characteristics for samples. The notion of a sample does not arise until we introduce the idea of statistical inference in Chapter 4. Hence, the reader may assume that in this chapter and the next we are discussing populations, or all observations, unless it is explicitly stated otherwise.

THE FREQUENCY DISTRIBUTION

The task of extracting information from data generally begins with the frequency distribution. It is something that may sometimes be overlooked by the researcher who prefers a quick, single descriptive measure, such as the mean or correlation coefficient. But the distribution, or shape, of data may reveal important information that these other measures cannot.

A frequency distribution can be defined simply as a listing of the number of observations[1] that fall into each of several categories or class intervals. It can be presented in tabular form, in a bar chart, or in the form of curves. Discrete variables, defined in Chapter 1, can be presented in tabular form or in a bar chart but not in the form of a curve. Continuous variables can be presented in tabular form, in a bar chart, and in the form of curves. We first consider the frequency distribution for discrete variables and then that for continuous variables.

Construction of Tabular Frequency Distributions for Nominal-scale (Discrete) Data

The construction of a frequency distribution for nominal-scale data is simply a matter of listing the categories and the number and percentage of observations that fall into each. Although the matter is simple, it is a place where sins are frequently committed, including such things as listing percentages in each category but not the number, bad captions, and the failure to include all information. Parl gives, as an example of the first, the newspaper story which reported that two-thirds of the female students at Johns Hopkins University had married faculty members after females were first admitted but which failed to state that only three females were involved.

A good deal of common sense is involved in the proper construction of a frequency distribution for nominal-scale data. The kinds of classifications, the manner of presentation, the direction in which the percentages are computed — these and other questions depend on what information the writer is trying to convey. Table 2-1, which is based on Benjamin Walter's study of political decision making in Arcadia,[2] illustrates the use of three frequency distributions in order to make a number of important points. The data show that the voter turnout in the 1954 municipal-building referendum was very low (9.8 percent for the entire city), and, except for the upper class, the majority of voters voted against the referendum. The failure of the voters to pass the referendum stimulated social and economic leaders to create an

---

[1] The word "observation" is used somewhat synonymously with the words "scores," "individuals," and "values" in this book. They refer to the basic unit being studied and the unit of measurement being used. For example, if we are studying cities and measuring properties such as their population size, each observation is the size of cities; the individuals are the cities, and the values are the different numbers the sizes take. "Scores" may also be the different values the variable may take, although it is more common to use the term "scores" for attitudinal variables. For example, if we are studying legislators and are measuring properties such as their role perception, the observations are the "scores" of each legislator on a series of questions designed to measure role perception. The word "observation" may thus be used in a more generic sense, and the other terms in a more specific sense.

[2] Benjamin Walter, "Political Decision Making in Arcadia," in F. Stuart Chapin, Jr., and Shirley F. Weiss (eds.), *Urban Growth Dynamics*, John Wiley & Sons, Inc., New York, 1962.

Table 2-1   Class and Voting in Arcadia City Referendums*

Type of Precinct	1954 Municipal Building Referendum			1956 Municipal Building Referendum			1956 Fluoridation Referendum		
	% Yes	Number of Votes	% Voting	% Yes	Number of Votes	% Voting	% Yes	Number of Votes	% Voting
Negro	46.8	77	2.5	24.0	288	8.2	27.3	286	8.2
Lower class	46.4	388	9.9	25.5	882	21.0	32.9	878	20.9
Negro and Lower Class Combined	46.5	465	6.6	25.1	1,170	15.2	31.5	1,164	15.2
Middle Class	47.7	1,084	9.2	36.0	3,789	23.3	53.5	3,794	23.3
Upper Class	50.1	809	15.7	45.3	2,281	36.3	70.3	2,295	36.5
City Totals	48.3	2,358	9.8	37.3	7,240	24.0	55.3	7,253	24.0

*Adapted from: Benjamin Walter, "Political Decision Making in Arcadia," in F. S. Chapin and Shirley Weiss (eds.), *Urban Growth Dynamics*. John Wiley & Sons, Inc., New York, 1962.

organization in an attempt to reverse the vote in 1956. Table 2-1 shows that they succeeded in getting a larger voter turnout (24 percent) but were unable to change the verdict of the voters, who again defeated the referendum. The frequency distribution of the 1956 fluoridation referendum shows that the voters did not vote against the municipal-building referendum "out of perennial pettiness and invincible perversity," because they did pass the fluoridation referendum.

Nominal-scale data, such as that in Table 2-1, are generally described in proportions, percentages, or ratios. A *proportion* is the number of observations falling into any category divided by the total number of observations. A *percentage* is a proportion multiplied by 100. For example, Table 2-1 shows that out of the 77 Negroes who voted in the 1954 referendum, 46.8 percent voted yes. Hence, the proportion who voted yes is 0.468 = 37/77, and the proportion who voted no, therefore, is 1 − 0.468 = 0.532 = 40/77, because, for proportions,

$$\frac{n_1}{N} + \frac{n_2}{N} + \cdots + \frac{n_k}{N} = 1.00$$

that is, the sum of proportions is equal to 1.

The way a percentage is constructed depends on what we are trying to illustrate. Turning again to Table 2-1, a column could be constructed that depicts what percentage of the total number each of the subclasses comprises. For example, in the 1954 referendum, 2,358 people voted. Of these, 809 are classified as upper class. Hence, about a third of all those who voted on this referendum are upper class. This leads us to conclude that the upper class turned out to vote in larger proportions than the other classes. These figures are not presented in Table 2-1, because the focus of the study was on the percentage voting yes on the referendum, not on the differences in social-class voting behavior.

Usually, percentages and proportions are sufficient descriptive indices for nominal-scale data. Under certain conditions, however, ratios and rates are very useful. A *ratio* is a means of comparing one group with another. For example, if a district has 5,000 Democrats and 4,000 Republicans, the ratio of Democrats to Republicans is 5,000/4,000 = 1.25. A *rate* is computed against a base, as in birth, death, and accident rates. Thus, the birth rate is the number of births per 1,000 population. The base can also be another quantity, as in computing income per factory or per person (per capita).

Construction of Tabular Frequency Distributions for Interval-scale Data

The construction of a frequency distribution for interval-scale data usually begins with an array in which the data are listed in ascending or descending order. The array in Table 2-2 contains the per capita gross national product (GNP) for 60 countries. In order to cast these into a frequency distribution, we must first decide how many class intervals to use. The *class interval* is the range of values for each class or grouping. For example, the array in Table 2-2 shows that several countries have a per capita GNP of between $50 and $60; three are in the next interval, from $61 to $70; and so on. There are several questions to be decided here. How many intervals should there be? How wide should they be? What should the lower and upper limits of each interval be?

Table 2-2 Per Capita GNP for 60 Countries in Descending Array*

$2,900	943	550	377	272	194	160	108	88	60
1,947	836	490	362	263	189	144	105	78	60
1,388	726	478	357	239	178	142	100	73	57
1,316	670	467	340	225	173	135	99	70	55
1,196	600	423	316	220	172	129	98	70	50
1,130	572	395	293	219	161	120	94	64	50

*Adapted from: Bruce M. Russett *et al., World Handbook of Political and Social Indicators,* Yale University Press, New Haven, Conn., 1964.

There are no exact answers to the first question. The rule of thumb is that the number of class intervals should not be so few as to obscure the data and not so many as to leave too few observations in each interval. In the large majority of cases, between 6 and 15 classes are sufficient. There are two purposes for constructing a frequency distribution. One is for conveying information to the reader, and the other is for use in computations. In the latter case, it does not matter how many classes are used except that computations will be more accurate if there is a large number of classes. In the former case, it is simply a matter of judgment. For the data in Table 2-2, 8 classes enable us to make the width of each exactly $400, as in Table 2-3, which shows that most countries have a relatively low per capita income, because 44 of the 60 fall into the first interval of $50 to $450. Selecting 8 rather than more or fewer intervals helps bring this out.

Table 2-3   Per Capita GNP in Frequency Distribution

Per capita GNP (Class interval)	Number of countries (Frequency)
$50– 450	44
450– 850	9
850–1,250	3
1,250–1,650	2
1,650–2,050	1
2,050–2,450	0
2,450–2,850	0
2,850–3,250	1
	60

Stated and True Class Intervals.   In this example it happened that none of the countries had a per capita GNP exactly equal to the upper or lower limits of the class intervals used. If a country had a per capita GNP of $1,650, there would be some ambiguity about the interval it should be placed into. In order to relieve this ambiguity, we derive true class intervals that are carried out one decimal place more than the precision of measurement of the data. In the data above, the per capita GNP of the countries measured is in whole numbers. Thus, if the class intervals ran from 49.5 to 449.5, 449.5 to 849.5, 849.5 to 1,249.5, and so on, there would be no doubt about where the observation 1,650 is to be placed. If the data above were measured to the first decimal (to tenths of a dollar), then the true class intervals could be extended another place: 49.95 to 449.95, 449.95 to 849.95, and so on.

The only other things to note about class intervals in frequency distributions are that the upper limit of one class interval corresponds with the lower limit of the next, and that the intervals are of equal width except for the last, which may be open-ended.

Charts and Curves

The tabular frequency distribution begins to give us some information that the arrayed data do not. In the example above we can see in Table 2-3 that most of the countries have relatively small per capita GNP. They fall into the two lower class intervals. We may also show this in terms of percentages to give a relative frequency distribution, as in Table 2-4. The use of percentages in a frequency distribution enables us to compare different frequency distributions. For example, we can compare the income distribution between countries with the income distribution within a country, using the relative frequencies.

**Table 2-4 Per Capita GNP of 60 Countries:
Relative Frequency Distribution**

Per capita GNP	Percent of countries
$50– 450	$73\frac{1}{3}$
450– 850	15
850–1,250	5
1,250–1,650	$3\frac{1}{3}$
1,650–2,050	$1\frac{2}{3}$
2,050–2,450	0
2,450–2,850	0
2,850–3,250	$1\frac{2}{3}$
	100

Additional information can be conveyed by casting the data in the form of a chart instead of a table. A *histogram*, or bar chart, is one such chart. In a histogram, we list the frequencies on the vertical or y axis and the class intervals on the horizontal or x axis. Figure 2-1 is a histogram for the GNP

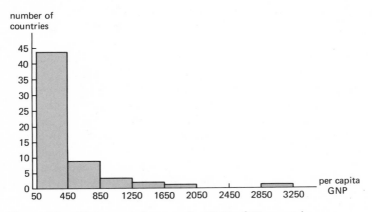

Figure 2-1. Histogram of per capita GNP of 60 countries.

of the 60 countries in Table 2-2. The class intervals are placed on the abscissa (horizontal axis) and the frequencies on the ordinate (vertical axis). In a histogram, the areas of the rectangles are proportional to the frequencies represented.

When we connect the midpoints of each rectangle with a straight line, we obtain a *frequency polygon*, as in Figure 2-2. The frequency polygon

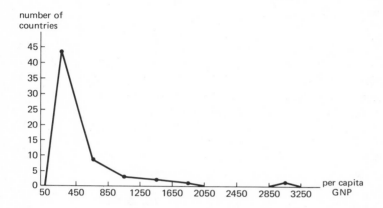

Figure 2-2. Frequency polygon of per capita GNP of 60 countries.

resembles the smooth curve most often used to describe continuous data. The methods for obtaining smooth curves are described later.

The frequency polygon and smooth curve have the advantage of conveying a good deal of information at a glance. After a little familiarization with frequency distributions, certain kinds of curves become recognizable as characteristic of empirical phenomena. The three smooth curves in Figure 2-3 represent frequency distributions of familiar phenomena. Figure 2-3*a* is similar to the polygon of per capita GNP for 60 countries and is the shape that

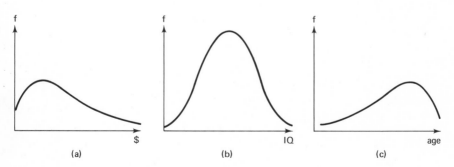

Figure 2-3. Three characteristic frequency curves.

almost all income-distribution data take, whether for one country, one city, or one organization. Most observations (countries, in this case) fall into the lower income intervals. Figure 2-3*b* is the shape that the distribution of intelligence scores generally takes. It is symmetrical, with most observations (people) falling near the middle and the number falling above the middle the same as that falling below. Figure 2-3*c* is the shape that mortality rates for chronic diseases, such as cancer and heart disease, take. Most deaths occur in the older age categories.

Cumulative Frequency Distribution or Ogive. Another useful descriptive device is the cumulative frequency distribution. We may want to answer questions such as "What percentage of families earns more than $15,000 a year" or "What percentage of countries has a per capita GNP of less than $450," and so on. Table 2-5 gives the cumulative frequency distributions for the data presented in Tables 2-3 and 2-4. Note that we can cumulate up or down. 73 1/3 percent of all countries have a per capita GNP of less than $450, and 26 2/3 percent have more; 88 1/3 percent have a per capita GNP of less than $850, and 11 2/3 percent have more; and so on.

Table 2-5   Cumulative Frequency Distribution of per Capita GNP

Per capita GNP	f	Percent	Countries having less than upper limit of class interval		Countries having as much or more than lower limit of class interval	
			N	Percent	N	Percent
$50– 450	44	$73^1/_3$	44	$73^1/_3$	60	100
450– 850	9	15	53	$88^1/_3$	16	$26^2/_3$
850–1,250	3	5	56	$93^1/_3$	7	$11^2/_3$
1,250–1,650	2	$3^1/_3$	58	$96^2/_3$	4	$6^2/_3$
1,650–2,050	1	$1^2/_3$	59	$98^1/_3$	2	$3^1/_3$
2,050–2,450	0	0	59	$98^1/_3$	1	$1^2/_3$
2,450–2,850	0	0	59	$98^1/_3$	1	$1^2/_3$
2,850–3,250	1	$1^2/_3$	60	100	1	$1^2/_3$
	60	100				

We can plot a cumulative frequency distribution on a chart. To do so, we locate the coordinates of each axis for each class interval. For example, because 73 1/3 percent of all countries have a per capita GNP of less than $450, the first set of coordinates is the point 73 1/3 percent (or 44 countries) and $450; the next is the point 88 1/3 percent and $850; the next is the point 93 1/3 percent and $1,250; and so on. The opposite cumulation is shown by the dotted line in Figure 2-4.

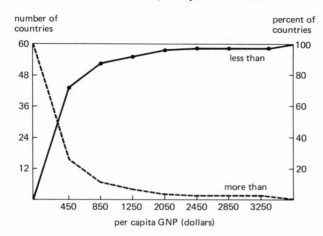

Figure 2-4. Cumulative frequency distribution of per capita GNP.

The Lorenz Curve

A particularly useful descriptive device is the Lorenz curve, which is a method for comparing two cumulative frequency distributions. By the Lorenz curve we can measure degrees of deviation from perfect equality. To illustrate, we use income-distribution figures for New York metropolitan area families for the years 1965 and 2000. Table 2-6 gives the frequency distributions, and Figure 2-5 the Lorenz curves.

In the Lorenz curve, both axes contain percentages. In the present example, the horizontal axis is the percentage of all families in the New York metropolitan area, and the vertical axis the percentage of total income. If we

Table 2-6 Family Income in New York Area for 1965 and Projected Income for 2000*

	1965		2000	
Income	Percent of families	Cumulative percent	Percent of families	Cumulative percent
≤ 5,000	31.5	31.5	12.8	12.8
5,000– 9,999	43.6	75.1	15.9	28.7
10,000–14,999	16.2	91.3	30.6	59.3
15,000–29,999	7.3	98.6	33.0	92.3
30,000 and over	1.4	100.0	7.7	100.0

*Adapted from: *The New York Times,* May 2, 1967.

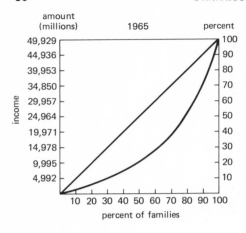

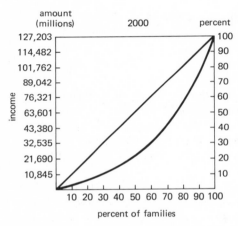

Figure 2-5. Lorenz curves of inequality of income for New York area residents, 1965 and 2000.

plot the coordinates of the two so that the resulting points form a straight line when connected, we have the line of perfect equality; that is, if 10 percent of all families have 10 percent of all income, 20 percent of all families have 20 percent of all income, ... , 90 percent of all families have 90 percent of all income, and so on, the line connecting all the coordinate points will cut across the graph in a perfect diagonal, running from the lower left-hand corner to the upper right-hand corner. This line represents perfect equality, because it represents a situation where the cumulative percentage of families equals the cumulative percentage of total income. We can also plot a curve representing the actual situation by picking several coordinates and connecting them with a smooth curve. To do so, we need more information than that given in Table 2-6. We need information on the cumulative percentage of income for the vertical axis. To obtain this, we must first get the total family income. This was not supplied in the source but may be computed from the data given in Table 2-6, along with information concerning the total number of families in the New York metropolitan area. The total number of families listed for the data in Table 2-6 is 5,854,600, and 31 1/2 percent of this is 1,844,200. Hence, 1,844,200 families fall into the class interval of $0–$5,000. We might assume that they all fall at the midpoint of this interval, which is $2,500, and, multiplying $2,500 times 1,844,200, we obtain a figure of $4,610,500,000 as the total income earned by these families for 1965. Doing the same for the other intervals, we get a total family-income figure for 1965 of $49,929,000,000.[3] We can now plot the cumulative percentage of total income on the vertical axis by taking convenient percentages of this total. Figure 2-5 lists the 10 percentage points; 10 percent of the total is $4,992,900,000; 20 percent is 9,995,800,000; and so on. To obtain a cumulative frequency curve, we pick several coordinate points and connect them with a smooth curve. For example, 31 1/2 percent of all families have a total income of $4,610,500,000, and this is 9.4 percent of the total. Hence, the first point is at $4,610,500,000 and 9.4 percent. This has been done for the years 1965 and 2000 in Figure 2-5.

The area between the line of perfect equality (the diagonal line) and the actual cumulative frequency is a measure of inequality. The larger the area, the more the inequality.[4] The Lorenz curves in Figure 2-5 show that income for the year 2000 is projected to be somewhat more equally distributed than that for the year 1965.

The Lorenz curve might be used in political science for computing inequality of representation. The cumulative percentage of total voters can be plotted on the horizontal axis and the cumulative percentage of total representatives on the vertical axis. The line of perfect equality represents the situation where 10 percent of all voters elect exactly 10 percent of all representatives, 20 percent of all voters elect exactly 20 percent of all representatives, and so on. As a simple illustration, suppose that in a particular state, 20 percent of all voters elect 15 percent of all representatives; 40 percent of all voters elect 30 percent of all representatives; 60 percent of all voters elect

---

[3] Treating all the observations as falling at the midpoint of the class interval produces some distortion. This problem and that of how to handle the open interval > $30,000 are considered below. At this point we simply note that the figures we are using are approximations and are sufficiently accurate for our purposes here.

[4] Computing the area of the difference between the line of perfect equality and the actual cumulative frequency line requires a knowledge of calculus. It can be measured by taking the definite integral of the function X as follows:

$$G = \frac{2 \int_1^{100} [X - f(X)]\, dX}{10,000}$$

This is called the Gini coefficient. Perfect equality produces a G of zero, and perfect inequality a G of one. The closer the G approaches one, the greater the inequality.

45 percent of all representatives; and 80 percent of all voters elect 70 percent of all representatives (from the opposite cumulation, this means that 20 percent of all voters can elect 40 percent of representatives, and so on). This is shown graphically in Figure 2-6, where curve *ADB* represents the actual situation; line *AB*, the line of perfect equality; and the area between the two, the amount of inequality.

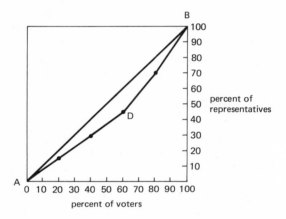

Figure 2-6. Hypothetical Lorenz curve for equality of representation.

MEASURES OF CENTRAL TENDENCY

The frequency distribution, in the form of a table or a chart, gives a basic summary of data. But if we want to compare different groups or variables, we need additional measures. For example, we might want to compare Western and Eastern nations with respect to GNP. These might be cast in frequency distributions, and one frequency distribution visually compared with the other. However, we can get more precise comparisons by using a number of other statistics.

1. Central tendency
2. Dispersion
3. Skewness
4. Kurtosis

We consider measures of central tendency in this section, and the rest in Chapter 3.

In every-day parlance we frequently talk about the typical or average person or observation. The meaning generally conveyed by these words is the person most like the greatest number of other individuals. The word "typical," however, also has stereotypic overtones. The typical American may be conceived of as a construct, or an ideal type, embodying all the qualities we think of when we think of an American. But the composite constructed as the sum of these qualities does not actually exist. The ideal type is obviously a useful analytical device, because it helps emphasize important qualities of what we are studying. But it can also be misleading, because it may exaggerate the importance of various characteristics. The statistical meaning of the typical or average is more precisely numerical. There are three commonly used statistical measures of central tendency: the mean, the median, and the mode.

The Arithmetic Mean

The mean is the most frequently used and most useful descriptive statistical concept. There are several different definitions of the mean, but the one used almost exclusively in statistical work is the arithmetic mean. It is defined as the sum of the individual observations divided by the total number of observations. In symbolic notation

$$\mu = \frac{\sum\limits_{i=1}^{N} X_i}{N} \tag{2-1}$$

This notation means to take the individual values, add them, and then divide by the total number of observations. The mean of the numbers 2, 5, 10, and 15 is thus $(2 + 5 + 10 + 15)/4 = 8$. The mean might best be conceived of as the "center of gravity" of a series of numbers, so that, if the numbers represented objects of uniform weight placed along a board, a fulcrum placed at the mean would balance the board perfectly.

The mean is defined so that the sum of the deviations of each value from the mean is equal to zero, or $\sum(X_i - \mu) = 0$. The sum of the distances of the numbers to the left of the mean are exactly offset by the sum of the distances of the numbers to the right of the mean:

X_i	$(X_i - \mu)$
2	-6
5	-3
10	2
15	7
	$\overline{0}$

A simple algebraic proof of this is as follows:

$$\sum(X_i - \mu) = 0$$
$$= \sum X_i - \sum \mu \qquad \text{by summation rule 1}$$
$$= \sum X_i - n\mu \qquad \text{by summation rule 3}$$
$$= \sum X_i - n\frac{\sum X_i}{n} \qquad \text{by the definition of } \mu$$
$$= \sum X_i - \sum X_i \qquad \text{canceling } n$$
$$= 0$$

We shall see in a moment how this fact can be used to simplify the computation of the mean.

The arithmetic mean allows us to say something about a frequency distribution with a single number and is a way of conveying a large amount of information very economically. For example, we can say that the mean family income of New York metropolitan area families for 1965 was $9,390, and this conveys a good deal of information in a single number. It would be much easier to compare this figure with the mean family income of all United States families, for example, than to compare overall frequency distributions or curves. But a word of caution should be added quickly, for the use of the mean results in a loss of information (about the overall distribution), and in some cases, the mean is not a good index of central tendency. An illustration of the latter is given shortly.

Geometric and Harmonic Means

There are other statistical definitions of the mean that are not used frequently in political science. The geometric mean μ_g can be defined as the Nth root of the product of N numbers:[5]

$$\mu_g = \sqrt[N]{(X_1)(X_2)\cdots(X_n)}$$

For example, the geometric mean of 2, 5, 10, and 15 is

$$\begin{aligned}
\mu_g &= \sqrt[4]{(2)(5)(10)(15)} \\
&= \sqrt[4]{1,500} \\
&= \sqrt[4]{38.73^2} \\
&= 6.22
\end{aligned}$$

The geometric mean can be used to average rates of change, percentages, and ratios. It can also be used to compute the central tendency of frequency distributions that are highly skewed,[6] such as income distribution, where the arithmetic mean is not a good description of central tendency. However, because it is the product of N numbers, it cannot be computed if any of the values in the distribution is zero or negative. Note that the geometric mean of the numbers 2, 5, 10, and 15 is less than the arithmetic mean. It is always less

---

[5] Another definition is

$$\log \mu_g = \frac{\sum_{i=1}^{N} \log X_i}{N}$$

where we simply compute the arithmetic mean of the logarithms of the numbers and then take the antilogarithm of this. This definition is easier for computational purposes. A logarithm to the base 10 of a number is the power to which the number ten must be raised in order to equal that number. Thus, $\log_{10}(100) = 2$, since $10^2 = 100$.

[6] Skewness is defined below. However, the reader should turn to Figure 3-7 at this point to get a visual idea of skewness.

than the arithmetic mean and is a better measure of central tendency for distributions that are positively skewed, because the arithmetic mean is pulled up by the small number of extremely high values. We return to this point in the discussion of the median.

The harmonic mean is defined as the reciprocal of the arithmetic mean of the reciprocals of X_i:[7]

$$\mu_h = \frac{N}{\sum 1/X_i}$$

The harmonic mean of the numbers 2, 5, 10, and 15 is

$$\frac{4}{1/2 + 1/5 + 1/10 + 1/15} = 4 \times \frac{150}{130} = 4.62$$

The harmonic mean is rarely used in political science and need not be discussed here.

Computation of the Mean

Equation (2-1) is the definition of the arithmetic mean. It is seldom worthwhile to compute the mean using the definition unless there are only a few observations involved. More often, we want to compute the mean directly from a frequency distribution. In doing so, we can greatly simplify the computation by taking advantage of the facts that the sum of the deviations of the individual observations from the mean is zero and that the observations within any interval are likely to be distributed evenly around the midpoint of the interval. Let us consider the latter point first. In the per capita GNP data presented in Table 2-3, there are nine observations in the class of interval $450–$850. We might assume that they all fall at the midpoint of the class interval, or $650. In fact, by the array in Table 2-2, we see that this is not so. The actual distribution of observations in this interval is $467, $478, $490, $550, $572, $600, $670, $726, and $836. However, if we treat them as falling at the midpoint, we shall not be too far off, because the number of observations falling below are somewhat offset by those falling above the midpoint for all the classes involved. In this instance six observations fall below and three above. The sum of the difference of those falling below is −$743, and the sum of the difference of those falling above is $282. Hence, for this interval our assumption is somewhat off. The reader should determine for these data how much the assumption is off for the other class intervals in the frequency distribution.

---

[7] A reciprocal of a number is 1 divided by the number; for example, the reciprocal of 2 is 1/2.

When we make the assumption that all observations fall at the midpoint of each class interval, we can write the equation for the mean as

$$\mu = \frac{\sum f_i M_i}{N} \qquad (2\text{-}2)$$

where M_i = midpoint of the class intervals

$$N = \sum f_i$$

The frequency in each class is multiplied by the midpoint of the interval. This is summed and finally divided by N. This is illustrated in Table 2-7.

Table 2-7   Mean per Capita GNP of 60 Countries

Per capita GNP	M_i	f_i	$f_i M_i$
$50– 450	$ 250	44	$11,000
450– 850	650	9	5,850
850–1,250	1,050	3	3,150
1,250–1,650	1,450	2	2,900
1,650–2,050	1,850	1	1,850
2,050–2,450	2,250	0	0
2,450–2,850	2,650	0	0
2,850–3,250	3,050	1	3,050
		60	27,800

$$\mu = \frac{27,800}{60} = \$463$$

Note that treating all the observations in the interval $450 to $850 as if they fell at the midpoint gives a total of $5,850 for this interval. When we add the actual values for the countries in this interval from the array in Table 2-2, we get $5,389. The overestimation for this interval is due to more observations falling below the midpoint than above. We also overestimate for the first, third, and eighth intervals, but we underestimate slightly for the fourth and fifth intervals, somewhat offsetting the overestimation. It can be seen that it is not sufficiently compensated, because the mean computed directly from the data is $396. Generally, the more skewed the data, the more likely we shall be off in computing a mean with Equation (2-2). When a distribution is skewed like that in Figure 2-2 (the present data), most of the observations within each interval behave like the overall frequency distribution; that is, most countries fall into the lowest class interval. Similarly, within each interval, there is a tendency for most observations to fall toward the lower end of the class interval, rather than around the midpoint. Hence, a mean computed for data skewed this way is likely to be too large. For practical purposes the amount is small. However, if precision is desired, Equation (2-1) should be used to compute the arithmetic mean for skewed data.

We can shorten the arithmetic chores even further by taking advantage of the fact that $\sum(X_i - \mu) = 0$. Because this is so, we can estimate a mean and correct for under or overestimating. Let us illustrate first by a simple example. Consider the series of numbers 1, 2, 3, 4, 5. The mean is 3. However, assume that we had to estimate the mean and that we estimated it to be 2. The sum of the deviations from 2 is not equal to zero, as shown below:

X_i	$(X_i - 2)$
1	-1
2	0
3	1
4	2
5	$\dfrac{3}{5}$

By estimating the mean of these numbers to be 2, we underestimate the mean each time by $5/5 = 1$, or by $\sum(X_i - \mu')/N$, where μ' represents the estimated mean. Thus, to get the actual mean, we have to add to our estimated mean, 2, the amount of the average underestimation, 1, to get $\mu = 2 + 1 = 3 = \mu' + \sum(X_i - \mu')/N$. This formula can be used to compute a mean from ungrouped data directly, but the reason for taking advantage of the fact that the sum of deviations from the mean is zero is to make the computation with grouped data easier. In doing so, we again treat all observations as if they fell at the midpoint of the interval and take the deviations from an estimated *midpoint*. The equation for the mean can then be written

$$\mu = \mu' + \frac{\sum_{i=1}^{N}(f_i d_i)}{N} \tag{2-3}$$

Table 2-8 illustrates computation by this equation.

In column 4 in Table 2-8 we have taken *step deviations* from the midpoint of the interval in which the mean has been estimated to lie. The deviations are in units of 1 rather than absolute deviations from each midpoint in order to shorten the computation even further. Note that, because we take the deviations in unit amounts, the width of the class interval has to be put back in the equation for computing the mean:

$$\mu = \mu' + \frac{\sum_{i=1}^{N}(f_i d_i')}{N} i \tag{2-4}$$

where d_i' represents step or unit deviations, and i the width of the class interval. In both cases, the estimated mean μ' is taken as the midpoint of the interval in which the mean is estimated to lie. Both equations yield the same mean and the same as that found by Equation (2-2) except for rounding error.

Table 2-8   Mean per Capita GNP as Computed by Using an Estimated Mean and Step Deviations

Per capita GNP	(Mid-point) M_i	(Deviations from mid-point of interval containing guessed mean) d_i	(Step deviations) d_i	(Frequency) f_i	$f_i d_i$	$f_i d'_i$
$50– 450	250	−400	−1	44	−17,600	−44
450– 850	650	0	0	9	0	0
850–1,250	1,050	400	1	3	1,200	3
1,250–1,650	1,450	800	2	2	1,600	4
1,650–2,050	1,850	1,200	3	1	1,200	3
2,050–2,450	2,250	1,600	4	0	0	0
2,450–2,850	2,650	2,000	5	0	0	0
2,850–3,250	3,050	2,400	6	1	2,400	6
				60	−11,200	−28

$$\mu = \mu' + \frac{\Sigma\,(f_i d_i)}{N} \qquad\qquad \mu = \mu' + \frac{\Sigma\,(f_i d'_i)}{N}\,i$$

$$= 650 + \frac{-11,200}{60} \qquad\qquad = 650 + \frac{(-28)}{60}\,(400)$$

$$= 650 - 186.6 \qquad\qquad = 650 + \frac{-11,200}{60}$$

$$= \$463.40 \qquad\qquad\qquad = \$463.40$$

Sometimes grouped data have an open-ended interval, such as that in the income data in Table 2-6. If we know the total for this interval, it is possible to compute the mean from these data; otherwise it is not, unless we estimate the average in the interval. For the data in Table 2-6, we know that the total income for all families is $49,929,000,000. We also know how much of this total is earned by all families up to the interval $\geq$ $30,000. The latter amount is $(1,844,200 \times 2,500) + (2,552,600 \times 7,500) + (948,400 \times 12,500) + (427,400 \times 22,500) = \$45,226,500,000$. Subtracting this from the total for all families gives $4,702,500,000. The average in this interval is thus $4,702,500,000/82,000 = \$57,347$. We can use the latter as the midpoint value for this interval and Equation (2-2) for computing the mean.

The Median

In some circumstances the median is a better description of central tendency than the mean. This is so if there are extremely large values on one side of the frequency distribution, that is, when the distribution is skewed. The median is a positional measure that designates the observation that is exactly halfway between the smallest and largest observation in the series. It divides the distribution into two equal parts. For example, the median of the numbers 2, 3, 5, 9, and 126 is 5. The median is a better description of central tendency in this case, because the mean would be highly inflated on account of one observation, 126. This is generally true of data, such as income distribution, that have very high values, with most values clustered at the lower end of the frequency distribution. The mean family income in the United States for 1963 was $7,510, but the median was $6,140. The latter figure is a more accurate representation of the typical income, because the mean is pulled up by the very large values for a small percentage of families. The median tells us that exactly 50 percent of all families earned less than $6,140 a year and exactly 50 percent earned more.

Because the median is a positional measure, it is found for ungrouped data by finding the middle observation. If the number of cases is odd, the median is the middle observation; if the number of cases is even, the median is half the distance between the middle two cases. For example, the median of the numbers 4, 6, 10, and 15 is $(6 + 10)/2 = 8$. The median is unaffected by the extreme cases; the mean takes account of all values of the set.

For grouped data, the median can be found by finding the interval within which the middle observation lies and going into this interval as far as necessary to reach the median observation. The equation for grouped data is

$$Md = l + \frac{N/2 - F}{f} i \qquad (2\text{-}5)$$

Table 2-9   Median per Capita GNP for 60 Countries

Per capita GNP	M_i	f_i	F
$50– 450	250	44	44
450– 850	650	9	53
850–1,250	1,050	3	56
1,250–1,650	1,450	2	58
1,650–2,050	1,850	1	59
2,050–2,450	2,250	0	
2,450–2,850	2,650	0	
2,850–3,250	3,050	1	60

$$Md = l + \frac{N/2 - F}{f} i$$

$$= 50 + \frac{30 - 0}{44} (400)$$

$$= \$322$$

where l is the lower limit of the interval in which the median lies, N is the number of observations in the distribution, F is the cumulative frequency up to the interval containing the median, and i is the size of the interval. Equation (2-5) simply tells us to go as far as necessary into the interval in which the median lies to get to the median case. Hence, we subtract from the middle case $(N/2)$ the cumulative frequency up to the interval in which the median lies. This can be illustrated by using the same data we used for the mean (see Table 2-9). Because there are 60 observations, the median lies between observations 30 and 31, which is somewhere in the interval 50 to 450. There are 44 cases in this interval and none before it. We must therefore go thirty forty-fourths into the interval to find the median case. Because the interval is 400 units wide, we want to go 70 percent into it, or to about point 322. If we check the array in Table 2-2, we find that the median is actually 219.5. The overestimation is due to the skewed nature of the data. Note, however, that the median for these data is much lower than the mean, as is the case with all income data. In this case, the median is a better description of the typical value than the mean.

We can use Equation (2-5), with appropriate adjustments, to find other positional values, such as the first quartile (the point below which 25 percent of all observations lie), the first decile (the point below which 10 percent of all observations lie), and so on. The median is actually the second quartile or the fifth decile. For the first quartile we have

$$Q_1 = l + \frac{1/4N - F}{f} i$$

where Q_1 represents the first quartile and the other symbols are the same as before. The first quartile for the data in Table 2-9 is

$$Q_1 = 50 + \frac{1/4(60) - 0}{44} (400) = 186$$

The first decile for these data is

$$D_1 = 50 + \frac{1/10(60) - 0}{44} (400)$$

$$= 104$$

The Mode

The mode is the least used of central-tendency measures. It is defined as the most frequent observation, or, for grouped data, as the midpoint of the class interval with the largest frequency. The series of numbers 1, 3, 4, 6, and 8 does not have a single mode. The series of numbers 1, 3, 3, 4, 6, and 7 has a mode of 3. The series of numbers 1, 3, 4, 4, 5, 6, 8, 8, 9, and 10 has two modes, or is bimodal. The mode for the data in Table 2-9 is 250.

Comparison of the Mean, Median, and Mode

There is no single answer to the question of which measure of central tendency should be used to describe the typical value in a particular case. It depends on the objective of the description. The mean is an artificial concept, not an actual value (unless an actual observation happens to coincide with it). Because it is computed by using every item in a distribution, it reflects all the values. But it may be affected too much by extreme values. For skewed data, it is misleading as a typical value. The median is a positional measure and is unaffected by extreme values, but it is affected by the number of observations in the frequency distribution. The mode is simply the most common value. The median and the mode are actual values (except in the case of the median where there are an even number of observations). It might also be noted that which measure should be used depends on the level of measurement. For nominal scales, only the modal class can be designated. For ordinal data the mode and the median can be found, and for interval or ratio scales, all three measures are possible.

The mean is the most useful measure in inferential statistics, because it is computed by an equation that lends itself to further algebraic treatment. Therefore, there are conventional symbolic distinctions for the mean of a sample $\bar{X}$ and the mean of a population μ. But there are no similar distinctions for the median and the mode, because these are seldom used in inferential statistics.

The definition of the mean and the method of computing it are exactly the same for samples as for populations except that different symbols are used to indicate whether it is a sample or the population that is being considered. In this chapter, we have used the symbolism for populations. The mean for a sample is defined in the same way:

$$\bar{X} = \frac{\sum_{i=1}^{n} X_i}{n} \tag{2-6}$$

Lowercase n is used to designate the sample size, and $\bar{X}$ indicates that a sample is involved.

READINGS AND SELECTED REFERENCES

Blalock, Hubert, Jr. *Social Statistics*, McGraw-Hill Book Company, New York, 1960, Chaps. 4–5.

Games, Paul, and George Klare. *Elementary Statistics*, McGraw-Hill Book Company, New York, 1967, Chaps. 2–4.

Parl, Boris. *Basic Statistics*, Doubleday & Company, Inc., Garden City, N.Y., 1967, Chaps. 6–7.

Spur, William A., and Charles P. Bonini. *Statistical Analysis for Business Decisions*, Richard D. Irwin, Inc., Homewood, Ill., 1967, Chaps. 4–5.

Wyatt, Woodrow, and Charles Bridges. *Statistics for the Behavioral Sciences*, D. C. Heath and Company, Boston, 1967, Chap. 2.

Yamane, Taro. *Statistics: An Introductory Analysis*, Harper & Row, Publishers, New York, 1967, Chaps. 2–3.

3

Basic Concepts:
Measure of Dispersion

In some cases a complete description of an empirical frequency distribution requires that we take account of the dispersion, or heterogeneity, of observations as well as typical values. As an illustration, consider the economic support scores computed by the *Congressional Quarterly* for each Congress.[1] Suppose that the mean economic support score for one Congress immediately following a presidential election is 55 and that for the Congress two years later after the off-year election is 52. We might assume that the two Congresses are almost identical in support. However, suppose that the scores of the first Congress are all closely bunched around the mean but those of the second are more widely scattered. The practical consequences of this wider dispersion of support in the second Congress could be very great, although the mean support for both Congresses is almost the same.[2]

Consider a second hypothetical example. Suppose that we are measuring the amount of disagreement in several state legislatures concerning the scope of legitimate action for government. Suppose further that this is done on a set of 20 questions each of which has a 0 to 10 choice range. Because we are measuring amounts of disagreement here rather than an attribute or property, we can treat the responses as an interval scale. The maximum score for any one respondent is 200 and the minimum is 0, where 200 represents a completely favorable attitude toward governmental activity and 0 represents a completely unfavorable attitude. Finally, suppose that four legislators in each of two states are interviewed, and the resulting scores are indicated in Table 3-1. It can be said that the mean attitude in each legislature is almost the same, but the dispersion in legislature A is obviously much greater. We might conclude that there is much more disagreement in legislature A concerning the legitimate scope of government.

[1] The economic support score is a measure of how much support a Congress gives to the administration's requests and ranges from 0 to 100.

[2] Another important use of measures of dispersion is in inferential statistics, which we consider in later chapters. Here we can note that, as the amount of variation in a set of scores increases, the amount of error that is likely in estimating a measure of central tendency like the mean also increases.

Table 3-1   Hypothetical Attitude Scores

Representative	Legislature A	Legislature B
1	20	70
2	80	75
3	90	90
4	130	95
	320	330
	$\bar{X} = 80$	$\bar{X} = 82.5$

These hypothetical examples illustrate that dispersion may be important as a research question. For example, we might want to look for the causes of variation. It would be theoretically useful to try to explain why the attitudes are more heterogeneous or homogeneous in different states.

There are various numerical alternatives for measuring how much dispersion there is in a set of observations. The *range* is the difference between the highest and lowest score. In the example above, it is as follows: range of legislature A: $130 - 20 = 90$; range of legislature B: $95 - 70 = 25$. The range gives us a rough idea of the amount of dispersion in each and shows that the level of disagreement in A is greater than that in B. It is almost four times as much by this measure. But if the number in each sample were different, it would no longer be meaningful to compare ranges, because the range can be a function of the number of observations. For example, the ranges of the two series of numbers (2, 5, 7, 9) and (2, 5, 6, 7, 9, 10, 12, 15) are 7 and 13 respectively. But it is not meaningful to compare them, because they depend on the number of values in each case.

One way to correct for this is to take the range between the first and third quarter scores, thus omitting the extreme values. This is called the *interquartile deviation*. The equation is $Q_i = (Q_3 - Q_1)/2$, where Q_3 and Q_1 are the third and first quartiles respectively. The interquartile deviation gives the amount of dispersion in the middle half of observations and is less affected by extreme observations. To illustrate the interquartile deviation, consider the data in Table 2-2 again. The range for these data is $2,850. The interquartile deviation is $(478 - 99)/2 = $189.50. Because of the extreme value $2,900, the range gives a false impression of the amount of dispersion in these data. The interquartile deviation tells us that the range of the middle 50 percent of all observations is $189.50. This is much smaller than the range.

Both the range and interquartile deviation as measures of dispersion are not too useful except as rough measures. To get a more precise and easily interpreted measure of dispersion, we can use the following reasoning. We begin by taking a stable central point of the frequency distribution and measuring how far each score deviates from it. A good stable central point in this case is μ. We could not simply measure deviations from μ, because $\sum (X_i - \mu) = 0$. One way to avoid getting a zero result is to take the *absolute*[3] sum of the deviation of each score from its mean: $\sum |X_i - \mu|$. Thus, for the data on the two legislatures we get

Legislature A	Legislature B
20 − 80 = 60	70 − 82 = −12
80 − 80 = 0	75 − 82 = −7
90 − 80 = 10	90 − 82 = 8
130 − 80 = 50	95 − 82 = 13
120	40

Using this measure, it turns out that legislature A has three times as much disagreement as legislature B, compared with the four times obtained by using the range. However, because in many cases the number of observations may not be the same for each case, it is necessary to take N into consideration:

$$\text{M.D.} = \frac{\sum_{i=1}^{N} |X_i - \mu|}{N} \tag{3-1}$$

This is called the *mean deviation*.[4] It measures how much on the average each individual score varies from its own mean. It might be a sufficient measure of dispersion except that, because the absolute value of the difference has been used, the resulting figure could not be used in further algebraic manipulation. Consequently it is not used very much as a measure of dispersion. In order to avoid this problem, we can square the deviation of each score from its mean, thereby eliminating the negative signs, and divide this by the number of observations in the sample, giving the following:

$$\sigma^2 = \frac{\sum (X_i - \mu)^2}{N} \tag{3-2}$$

This measure is called the *variance*.

The variance gives us the average squared deviation of the scores from

---

[3] An operation enclosed in straight lines rather than parentheses tells us to ignore the signs when summing.

[4] We can also compute the mean deviation for a sample in the same way by

$$\text{M.D.} = \frac{\sum_{i=1}^{n} |X_i - \overline{X}|}{n}$$

their own mean and can be used to compare the amount of dispersion in two groups of scores. The data for the previous example gives

$$
\begin{array}{cc}
\text{Legislature } A & \text{Legislature } B \\
(X_i - \mu)^2 & (X_i - \mu)^2 \\
3{,}600 & 144 \\
0 & 49 \\
100 & 64 \\
\underline{2{,}500} & \underline{169} \\
6{,}200 & 426
\end{array}
$$

$$\sigma^2 = \frac{6{,}200}{4} = 1{,}550 \qquad\qquad \sigma^2 = \frac{426}{4} = 106.5$$

Now we might say that the amount of disagreement about the legitimate scope of government is about fourteen times as great in legislature A as in legislature B. The scale originally used to measure disagreement in this case had minimum and maximum values of 0 to 200. In order to get the result above back to its original scale — we squared the deviations to compute the variance — we simply extract the square root of the final product, arriving at the definition of the *standard deviation;*

$$\sigma = \sqrt{\frac{\sum (X_i - \mu)^2}{N}} \tag{3-3}$$

Completing the example of disagreement within two state legislatures, we get 39.37 and 10.32 respectively. We finally arrive at the conclusion that the amount of disagreement in legislature A is almost four times that in legislature B. The standard deviation might be interpreted as the square root of the mean amount of deviation of each score from its own mean, or the *root mean deviation*, as it is sometimes called.

COMPUTATION OF VARIANCE AND STANDARD DEVIATION

Equations (3-2) and (3-3) define the variance and the standard deviation and could be used to compute these measures of dispersion. However, if the number of observations is large or if we are working with grouped data, easier methods of computing the variance and the standard deviation are available. Several alternative equations can be derived with simple algebra:

$$\sigma^2 = \frac{\sum (X_i - \mu)^2}{N}$$

$$= \frac{\sum (X_i^2 - 2\mu X_i + \mu^2)}{N}$$ by carrying out the multiplication of $(X_i - \mu)^2$

$$= \frac{\sum X_i^2}{N} - 2\mu \frac{\sum X_i}{N} + \frac{N\mu^2}{N}$$ by rules 1 and 2 of summation

$$= \frac{\sum X_i^2}{N} - 2\mu(\mu) + \mu^2$$ by substitution and canceling

$$= \frac{\sum X_i^2}{N} - \mu^2$$ since we are subtracting two squared means and adding one

Therefore

$$\sigma^2 = \frac{\sum X_i^2}{N} - \mu^2 \qquad (3\text{-}4)$$

and the standard deviation is

$$\sigma = \sqrt{\frac{\sum X_i^2}{N} - \mu^2} \qquad (3\text{-}5)$$

Additional equations can be derived with simple algebra as follows:

$$\sigma = \sqrt{\frac{\sum X_i^2}{N} - \mu^2}$$

$$= \sqrt{\frac{\sum X_i^2}{N} - \left(\frac{\sum X_i}{N}\right)^2}$$ substituting the definition of μ

$$\qquad (3\text{-}6)$$

$$= \sqrt{\frac{\sum X_i^2 - (\sum X_i)^2/N}{N}}$$ multiplying and dividing by N (3-7)

$$= \frac{\sqrt{\sum X_i^2 - (\sum X_i)^2/N}}{\sqrt{N}}$$ by the division property of (3-8) radicals

$$= \frac{1}{N}\sqrt{N\sum X_i^2 - (\sum X_i)^2}$$ multiplying by N and taking (3-9) the square root of $\dfrac{1}{\sqrt{N^2}}$

Any of Equations (3-5) to (3-9) give the same standard deviation and, removing the radical, give the variance. The last (3-9) is obviously the simplest, because it involves finding only two quantities: $\sum X_i^2$ and $(\sum X_i)^2$.

For grouped data we can follow the same reasoning except that we substitute $\sum f_i d_i^2$ for $\sum (X_i - \mu)^2$, where d_i^2 is the squared deviation from the midpoint of the class interval with the estimated mean (or $d_i'^2$ can be used if we take step deviations). The following three equations for grouped data are derived in the same manner as their equivalents for ungrouped data. Each is mathematically equivalent:[5]

$$\sigma = i\sqrt{\frac{\sum f_i d_i'^2}{N} - \left(\frac{\sum f_i d_i'}{N}\right)^2} \tag{3-10}$$

$$= i\sqrt{\frac{\sum f_i d_i'^2 - \dfrac{(\sum f_i d_i')^2}{N}}{N}} \tag{3-11}$$

$$= \frac{i}{N}\sqrt{N\sum f_i d_i'^2 - (\sum f_i d_i')^2} \tag{3-12}$$

The computation of the standard deviations for grouped data, using Equation (3-12), is illustrated in Table 3-2.

Table 3-2   Standard Deviation of per Capita GNP

Per capita GNP	f_i	d_i'	$f_i d_i'$	$f_i d_i'^2$
$50– 450	44	−1	−44	44
450– 850	9	0	0	0
850–1,250	3	1	3	3
1,250–1,650	2	2	4	8
1,650–2,050	1	3	3	9
2,050–2,450	0	4	0	0
2,450–2,850	0	5	0	0
2,850–3,250	1	6	6	36
	60		−28	100

$$\sigma = \frac{i}{N}\sqrt{N\sum f_i d_i'^2 - (\sum f_i d_i')^2}$$

$$= \frac{400}{60}\sqrt{60(100) - (-28)^2}$$

$$= \frac{400}{60}\sqrt{5216}$$

$$= \frac{400}{60}(72.2)$$

$$= 480.9$$

[5] If step deviations cannot be used for grouped data because the intervals are of unequal width, then the i in these formulas changes to 1 and d' to d. In this case, we take deviations from the midpoint of the interval in which we estimate the mean to lie in actual amounts rather than in steps.

In a distribution that is symmetrical, that is, not skewed to the right or left, there is a constant relationship among the standard deviation, the mean deviation, and the interquartile deviation. The interquartile deviation is approximately $4/5\sigma$, and the mean deviation approximately $2/3\sigma$. These facts are useful for estimating σ when we do not know it and for checking the accuracy of a calculated σ.

The former point is emphasized in Chapter 5. In statistical inference we sometimes need to estimate σ in order to decide how large a sample is needed for a given level of error. No other information about the population values may be available than its range and an idea of the interquartile deviation. If we are dealing with data that are normally distributed, we can estimate σ by using the fact that

$$Q_i = 4/5\sigma$$

$$5/4Q_i \doteq \sigma$$

Thus we can take $5/4$ of the interquartile deviation as an estimate of σ. This is covered in Chapter 5.

Sample and Population Dispersion

Thus far we have considered the definition and methods of computing the variance and the standard deviations for populations. For samples, a slight change must be made. The quantity used in the denominator is $n - 1$ rather than n, because s consistently underestimates σ, especially for small samples. Dividing by $n - 1$ helps correct for this. We consider this in more detail in Chapters 8 and 11. The equation for the standard deviation of a sample, therefore, is

$$s = \sqrt{\frac{\sum_{i=1}^{n} (X_i - \bar{X})^2}{n - 1}} \tag{3-13}$$

and for the variance we simply remove the radical.

For example, suppose that the data in Table 3-2 are a sample of $n = 60$ rather than the population. When we compute s using the correction factor $n - 1$ in the denominator, we get $s = 489.5$ in contrast to $\sigma = 480.9$. If the sample is large, the correction factor will make almost no difference and can safely be ignored.

Variation and Sum of Squares

Although not useful by itself, two additional aspects of variance should be considered here because of their importance in measures of association and more advanced statistics. The first of these, called *variation*, is simply the sum of the deviations of the observations from the mean, that is, $\sum_{i=1}^{n} (X_i - \bar{X})$. This, as we said above, is equal to zero when standing by itself, but when multiplied by another series of values, such as the deviations of another variable from its mean, it becomes a very useful measure; that is,

$$\sum_{i=1}^{n} (X_i - \bar{X})(Y_i - \bar{Y})$$

is a part of the measure of correlation, as we shall see in Chapter 10.

We also note at this point that it is conventional to designate $\sum_{i=1}^{n} (X_i - \bar{X})$ by shorthand notation as $\sum_{i=1}^{n} x$, where the lowercase x is simply used in place of the longer statement. We follow this convention in this book. Lowercase x means $(X_i - \bar{X})$ wherever it appears. Similarly, the sum of the squared deviations from the mean $\sum_{i=1}^{n} (X_i - \bar{X})^2$ may be written $\sum_{i=1}^{n} x^2$, and it means the same thing. It is conventionally called the *sum of squares* and is used frequently in various statistical procedures. Because of its frequent use, short-form computing equations are useful and can be derived in a similar manner to those for variance and standard deviation given in Equations (3-5) to (3-9). We show a few here:

$$\sum x^2 = \sum (X_i - \bar{X})^2$$
$$= \sum X_i^2 - \sum \bar{X}^2$$
$$= \sum X_i^2 - n\bar{X}^2$$
$$= \sum X_i^2 - \frac{(\sum X_i)^2}{n}$$

The last two provide the simplest computational equations for $\sum x^2$. The same holds for populations, and we could thus substitute μ and N for $\bar{X}$ and n in each equation.

Using the shorthand notation for the sum of squares, we note that the variance may be written $\sigma^2 = \sum x^2/N$; that is, it is the mean of the squared deviations. Like the mean, therefore, it is affected by large values of x^2 and is large for large values of x^2. Hence, if one observation is added to a frequency distribution that is an extreme score, x^2 will be increased, because we take the square of this deviation, although at the same time N increases by only 1. However, if an observation that is close to the mean is added, the variance will be reduced. It also follows that, if an observation is added whose squared deviation from the mean is close to the average squared deviations of other observations, the variance will not be affected much.

Relative Dispersion: The Coefficient of Variation

The standard deviation measures the absolute dispersion of a distribution in terms of the units of the original data. But if we wanted to compare two distributions measured in different units, for example, one in dollars and the other in years, or two distributions measured in the same units but with different means, we could not use σ. To get a measure of the relative variation we take the standard deviation of a distribution as a proportion of its own mean and multiply by 100 to get a percentage, giving

$$V = 100\,\frac{\sigma}{\mu} \tag{3-14}$$

which is called the *coefficient of variation*.

As an illustration, suppose that the mean age of United States senators is 55, with a standard deviation of 5, and the mean age of United States representatives is 40, with a standard deviation of 4. It appears that there is more dispersion in the former than the latter. However, the coefficient of variation for each is

$$V_1 = 100 \times 5/55 = 9 \text{ percent}$$
$$V_2 = 100 \times 4/40 = 10 \text{ percent}$$

which shows that there is slightly more dispersion or variation in the second series of data than the first, although the standard deviation of the first is greater.

SKEWNESS AND KURTOSIS

Generally speaking, if we have described a set of data by casting it in a frequency distribution, plotting it on a chart, indicating proportions or percentages, and giving the mean, median, and standard deviations where appropriate, we have completed the descriptive work, provided that we are not looking for relationships. We might go on to describe the skewness and how peaked the distribution is, but normally these two measures are more useful for describing probability distributions and for comparing empirical frequency distributions with probability distributions.

Skewness

In a perfectly symmetrical distribution (the full properties and importance of the symmetrical distribution are considered below) there is one mode; the mean, median, and mode all fall at exactly the same place; and one half of the distribution is the mirror image of the other half (see Figure 3-1). Skewness

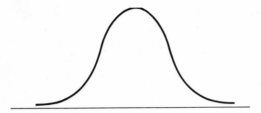

Figure 3-1. Symmetrical distribution.

is defined in terms of deviations from perfect symmetry. If the distribution bunches to the left of center, the mean will be larger than the median and mode. Such a distribution is said to be positively skewed, defined by the direction of the smaller tail. If the distribution bunches to the right, the mean will be smaller than the mode, and the distribution is said to be negatively skewed (see Figure 3-2). Given these facts, it is apparent that we may measure

positive skewness   negative skewness

Figure 3-2. Skewness.

skewness by taking the difference between the mean and mode and dividing by s in order to get a relative rather than an absolute measure; that is, $SK = (\bar{X} - Mo)/\sigma$. If the resulting figure is positive, the distribution is positively skewed; if negative, the distribution is negatively skewed. For the perfectly symmetrical distribution $\bar{X} - Mo = 0$. The upper limit for positive skewness is 3, and the lower limit for negative skewness is -3.[6]

A more accurate measure of skewness can be made in terms of *moments* of a distribution. The concept of moment is important in statistics, but a precise definition would require more advanced mathematics than assumed for this book. A moment is usually defined as the arithmetic mean of the power of squared deviations of X_i from the mean. Let us symbolize moments

---

[6] We may also measure skewness by $SK = [3(\bar{X} - Md)]/\sigma$. This has the same upper and lower limits as the definition just given.

as M_1, M_2, ..., M_i, representing the first, second, and ith moment respectively. The first three moments of a distribution are

$$M_1 = \frac{\sum (X_i - \bar{X})}{N}$$

$$M_2 = \frac{\sum (X_i - \bar{X})^2}{N}$$

$$M_3 = \frac{\sum (X_i - \bar{X})^3}{N}$$

Let α_3 (alpha three) represent skewness. We can measure it by the following:

$$\alpha_3 = \frac{M_3}{\sigma^3} = \frac{(1/N) \sum_{i=1}^{N} (X_i - \bar{X})^3}{\sigma^3}$$

In this equation, the mean of the cubed deviations is divided by the third power of the standard deviation. If $\alpha_3 = 0$, the distribution is symmetrical; if $\alpha_3 > 0$, the distribution is positively skewed; and if $\alpha_3 < 0$, the distribution is negatively skewed.

Kurtosis

Kurtosis is defined in terms of how much of the total area under the curve lies within plus or minus one, two, and three standard deviations from the mean. This also defines the characteristics of the normal curve, something that is discussed in Chapter 4, because its importance pertains to probability distributions. For the present, we simply state that, when a distribution is perfectly normal, it is said to be *mesokurtic*; when it is peaked, it is said to be *leptokurtic*; and when it is flat, it is said to be *platykurtic* (see Figure 3-3).

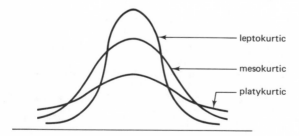

Figure 3-3. Kurtosis of frequency distribution.

If we let α_4 represent kurtosis, we can measure it in terms of moments by the equation

$$\alpha_4 = \frac{M_4}{\sigma^4} = \frac{(1/N) \sum_{i=1}^{N} (X_i - \bar{X})^4}{\sigma^4}$$

If the distribution is normal, $\alpha_4 = 3$; if it is leptokurtic, $\alpha_4 > 3$; and if it is platykurtic, $\alpha_4 < 3$.

The concepts discussed in this chapter and Chapter 2 are the basic instruments for describing singular characteristics of data. There are, in addition, many statistical concepts for describing relationships and trends in data. However, rather than discussing them all as descriptive tools and then turning to the question of how they may be used to make inferences to populations, it is better first to introduce the basic ideas needed for making statistical inferences. Later in the book we note that many of the measures described there concerning relationships between variables can be used either in a strictly descriptive manner or in an inferential manner if certain assumptions are met.

Generally, in research, we first describe data in terms of frequency distributions, curves, percentages, measures of central tendency, and measures of dispersion, for which the concepts discussed in this chapter and the last are sufficient. Most often, however, we wish to go beyond the particular data we have and to say something about a population based on a sample. To do so, we need to use the mathematical theory of probability and its applications in various sampling distributions. It is to this that we now turn.

READINGS AND SELECTED REFERENCES

Blalock, Hubert, Jr. *Social Statistics*, McGraw-Hill Book Company, New York, 1960, Chap. 6.

Freund, John. *Modern Elementary Statistics*, Prentice-Hall, Englewood Cliffs, N.J., 1960, Chaps. 4–5.

Games, Paul, and George Klare. *Elementary Statistics*, McGraw-Hill Book Company, New York, 1967, Chap. 5.

Hammond Kenneth, and James Householder. *Introduction to the Statistical Method*, Alfred A. Knopf, Inc., New York, 1963, Chap. 4.

Parl, Boris. *Basic Statistics*, Doubleday & Company, Inc., Garden City, N.Y., 1967, Chaps. 8–9.

Spurr, William A., and Charles P. Bonini. *Statistical Analysis for Business Decisions*, Richard D. Irwin, Inc., Homewood, Ill., 1967, Chap. 6.

4

Probability Theory and Statistical Inference

Statistics is useful not only in describing certain characteristics of data economically but also in saying something about unknown population parameters on the basis of known statistics. Statistics such as the mean or variance are computed from samples drawn from a population and are used as estimates of the same characteristics of the population. The mathematics of probability makes it possible to make this inferential leap. With probability theory, *sampling distributions*, or hypothetical frequency distributions, are used to calculate the probability of error when something about populations is inferred on the basis of a sample. The concept of sampling distributions is central to inferential statistics. Chapter 5 is devoted entirely to a discussion of three kinds of sampling distributions; in the present chapter it is necessary to discuss the concept generally in order to introduce probability theory and its function in applied statistics.

Statistical inference deals with three kinds of distributions: (*a*) the population frequency distribution, which is unknown or assumed to be unknown; (*b*) the empirical frequency distribution, obtained by drawing a probability sample; and (*c*) the sampling distribution of the particular statistic we want to say something about. A simple example illustrates these points; it is developed in subsequent chapters. Suppose that we want to learn the mean income of all present state legislators (a finite and limited universe). The unknown population parameter in this case is μ. Suppose that we are not able to ask all legislators their income, so that we draw a random[1] sample of 50 from this universe. Suppose that the sample mean $\bar{X}$ is $11,500. If the reporting is completely honest and accurate, may we say that the population parameter is $11,500? We may, but this is likely to be wrong; if we draw another sample from the same population, the mean for the second sample is not likely to be $11,500 and may be as little as $9,000 or as much as $13,000. If we continue to draw samples of the same size from this population many times, we observe that each time the mean is a little different. Thus, if we assert on the first sample that the parameter value is $11,500, we run the risk of being wrong. In order to measure by how much

[1] Random sampling is defined in general terms in Chapter 5 and in detail in Chapter 15.

we are likely to be wrong and the probability of being wrong, we construct a sampling, or probability, distribution of all the possible means we may get by repeating the sampling process many times; that is, the sampling distribution of means is the frequency distribution obtained if we repeat the process of drawing samples of the same size, computing the mean each time and then casting the means in the form of a frequency distribution like those discussed in Chapters 2 and 3. This probability distribution is used to answer questions about how far we are likely to be off in saying something about μ on the basis of $\bar{X}$ and the risk we run of being wrong. Actually, as we shall see in Chapter 5, it is not necessary to construct a sampling distribution, because general theorems and the labor of others save us from this task in most instances. But if we have to construct a distribution or if we want to understand what is done when using a distribution constructed by others, it is necessary to understand something about elementary probability theory.

ELEMENTARY PROBABILITY THEORY

There are two main schools of thought about probability: the subjectivist and the objectivist. The subjectivist school looks at probability from the point of view of personal belief that an event[2] will occur. The objectivist considers probability in terms of the relative frequency of occurrence of events or expected relative frequencies. In the present chapter we consider the objectivist definition of probability; we return to the subjectivist definition in Chapter 14.

The relative-frequency definition of probability states that the laws of probability apply only to events that can be repeated many times under relatively similar conditions. Two examples follow. A machine that punches out bottle caps produces 10 defective caps out of 1,000; the probability of its producing a defective cap is $10/1,000 = 0.01$. Of 100,000 nineteen-year-olds, 754 died during their nineteenth year; the probability of dying at the age of nineteen is $754/100,000$, or 0.00754. These examples illustrate the *statistical* or relative-frequency definition of probability. We shall see in a moment that this definition can be distinguished from the *mathematical* one, which is a priori rather than a posteriori. Let us note first that, given the statistical definition, nothing can be said about the probability of events such as whether the next president of the United States will be a Republican or a Democrat. Nor can anything be said about the probability of a particular bill's being passed by Congress. These events cannot be repeated many times under similar conditions.

In order to state the mathematical or a priori definition of probability, we must first introduce the ideas of events and sample space. If we toss a

---

[2] The word "event" has a precise meaning in statistics. It is considered below.

die and it turns up 4, this is called a *simple event*. If we toss two dice and a 3 turns up on one and a 4 on the other, the result (3,4) is called a *compound event*. A compound event can be broken down into simple events. In measurements taken of people, such as their beliefs, income, age, and so on, each observation for each person is a simple event. For example, the observation that person *A* has an income of $4,200 is an event. Hence, we may use the words "observation" and "event" interchangeably. Similarly, the outcomes of an experiment are events. If we time how long it takes a group of three students to solve a problem, the outcome is an event. We may therefore use the words "outcome," "observation," and "event" interchangeably. If we list all the possible outcomes of a particular experiment (or all possible events), we have a *sample space*. It may also be called a *set*. For example, if we toss a coin once, it may turn up *H* or *T*. Each is a simple event. The sample space, or set, is (*H*,*T*). If we toss the same coin twice, it can turn up (*H*,*H*;*H*,*T*;*T*,*H*;*T*,*T*), that is, heads on the first toss and heads on the second; heads on the first toss and tails on the second, and so on. The four compound events comprise the sample space. Each compound event consists of two simple events. A particular *sample point* might be the compound event (*H*,*H*). The sample space in this case is finite. There is a limited number of possible outcomes. If we listed all the possible weights of a group of 20 students, we should have an infinite sample space, because there is an infinite number of possible weights that may be listed even if we restrict the range to 100 to 300 pounds, because there is an infinite number of fractions between any two whole numbers.

Sample spaces may be represented in set notation, as Venn or Euler diagrams, or as cartesian coordinates. In set notation events are listed in braces, as below. A Venn or Euler diagram is the representation of sample spaces and events in the form of circles and rectangles, as in Figures 4-2 and 4-3. A set of cartesian coordinates is the representation of events as intersecting points plotted on two coordinate axes (straight lines intersecting at right angles), as in Figure 4-4 .

Consider a set consisting of two white balls and three black balls in a box. The set can be written

$$I = \{W_1, W_2, B_1, B_2, B_3\}$$

Suppose now that we gamble with these balls. The gamble is a wager about what is likely to occur if one of the balls is drawn from the box at random. We know that there are five possible outcomes, represented as $\{W_1, W_2, B_1, B_2, B_3\}$; that is, we might draw the first white ball, the second white ball, the first black ball, and so on. Each of these events is equally likely to occur if we draw at random and the balls are all the same size. The five possible outcomes are the sample space. Let $n(I)$ represent this universe of possible outcomes. The number of elements in this sample space is $n(I) = 5$. Suppose that the wager is whether a white ball will be drawn on a single draw. This outcome of the experiment in which a particular characteristic may be present is an event, denoted by A. The event A is a subset of the universe I. The number of elements in the event A is denoted by $n(A)$. In the particular wager we are concerned with, the number of elements in the event A, that is, the ball drawn has the characteristic of white, is $n(A) = 2$. Finally, the probability of event A is denoted by $p(A)$. The probability of event A can now be defined:

$$p(A) = \frac{n(A)}{n(I)} \tag{4-1}$$

that is, the probability of event A is equal to the number of points in event A divided by the number of points in the sample space. For A equals drawing a white ball from this box, the probability is

$$p(A) = \frac{n(A)}{n(I)} = \frac{2}{5}$$

The chance of drawing a white ball from the box at random is $2/5$. Hence, if we make a wager concerning this event and bet that it will occur, the person betting that it will not occur has a better chance of winning, because it is more likely that a white ball will not be drawn. Because the probability of drawing any ball from the box in one draw is 1, that of not drawing a white ball, denoted by $p(A')$, is $1 - p(A) = 1 - 2/5 = 3/5$. To make this bet fair, the person betting on event A' should give odds. The odds are the ratio of event A' to event A, which is $3:2$ in this case. The person betting on event A' should put up \$3 to the \$2 put up by the person betting on event A. We shall return to these questions later; at present, let us consider some of the important characteristics of this mathematical definition of probability.

The mathematical definition of probability does not say that, if we draw five balls from the box, replacing the one drawn each time and shaking the box thoroughly before drawing again, exactly two balls will be white and three black. It says that in an infinite number of draws the proportion of white balls approaches $2/5$ and that of black balls $3/5$. Thus, for 100 draws, the expected number of white balls is $2/5 \times 100 = 40$. But an actual experiment does not yield exactly this result. If we perform the experiment of drawing 100 balls at random from the box, we can get 38 white and 62 black balls, or 45 white and 55 black, and so on. The mathematical definition of probability is a statement of what to expect under ideal conditions in the long run, or in an infinite number of trials, not of what will actually occur. Thus, we can distinguish the statistical definition from the mathematical definition of probability. The statistical definition states that the probability of event A is equal to the relative proportion of occurrences of the event to the total number of trials. It can be defined as $p(A) = m/n$, where m

is the number of outcomes of event A and n the total number of outcomes.[3] The mathematical definition is a priori. It is what may be expected in the long run if the sample space has the characteristics defined and if the sampling process is perfectly random. The two definitions do come together; that is, the statistical definition leads us to expect that in the long run, if we could observe an infinite number of events,

$$p(A) = \frac{m}{n} \rightarrow \frac{n(A)}{n(I)} \qquad \text{as } n \rightarrow \infty$$

where $\rightarrow$ means "approaches." For example, assume that we toss a coin 10 times. We may get 8 heads and 2 tails, or a proportion of heads of $8/10 = 0.8$. If we toss the same coin 100 times, we may get 60 heads and 40 tails, or a proportion of heads of 0.6. As we increase the number of tosses, the proportion of heads approaches 0.5, although it may never be exactly equal to this. This is shown in Figure 4-1. The expected long-run consequence is

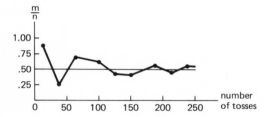

Figure 4-1.

not something that can be practically proved each time probability is used. Instead, the mathematical models of probability are used to solve probability questions without having to carry out repeated experiments.

$P(A)$ is called a *probability function* with a range of $0 \geq p(A) \leq 1$. If there are no points in the sample space, the probability of the event is 0. For example, the probability of drawing a green ball from the box in the example above is

$$p(A) = 0/5 = 0$$

because there are no green balls in the box. The probability of drawing a ball from the box regardless of color, therefore, is

$$p(A) = 5/5 = 1$$

because the sample space contains nothing but balls.

[3] Note that probability defined as relative frequency is the same as that which was defined in Chapter 2 as a proportion.

Properties of Probability

Given the definition of probability in Equation (4-1), we can deduce several properties of probability. The following notation will be used in this discussion (see Figure 4-2):

$$A' = \text{compliment of } A, \text{ that is, the event ``not } A\text{''}$$
$$A \cap B = \text{intersection of events } A \text{ and } B$$
$$A \cup B = \text{union of events } A \text{ or } B$$

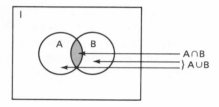

Figure 4-2.

Because we have defined the probability of event A as

$$p(A) = \frac{n(A)}{n(I)}$$

and because $n(A)$ must be $\leq n(I)$, it follows that $p(A') = 1 - p(A)$. The proof of this is simple. By definition,

$$n(A') = n(I) - n(A)$$

dividing each member by $n(I)$,

$$\frac{n(A')}{n(I)} = \frac{n(I)}{n(I)} - \frac{n(A)}{n(I)}$$

substituting definitions, we have

$$p(A') = 1 - p(A)$$

The definition of probability for compound events is a straightforward extension of that for simple ones. There are two different ways to consider a compound event: (*a*) either one or the other event may occur, and we speak of the probability of A or B, or $p(A \cup B)$ in set notation; (*b*) both events may occur, and we speak of the probability of A and B, or $p(A \cap B)$. As an illustration, the following data from Prothro and Matthews[4] will be

---

[4] See Donald R. Matthews and James W. Prothro, "The Concept of Party Image and Its Importance for the Southern Electorate," in M. Kent Jennings and L. Harmon Zeigler, *The Electoral Process*, Prentice-Hall, Inc., Englewood Cliffs, N. J., 1966.

used (the numbers have been rounded for arithmetic clarity):

$$I = 1{,}312 \text{ Southern voters}$$
$$A = 618 \text{ Negroes}$$
$$B = 694 \text{ whites}$$
$$C = 374 \text{ strong Democrats}$$
$$A \cap C = 173 \text{ Negroes who are strong Democrats}$$
$$B \cap C = 201 \text{ whites who are strong Democrats}$$
$$A \cap B = 0$$

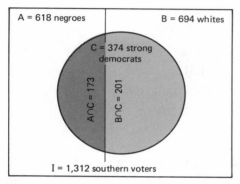

Figure 4-3.

This may be represented in diagrammatic form as in Figure 4-3. Assume that we draw a person at random from this population. The probabilities of a few simple events are

$$p(A) = \frac{n(A)}{n(I)} = \frac{618}{1{,}312} \doteq 0.47$$

$$p(B) = \frac{n(B)}{n(I)} = \frac{694}{1{,}312} \doteq 0.53$$

$$p(C) = \frac{n(C)}{n(I)} = \frac{374}{1{,}312} \doteq 0.28$$

The probability that the person selected is a Negro is 0.47; the probability that he is a white is 0.53; and the probability that he is a strong Democrat is 0.28. Let us consider the compound event A or C, the probability that a person drawn from I is either a Negro or a strong Democrat. This may be stated in the form of the following theorem:

$$p(A \cup C) = p(A) + p(C) - p(A \cap C) \tag{4-2}$$

The probability of A or C occurring is equal to the sum of the probabilities of A plus C minus the probability of both A and C. We prove this theorem first with simple algebra and then in terms of the specific example:

$$n(A \cup C) = n(A) + n(C) - n(A \cap C) \qquad \text{by definition}$$

$$\frac{n(A \cup C)}{n(I)} = \frac{n(A)}{n(I)} + \frac{n(C)}{n(I)} - \frac{n(A \cap C)}{n(I)} \qquad \text{dividing by } n(I)$$

therefore

$$p(A \cup C) = p(A) + p(C) - p(A \cap C)$$

Now, using the data in the specific example,

$$n(A \cup C) = 618 + 374 - 173$$

$$\frac{819}{1,312} = \frac{618}{1,312} + \frac{374}{1,312} - \frac{173}{1,312}$$

$$.62 = .47 + .28 - .13$$

therefore

$$p(A \cup C) = .62$$

The probability that one person selected at random from this population is a Negro or a strong Democrat is 0.62. This is also the proportion of the total 1,312 voters who are either Negro or strong Democrats. To find this proportion, we subtracted the intersection of events A and C, because some of the individuals were both Negro and strong Democrats and, if they were not subtracted, we should count them twice. However, if the events are mutually exclusive, such as events A and B in Figure 4-3, then $p(A \cup B) = p(A) + p(B)$. This is because $(A \cap B) = 0$ and there is no need to subtract it. In this example, the probability of drawing either a Negro or white voter at random is

$$p(A \cup B) = 0.47 + 0.53$$

$$= 1.00$$

Consider next the probability of both events occurring, the probability of A and C, or $p(A \cap C)$. To find this probability, we use the following:

$$p(A \cap C) = p(A)p(C|A) \tag{4-3}$$

Equation (4-3) states that the probability of events A and C occurring is equal to the probability of A times the probability of C given that A occurs. The latter is called the *conditional* probability of C given that A occurs and is defined in the following manner:

$$p(C|A) = \frac{n(C \cap A)}{n(A)} \tag{4-4}$$

The probability of C given A is equal to the number of elements in the intersection of C and A divided by the number of elements in A. Note that, if there are no elements in the intersection of C and A, the two events are independent, and $p(C|A) = p(C)$. Notice also that $(C|A)$ is not a symbol for a set. If $n(C \cap A) = 0$, $p(C|A) = p(C)$ rather than 0. For example, what is the probability that a person selected at random from the sample space of 1,312 Southern voters is a strong Democrat given that he is a Negro? There are 374 strong Democrats in the sample, of which 173 are Negroes, or $n(C \cap A) = 173$. It is obvious that the proportion of Negroes who are strong Democrats (or the probability of a person being a strong Democrat given that he is a Negro) is $173/618 \doteq 0.28$. Thus, $p(C|A) = 0.28 = n(C \cap A)/n(A) = 173/618$. Let us return now to the joint event, the probability of A and C. What is the probability that a person selected at random is both a Negro and a strong Democrat? According to Equation (4-3),

$$p(A \cap C) = p(A)p(C|A)$$
$$= (0.47)(0.28)$$
$$\doteq 0.13$$

We might look at this question in another manner. The probability of a person's being both a Negro and a strong Democrat is, by definition,

$$\frac{\text{Number in event}}{\text{Number in sample space}}$$

In this case it is $173/1,312 = 0.13$, or the same as the previous answer. This also says that 13 percent of the total sample are Negroes who also are strong Democrats. The multiplicative property of probability can also be proved easily by means of simple algebra:

$$p(A \cap C) = p(A)p(C|A)$$

$$= \frac{n(A \cap C)}{n(I)} \qquad \text{by definition}$$

$$= \left(\frac{n(A)}{n(I)}\right)\left(\frac{n(A \cap C)}{n(A)}\right) \qquad \begin{array}{l}\text{multiplying numerator and}\\\text{denominator by } n(A)\end{array}$$

therefore

$$p(A \cap C) = p(A)p(C|A)$$

In this example, the two events "Negro voter" and "strong Democrat" are dependent. Whether a person selected at random is a strong Democrat or not depends on whether he is a Negro or white voter. If the two events are independent, then Equation (4-3) reduces to $p(A \cap B) = p(A)p(B)$. In this example, the probability of drawing a Negro voter and a white voter in two draws from the universe with replacement of the first person drawn is

$$p(A \cap B) = p(A)p(B) = (0.47)(0.53) \doteq 0.25$$

However, if the first person drawn is not replaced after he is drawn, then the two events are dependent, and

$$p(A \cap B) = p(A)p(B|A) = 618/1{,}312 \times 694/1{,}311 \doteq 0.2493$$

also

$$p(A \cap A) = p(A)p(A|A) = 694/1{,}312 \times 693/1{,}311 \doteq 0.2796$$

We may conclude that, if $p(A|B) = p(A)$, the two events are statistically independent. Let us illustrate this point. Suppose that we have a sample of 100 persons and place them in categories in accordance with whether they are male or female and whether they voted or did not in the previous election, as in Table 4-1. If we select one person at random from this group, what is

Table 4-1   Voting Behavior of Men and Women

	Men (M)	Women (W)	Total
Voted (V)	50	20	70
Did not vote (V')	10	20	30
Total	60	40	100

the probability that he voted? $P(V) = n(V)/n(I) = 0.70$. What is the probability that he voted given that he is a man; that is, what is $p(V|M)$? $P(V|M) = n(V \cap M)/n(M) = .83\ 1/3$. Hence, $p(V|M) \neq p(V)$, and we conclude that the two events are statistically dependent. Knowledge of one event helps predict the other.

Given this definition of statistical independence, we say that, if we sample from a universe with replacement, the two events are statistically independent; if we sample without replacement, the two events are statistically dependent. However, if the universe is large, for all practical purposes sampling without replacement is the same, and the probability of the second event is not dependent on the first.

The additive and multiplicative properties of probability can be made clearer by illustrations. The following examples illustrate several characteristics.

Example. What is the probability of drawing an ace and a king on two successive draws from an ordinary playing deck if the first card is replaced after it is drawn? Let A represent "draw an ace" and B represent "draw a king":

$$p(A) = \frac{n(A)}{n(I)} = \frac{4}{52} = \frac{1}{13}$$

$$p(B) = \frac{n(B)}{n(I)} = \frac{4}{52} = \frac{1}{13}$$

Because the first card is replaced after being drawn, $p(B|A) = p(B)$; therefore

$$p(A \cap B) = p(A)p(B) = 1/13 \times 1/13 = 1/169 \doteq 0.0059$$

or only once in 169 draws does such an event occur in the long run.

Example. The same problem as in the first example, but now the first card is not replaced after being drawn. In this case, $p(B|A) \neq p(B)$, because there is one card less in the deck after the first card is drawn. Therefore, $p(B|A) = 4/51$, and

$$p(A \cap B) = p(A)p(B|A) = 4/52 \times 4/51 = 4/663 \doteq 0.0060$$

Example. What is the probability of getting an ace or a king on one draw from an ordinary playing deck? Because $p(A \cap B) = 0$,

$$p(A \cup B) = p(A) + p(B) = 1/13 + 1/13 = 2/13 \doteq 0.15$$

Example. What is the probability of getting a red and a black card on two draws from an ordinary playing deck if the first card is not replaced?

$$p(R) = 1/2$$
$$p(B|R) = 26/51$$
$$p(R \cap B) = 1/2 \times 26/51 \doteq 0.25$$

Example. What is the probability of getting a red or a black card on two draws from an ordinary playing deck if the first card is replaced?

$$p(R \cup B) = p(R) + p(B) = 1/2 + 1/2 = 1$$

Summary of Elementary Properties of Probability for Finite Sample Spaces

Elementary properties of probability can be summarized under two major headings and two subheadings each:

1. Additive Property
   a. For nonmutually exclusive events,
   $p(A \cup B) = p(A) + p(B) - p(A \cap B)$.
   b. For mutually exclusive events, $p(A \cup B) = p(A) + p(B)$.
2. Multiplicative property
   a. For independent events, $p(A \cap B) = p(A)p(B)$.
   b. For dependent events, $p(A \cap B) = p(A)p(B|A)$.

Counting Properties

If an event can occur in more than one way, we have first to determine how many different ways it can occur before we can compute the probability of its occurring. For example, if we ask the probability of getting three heads on three flips of a coin, we have

$$p(H \cap H \cap H) = p(H)p(H)p(H)$$
$$= 1/2 \times 1/2 \times 1/2 = 1/8$$

But suppose that we ask the probability of getting two heads and one tail in three flips of a coin? In this case, the multiplicative rule alone does not give the answer, because the event $(2H,1T)$ can occur in more than one way. There are three different sequences that yield two heads and one tail:

$$(H,H,T) \qquad (H,T,H) \qquad (T,H,H)$$

that is, we may get two heads followed by a tail; a head, a tail, and a head; or a tail, followed by two heads. Any of these three sequences conforms to the event $(2H,1T)$. The probabilities of the three different sequences are

$$p(H \cap H \cap T) = 1/2 \times 1/2 \times 1/2 = 1/8$$
$$p(H \cap T \cap H) = 1/2 \times 1/2 \times 1/2 = 1/8$$
$$p(T \cap H \cap H) = 1/2 \times 1/2 \times 1/2 = 1/8$$

What is the probability of getting two heads and one tail on three flips of a fair coin? This implies the probability of getting sequence 1 or sequence 2 or sequence 3, or

$$p(\text{sequence } 1 \cup \text{sequence } 2 \cup \text{sequence } 3) = 1/8 + 1/8 + 1/8 = 3/8$$

In general, the probability of a compound event that can occur in more than one way is the ratio of the number of ways it can occur to all possible outcomes. This can be represented by a tree diagram:

First flip	Second flip	Third flip	Possible outcomes	Probability
		H	HHH	1/8
	H	T	HHT	1/8
H		H	HTH	1/8
	T	T	HTT	1/8
		H	THH	1/8
	H	T	THT	1/8
T		H	TTH	1/8
	T	T	TTT	1/8

$$8/8 = 1$$

There are eight different possible outcomes when we flip a fair coin three times. Three of these different outcomes qualify for the event "two heads and one tail." Because each outcome is equally probable, the probability of event A, defined as two heads and one tail, is

$$p(A) = \frac{n(A)}{n(I)} = \frac{3}{8}$$

The counting properties of probability provide a shortcut means of determining the possible number of outcomes under different conditions. We consider three counting properties: (*a*) all possible arrangements, (*b*) permutations, and (*c*) combinations.

All Possible Arrangements. Suppose that we have three cards on which the letters *A*, *B*, and *C* are marked. Assume that we select samples of two from this deck of three cards, selecting one card at a time, replacing it, and shuffling the deck before selecting the second. How many arrangements (samples, sets, or outcomes) of two cards are possible? For example, we might select the card marked *A* twice in a row and get a sample of *A*, *A*. We might also select the card *A* on the first draw and the card *B* on the second and get a sample of *A*, *B*. In order to determine how many possible samples of three can be selected from this deck, we construct a tree diagram as follows:

Member selected first	Member selected second	All possible arrangements
	A	AA
A	B	AB
	C	AC
	A	BA
B	B	BB
	C	BC
	A	CA
C	B	CB
	C	CC

It can thus be seen that there are nine possible samples of two cards each that can be selected. The number of possible samples can also be found by the following rule: k^n, where k is the total number of elements in the population and n the number in the sample to be selected. In the example above, the total number of possible sets of two that can be taken from a population of three is $3^2 = 9$, as we have seen. Using this rule, we can answer questions such as "How many outcomes (sets, samples) are possible on three flips of a coin?" The population here is two, because a coin can turn up either a head or a tail, and we are selecting sets (or samples) of three each. Hence, we have $2^3 = 8$ possible outcomes, as the tree diagram above shows. Consider another example. If a state uses combinations of six letters and no numbers for license plates, how many possible license plates can be constructed? The answer is $26^6 = 308,915,776$. Of course, in this example, the same letter can appear more than once in the same license plate.

Permutations. A permutation is an event in which order is distinguished and the same member cannot appear more than once. Assume a universe consisting of the letters *A*, *B*, and *C*. If we take each member of the universe one at a time without replacing it before selecting the next, the number of possible permutations of the three members is

Member selected first	Member selected second	Member selected third	Different permutations
A	B	C	ABC
	C	B	ACB
B	A	C	BAC
	C	A	BCA
C	A	B	CAB
	B	A	CBA

Any one of the three members may be selected first (n). Once the first member has been selected, either one of the two remaining members may be selected next ($n - 1$). Once the second number has been selected, only one member remains to be selected ($n - 2$). Therefore, the total number of permutations possible from this set is $3(3 - 1)(3 - 2) = 6$. In general notation we have

$$_nP_n = n(n - 1)(n - 2)\cdots[n - (n - 1)] = n!$$

This should be read, the number of possible permutations of n things when all n are being selected is equal to n factorial.[5]

Example. How many different permutations are possible of a set consisting of six books on a shelf?

$$_6P_6 = 6! = 6 \times 5 \times 4 \times 3 \times 2 \times 1 = 720$$

[5] n factorial ($n!$) is a number multiplied by itself, each time reducing the number by 1 until 1 is reached. Thus, 4! is $4 \times 3 \times 2 \times 1 = 24$. Note that, by convention, $0! = 1$.

Suppose instead that we want to calculate the number of different possible permutations if we take three books from a total of four. A tree diagram shows that there are 24 possible permutations:

First book	Second book	Third book	Permutations

The first book may be any one of the four, the second any one of the remaining $n - 1 = 3$, and the third either of the remaining $n - 2 = 2$. Therefore

$$_nP_r = n(n - 1)(n - 2)\cdots n - (r + 1) = \frac{n!}{(n - r)!}$$

where r represents the number of things taken from n; in the specific case,

$$_4P_3 = \frac{4!}{(4 - 3)!} = \frac{4 \times 3 \times 2 \times 1}{1} = 24$$

Some further illustrations:

$$_8P_2 = 8 \times 7 = 56$$

$$_{10}P_2 = 10 \times 9 = 90$$

$$_7P_3 = 7 \times 6 \times 5 = 210$$

Combinations. In some cases we may want to consider an arrangement such as *ABC* as identical to *ACB*. In this case, the order in which the same members appear is not distinguished. Consider again the set consisting of *A*, *B*, and *C*. The number of possible permutations of two things taken from this set is $_3P_2 = 3 \times 2 = 6$. Some of the permutations contain the same members in opposite order. If we consider the subset *AB* to be the same as the subset *BA*, then we have only three possible combinations and six permutations:

Permutations   Combinations

$$
\begin{array}{ll}
A \left< \begin{array}{l} B \longrightarrow AB \\ C \longrightarrow AC \end{array} \right. & \longrightarrow AB \\
B \left< \begin{array}{l} A \longrightarrow BA \\ C \longrightarrow BC \end{array} \right. & \longrightarrow AC \\
C \left< \begin{array}{l} A \longrightarrow CA \\ B \longrightarrow CB \end{array} \right. & \longrightarrow BC \\
\end{array}
$$

Let $_nC_r$ represent the number of combinations of n things taken r at a time. Each combination has $r!$-permutations. Hence

$$r! \times _nC_r = _nP_r$$

$$_nC_r = \frac{_nP_r}{r!}$$

$$= \frac{n!/(n-r)!}{r!}$$

$$= \frac{n!}{(n-r)!} \times \frac{1}{r!}$$

$$= \frac{n!}{r!(n-r)!}$$

Therefore, the number of different combinations of three things taken two at a time is

$$_3C_2 = \frac{3!}{2!(3-2)!} = \frac{3 \times 2 \times 1}{(2 \times 1)(1)} = 3$$

In general, we have

$$_nC_r = \frac{n!}{r!(n-r)!} \qquad (4\text{-}7)$$

Some further illustrations:

$$_{10}C_2 = \frac{10 \times 9}{1 \times 2} = 45$$

$$_{8}C_3 = \frac{8 \times 7 \times 6}{1 \times 2 \times 3} = 56$$

$$_{7}C_1 = \frac{7}{1} = 7$$

Computing Probabilities

Consider the set $I = \{5W, 4B, 6R\}$, in which W is a white ball, B a black ball, and R a red ball. The situation represents a box containing five white, four black, and six red balls. If we draw one ball at random from the box, what is the probability that it will be white? Let A represent the event "draw a white ball." Then, $n(A) = 5$, $n(I) = 15$, and

$$p(A) = \frac{n(A)}{n(I)} = \frac{5}{15} = \frac{1}{3}$$

Now, if we draw three balls at random from the box without replacement, what is the probability that we shall get three white balls? Because we do not replace balls after each draw, the events are dependent. By the multiplicative property for dependent events, we have

$$p(W \text{ and } W \text{ and } W) = p(W) \times p(W|W) \times p(W|W \text{ and } W)$$

$$= 5/15 \times 4/14 \times 3/13$$

$$= 2/91$$

By the counting property for combinations, we have

$$p(A) = \frac{_5C_3}{_{15}C_3} = \frac{(5 \times 4 \times 3)/(1 \times 2 \times 3)}{(15 \times 14 \times 13)/(1 \times 2 \times 3)} = \frac{5 \times 4 \times 3}{15 \times 14 \times 13} = \frac{2}{91}$$

The probability of event A is equal to all possible combinations of three things taken from five (white balls) as a proportion of all possible combinations of three things taken from fifteen (all balls).

What is the probability of drawing two white and one red ball from the same box in three random draws without replacement? There are three different sequences in which the event "two white balls and one red ball" may occur: (W,W,R), (W,R,W), and (R,W,W). By the multiplicative rule, we have

$$
\begin{aligned}
p(W \text{ and } W \text{ and } R) &= p(W) \times p(W|W) \times p(R|W \text{ and } W) \\
&= p(W) \times p(R|W) \times p(W|W \text{ and } R) \\
&= p(R) \times p(W|R) \times p(W|R \text{ and } W) \\
&= 5/15 \times 4/14 \times 6/13 \\
&= 4/91
\end{aligned}
$$

Each of the three sequences has the same probability, 4/91. Our question asks the probability of getting

$$
p(W,W,R) \text{ or } (W,R,W) \text{ or } (R,W,W) = 4/91 + 4/91 + 4/91 = 12/91
$$

The counting rule for combinations yields the same result by a different route:

$$
p(A) = \frac{{}_5C_2 \times {}_6C_1}{{}_{15}C_3} = \frac{(5 \times 4)/(1 \times 2) \times 6/1}{(15 \times 14 \times 13)/(1 \times 2 \times 3)} = \frac{5 \times 4 \times 6 \times 3}{15 \times 14 \times 13} = \frac{12}{91}
$$

Note that here the numerator changes to take into account that we are interested in getting all the combinations of two white balls taken from five and one red ball taken from six.

A few additional examples will illustrate the use of the three different counting rules. (*a*) What is the probability of getting an ace and a king on two draws from an ordinary bridge deck if the first card is *not* replaced before the second card is drawn? Since we do not replace the first card drawn the two events are dependent, and we have

$$
p(A \cap K) = p(A)p(K|A) = \frac{{}_4C_1 {}_4C_1}{{}_{52}C_2} = \frac{32}{2652}
$$

(*b*) What is the probability of getting an ace and a king on two successive draws from an ordinary bridge deck if the first card *is* replaced before the second is drawn? Since the events are independent we have

$$
p(A \cap K) = p(A)p(K) = \frac{{}_4C_1 {}_4C_1}{52^2} = \frac{16}{2604}
$$

Notice here that since the first card is replaced the same card can be drawn twice and therefore the denominator is k^n rather than ${}_nC_r$. (*c*) What is the probability of getting an ace and a king on two draws in that order if the first card is not replaced?

$$
p(A \text{ followed by a } K) = \frac{{}_4P_1 {}_4P_1}{{}_{52}P_2} = \frac{16}{2652}
$$

THE BINOMIAL THEOREM

We can now take advantage of the counting rules and properties of probability to formulate a general theorem that is more useful in statistics than the elementary properties of probability. Let $p(r,n;P)$ represent the probability of getting r successes in n trials, where the probability of any one success is P. The combination rule, and the multiplicative property of probability, tell us that the probability of r successes in n trials is equal to all of the possible combinations of the event, times the probability of its occurrence each time, times the probability of its nonoccurrence on each of the remaining $n - r$ times. In symbolic notation, then, the binomial theorem can be stated as follows:

$$p(r,n;P) = {}_nC_rP^r(Q)^{n-r} \qquad (4\text{-}8)$$

where $Q = 1 - P$. This theorem states that there are ${}_nC_r$ combinations in which r events A may occur in n trials, and each combination is an event with a probability of $p(A)p(A')$. If $p(A) = P$ and $p(A') = 1 - P = Q$, the resulting equation is (4-8). This can be illustrated with the following examples.

What is the probability of getting two heads in three tosses of a fair coin?

$$p(2,3;1/2) = {}_3C_2(1/2)^2(1/2)$$
$$= \frac{3 \times 2}{1 \times 2}\left(\frac{1}{4}\right)\left(\frac{1}{2}\right)$$
$$= 3/8$$

What is the probability of getting one head in three tosses of a balanced coin?

$$p(1,3;1/2) = {}_3C_1(1/2)(1/2)^2$$
$$= 3/8$$

If the probability of a person's voting in a presidential election is 0.6, what is the probability that four out of five voters selected at random will vote in the next election?

$$p(4,5;0.6) = {}_5C_4(0.6)^4(0.4)$$
$$= 0.259$$

It can be seen from these three examples that the binomial theorem can be applied in those cases where the outcome of the event can be dichotomized, such as head or tail, and vote or not vote. It is called the binomial theorem because it is the general term of the expansion of the powers of the term $(a + b)$. The number of different combinations in which r events A may occur are the coefficients in the expansion, and may be found by Pascal's triangle. Consider the following expansion:

$$(a + b)^3 = (a + b)(a + b)(a + b)$$
$$= aaa + baa + aab + aba + bba + bab + abb + bbb$$

If we combine all the similar terms, we have

$$(a + b)^3 = a^3 + 3a^2b + 3ab^2 + b^3$$

That is, there is one way of getting three a's, three ways of getting two a's and one b, three ways of getting two b's and one a, and one way of getting three b's. The number of ways of getting each combination are the binomial coefficients. We may write

$$(a + b)^3 = {}_3C_0a^3 + {}_3C_1a^2b + {}_3C_2ab^2 + {}_3C_3b^3$$

or, in general,

$$(a + b)^n = {}_nC_0a^n + {}_nC_1a^{n-1}b + {}_nC_2a^{n-2}b^2 + \cdots + {}_nC_nb^n$$

We shall see in Chapter 5 how this regular progression of coefficients in the binomial expansion shows up in constructing sampling distributions for events that have only two possible outcomes.

RANDOM VARIABLES

One more concept needs to be considered before turning to sampling distributions. A random variable may be thought of as all the possible outcomes of an event in a sample space. Consider the experiment of tossing a die. There are six possible outcomes, and the sample space is $\{1,2,3,4,5,6,\}$. If the experiment of tossing a die is thought of as a variable X, then there are six possible values this variable may have, and they can be considered the values X_i of the random variable X. The variable is random because we do not know what value it will have on any particular toss, but only the probability that it will have this value. In this example, all the possible outcomes in the toss of a die are the population or sample space. Consider populations of N elements we are more familiar with, such as all registered voters in the United States or all cities in the United States. Associated with each of the elements in these populations are values of one or more variables. For the population of all registered voters, the elements are the individual voters, and associated with each voter are values of such variables as age, beliefs, religion, and intelligence. In the case of cities, the elements are the cities, and associated with each are values of variables such as size, number of registered Republicans, and total local taxes. These variables X are random if they have a finite number of possible values and if associated with each possible value there is a probability p_i that $X = X_i$ (the ith value) if an element is selected at random.

Consider next the experiment of tossing two dice. There are 36 possible outcomes, which are given in Figure 4-4. Suppose that we are interested in

die number 1

6	6,1	6,2	6,3	6,4	6,5	6,6
5	5,1	5,2	5,3	5,4	5,5	5,6
4	4,1	4,2	4,3	4,4	4,5	4,6
3	3,1	3,2	3,3	3,4	3,5	3,6
2	2,1	2,2	2,3	2,4	2,5	2,6
1	1,1	1,2	1,3	1,4	1,5	1,6

die number 2

| 1 | 2 | 3 | 4 | 5 | 6 |

Figure 4-4.

only a part of this sample space, the event "sum of two dice." There are 11 different compound events in this subset: (2,3,4,5,6,7,8,9,10,11,12). If we think of a random variable X as the event "sum of two dice," then each event in the subset is a value X_i of the random variable X. Suppose now that we are interested in the mean $\bar{X}$ of the two numbers that may turn up if we toss the two dice once. Thus, if a 1 turns up on the first die and a 1 on the second, the mean $\bar{X}$ is 1; if we obtain either (2,1) or (1,2), the mean $\bar{X}$ is $1\frac{1}{2}$, and so on. The possible values of this random variable $\bar{X}$ are $\{1,1\frac{1}{2}, 2,2\frac{1}{2}, 3,3\frac{1}{2}, 4,4\frac{1}{2}, 5,5\frac{1}{2}, 6\}$. Note in Figure 4-4 that there is one sample point that satisfies the first possible value of the random variable $\bar{X} = 1$: the point (1,1). There are two sample points that satisfy the value of the random variable $\bar{X} = 1\frac{1}{2}$: (2,1) and (1,2), and so on. Because each of the 36 sample points in the sample space in Figure 4-4 is equally probable, the probability of any particular sample point is $1/36$. The probability of a compound event, such as $\bar{X} = 4$, is the sum of the probabilities of the sample points comprising it. Because there are five sample points by which it is possible to get $\bar{X} = 4$, the probability of observing this value of the random variable $\bar{X}$ is

$$1/36 + 1/36 + 1/36 + 1/36 + 1/36 = 5/36$$

Similarly, the probability that the random variable $\bar{X} = 2$ is $p(\bar{X} = 2) = 3/36$. When we compute the probability of each possible value of the random variable $\bar{X}$, we have a *probability distribution*. This may be portrayed in the following manner:

$\bar{X}$	1	1 1/2	2	2 1/2	3	3 1/2	4	4 1/2	5	5 1/2	6
$p(\bar{X})$	$\frac{1}{36}$	$\frac{2}{36}$	$\frac{3}{36}$	$\frac{4}{36}$	$\frac{5}{36}$	$\frac{6}{36}$	$\frac{5}{36}$	$\frac{4}{36}$	$\frac{3}{36}$	$\frac{2}{36}$	$\frac{1}{36}$

This probability distribution may also be portrayed in the form of a bar chart, as in Figure 4-5. Because this is a frequency distribution, we may

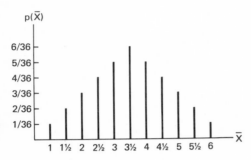

Figure 4-5.

talk about the mean, standard deviation, skewness, and kurtosis of this probability distribution. The mean of a probability distribution is

$$\mu = \sum_{i=1}^{N} p_i X_i \qquad (4\text{-}9)$$

This also is the *expected value* of the random variable. The expected value of a random variable is its average long-run value. It is found by multiplying each possible value by its corresponding probability and summing the products. The expected value of a random variable is designated generally by the capital English letter E followed by the variable in parentheses. The expected value of the random variable $\bar{X}$ in the example above is

$$E(\bar{X}) = \sum_{i=1}^{N} p_i X_i = 3\ 1/2$$

We are saying here, in effect, that the value we expect when we toss two dice and compute the mean of the values that turn up is 3 1/2. The expected value is called the *first moment* of a probability distribution and can be conceived of as a weighted mean when we are discussing discrete distributions. The variance of a probability distribution, called the second moment, is defined for a discrete distribution as

$$\sum [X - E(X)]^2 \cdot p(X)$$

for example, the variance for the probability distribution given above is

$$\sigma^2 = \sum [\bar{X} - E(\bar{X})]^2 \cdot p(\bar{X}) = (1 - 3\ 1/2)^2(1/36) + (1\ 1/2 - 3\ 1/2)^2(2/36)$$
$$+ \cdots + (6 - 3\ 1/2)^2(1/36)$$
$$= 1.46$$

The mean of the probability distribution of the random variable $\bar{X}$ is at the point of highest density, because this distribution is perfectly symmetrical. Its probability is higher than that of any other value. For example, the probability that the random variable $\bar{X} = 1$ is $1/36$. If we throw two perfectly balanced dice and compute the mean of the numbers that appear on the face, it is not so likely that we shall find a mean of 1 as it is that we shall find a mean of 3 1/2.

The probability distribution shown in Figure 4-5 is discrete, because the random variable $\bar{X}$ can have only specific values. A continuous probability distribution, however, appears in a chart as a smooth curve, and probabilities are represented by areas under the curve. A continuous curve can be represented by a mathematical equation, and vice versa. For example, the mathematical equation $P(X) = 0.08X^2$ is graphed in Figure 4-6 for values of the random variable X between 0 and 10. The curve is symmetrical, as is the

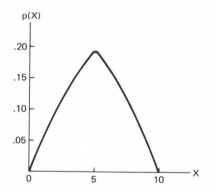

Figure 4-6. Continuous probability distribution for $p(X) = 0.08X - 0.008X$, $0 < X < 10$.

distribution in Figure 4-5, and its mean is at the point of highest density. The expected value of this probability function is 5.

In classical statistical inference there are several general probability distributions like the one in Figure 4-6 that are used for the majority of inference problems. The four that are used most often are called the normal, the t, the F, and the chi-square distributions. These, along with a few others, are described in this book. We now turn to a detailed consideration of three of these probability distributions.

READINGS AND SELECTED REFERENCES

Edwards, Allen L. *Expected Values of Discrete Random Variables and Elementary Statistics*, John Wiley & Sons, Inc., New York, 1964, Chap. 4.

Hays, William. *Statistics for Psychologists*, Holt, Rinehart and Winston, Inc., New York, 1965, Chaps. 2–4.

Mosteller, Frederick, Robert Rourke, and George Thomas. *Probability and Statistics*, Addison-Wesley Publishing Company, Inc., Reading, Mass., 1961.

Spurr, William A., and Charles P. Bonini. *Statistical Analysis for Business Decisions*, Richard D. Irwin, Inc., Homewood, Ill., 1967, Chap. 7.

Western, D. W., and V. H. Haag. *An Introduction to Mathematics*, Holt, Rinehart and Winston, Inc., New York, 1959, Chap. 4.

Yamane, Taro. *Statistics: An Introductory Analysis*, Harper & Row, Publishers, 1967, Chap. 5.

5

Sampling Distributions

We noted in Chapter 4 that the objective of statistical inference is to say something about an unknown population parameter on the basis of a sample statistic. The usual procedure is to draw a simple random sample[1] from a population and make an inference about the population based on the results of the sample. For example, we may infer something about μ on the basis of $\bar{X}$, but we may not simply take the sample result as being true of the population, because sampling error can produce a bad estimate of the population parameter. If more than one sample is drawn from a population, each sample will produce a different result. In order to know how accurate the result of a sample is likely to be, we can construct a *probability distribution* of all the possible results we might get if we repeated the sampling process an infinite number of times. The probability distribution of all possible sample values for a particular statistic is called a *sampling distribution.*

It is called a probability distribution because if all the sample results are plotted on the x axis and the frequency of their occurrence on the y axis the resulting curve may be used to determine the probability of observing a particular outcome. It is called a sampling distribution when it refers to all the possible sample results. It is possible to have sampling distributions of $\bar{X}$, s, s^2, and so on. To fix this very important point firmly in your mind, think of the following process, repeated an infinite number of times. A sample of a given size is drawn from a population. The $\bar{X}$ for this sample is computed using the equation described in Chapter 2. The elements are returned to the population, and a second sample is drawn. The $\bar{X}$ is computed for this sample. Note that the value of $\bar{X}$ is different the second time. This process is repeated an infinite number of times, and the resulting values of $\bar{X}$ are cast in the form of a frequency distribution that is the sampling distribution of the statistic $\bar{X}$.

With the aid of probability theory, we may construct a sampling distribution of all possible sample results and get the same thing as if we actually repeated the sampling process an infinite number of times. For example,

[1] Simple random sampling is defined briefly at the beginning of Chapter 6 and discussed in more detail in Chapter 15.

if we want to say something about the population mean on the basis of a sample mean $\bar{X}$, we can use probability theory to construct a sampling distribution of all possible sample means and use it as a basis for determining how accurate the particular sample mean is. Usually this is not necessary, because certain kinds of distributions can be used as approximations of the sampling distribution we should get if we constructed one using probability theory or even if we constructed one by repeating the process of sampling many times.

We consider three of these general kinds of probability distributions in this chapter: the binomial, the normal, and the Poisson. For the moment, let us describe in general terms the method of using the properties of probability described in Chapter 4 to construct sampling distributions.

Recall that all the possible samples of n that can be selected from a population are given by N^n. Suppose that we had a universe consisting of three people: a Republican, a Democrat, and an Independent. Supposing that we take samples of size two from this population, each time replacing those drawn; there are $3^2 = 9$ possible samples that may be obtained:

$$
\begin{array}{ccc}
\text{RR} & \text{RD} & \text{RI} \\
\text{DR} & \text{DD} & \text{DI} \\
\text{IR} & \text{ID} & \text{II}
\end{array}
$$

that is, the Republican may be drawn twice, the Democrat twice, or the Independent twice, although it is more likely that we shall get a mixed sample (2/3 to 1/3). To be more realistic, we should consider sampling from this universe without replacement. Then, the first person selected is any one of the three, and the second is any one of the remaining two. There are therefore six possible samples that can be drawn without replacement:

These are the number of *permutations* of two things selected from three, as defined by Equation (4-6):

$$
{}_3P_2 = \frac{3!}{1!} = 6
$$

Of course, normally we consider a sample of RD the same as a sample of DR, because they are the same two people. Therefore, we are normally interested in the number of different *combinations* of size two that can be drawn from this universe. This is given by Equation (4-7):

$$_3C_2 = \frac{3!}{2!} = 3$$

That is, there are three possible combinations of two different people: RD, RI, and DI. If we use simple random sampling, defined in Chapter 6, each of the three possible outcomes is equally probable. For example, in sampling with replacement in which we distinguish each outcome, the probability of drawing a sample of a Republican and a Democrat is 2/9. If we sample without replacement and are concerned with samples of two different persons, the probability of drawing a sample of a Republican and a Democrat is 1/3.

Now consider a slightly better example. Suppose that the population consists of the numbers (4,5,6,7,8,9). The mean is $\mu = 6\,1/2$. We do not know anything about the population, and we draw a sample of $n = 3$ from it. There are $_6C_3 = 20$ different samples that can be drawn from the population:

(4,5,6)	(4,6,7)	(4,7,9)	(5,6,9)	(6,7,8)
(4,5,7)	(4,6,8)	(4,6,9)	(5,7,8)	(6,7,9)
(4,5,8)	(4,6,9)	(5,6,7)	(5,7,9)	(6,8,9)
(4,5,9)	(4,7,8)	(5,6,8)	(5,8,9)	(7,8,9)

The means of each of these samples are 5, 5 1/3, 5 2/3, 6, 5 2/3, 6, 6 1/3, 6 1/3, 6 2/3, 7, 6, 6 1/3, 6 2/3, 6 2/3, 7, 7 1/3, 7, 7 1/3, 7 2/3, and 8 respectively. Some of the samples have the same mean. The means of all 20 possible samples can be shown in the form of a frequency distribution that is called the *sampling distribution of means*, as follows:

$\bar{X}$	5	5 1/3	5 2/3	6	6 1/3	6 2/3	7	7 1/3	7 2/3	8
$p(\bar{X})$	$\frac{1}{20}$	$\frac{1}{20}$	$\frac{2}{20}$	$\frac{3}{20}$	$\frac{3}{20}$	$\frac{3}{20}$	$\frac{3}{20}$	$\frac{2}{20}$	$\frac{1}{20}$	$\frac{1}{20}$

Each of the 20 samples is equally likely. The probability that any one of the samples will be selected is 1/20, but because some of the samples have the same means, the probability of obtaining a sample mean $\bar{X} = 6$ is more likely than that of obtaining an $\bar{X} = 5$. It can be seen that the most likely sample means are close to the true population mean. The farther away from the true population mean we get, the less likely it is that we shall draw the sample. We can use a sampling distribution of means to say something about μ on the basis of $\bar{X}$. If $\bar{X}$ is used as the estimate of μ, we run the risk of making an error, because there is a chance that a sample with the wrong mean will be drawn. This is because we allow chance (random sampling) to determine which members of the population will be included in the sample. In order to determine what the chances of error are, we use a sampling distribution of means. We discuss this in more detail in Chapter 6 and show that we do not have to construct a sampling distribution each time, because we can rely on some general families of distributions in most cases. At this point we describe some of these general families of sampling distributions.

THE BINOMIAL DISTRIBUTION

We begin this discussion with a simple example of coin flipping and proceed to more realistic examples. Suppose that a balanced coin is flipped five times and one head turns up. This seems to be an unusual result. In order to determine how unusual, we can construct a probability distribution of all possible outcomes in five flips of a balanced coin. To do so, we use the binomial theorem developed in Chapter 4:

$$p(0,5;1/2) = {}_5C_0(1/2)^0(1/2)^5 = 1/32$$
$$p(1,5;1/2) = {}_5C_1(1/2)^1(1/2)^4 = 5/32$$
$$p(2,5;1/2) = {}_5C_2(1/2)^2(1/2)^3 = 10/32$$
$$p(3,5;1/2) = {}_5C_3(1/2)^3(1/2)^2 = 10/32$$
$$p(4,5;1/2) = {}_5C_4(1/2)^4(1/2)^1 = 5/32$$
$$p(5,5;1/2) = {}_5C_5(1/2)^5(1/2)^0 = 1/32$$

This is called a binomial sampling distribution for $n = 5, p = 1/2$. It may be shown as follows:

Number of heads	0	1	2	3	4	5
P	1/32	5/32	10/32	10/32	5/32	1/32

or as raised ordinates, as in Figure 5-1.

The sampling distribution shows that, if the coin is perfectly balanced,

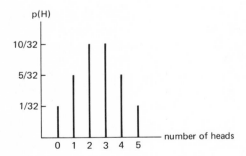

Figure 5-1.

we may expect to get 1 head in 5 flips only about 5 times out of every 32 flips in the long run. The probability of getting only 1 head in 5 flips is therefore $5/32 = 0.16$. Note that all we have said here is that, if the coin is perfectly balanced and there is no bias in its landing, we may not expect to find 1 head in 5 flips very often. In the long run, if the same experiment is repeated many times, we may expect to get 1 head in 5 flips only about 16 percent of the time. More often, we may expect to get 2 or 3 heads in 5 flips. The exact number of heads we may expect, and how often we may observe deviations from this, is considered below. First, let us take a more realistic example.

Suppose that an advisor to the mayor of a large city suggests that the mayor recommend subway fares be increased so as to raise more revenue for the city. A second advisor suggests that, if he does so, he will lose votes, because at least 8 out of 10 voters would vote against him in the next election.

Table 5-1   Binomial Probability Distribution for $N = 10$, $p = .08$

Number of Successes (X)	0	1	2	3	4	5	6	7	8	9	10
Probability $p(X)$	.0000001	.0000041	.000074	.00079	.006	.026	.088	.201	.302	.268	.107

A third advisor doubts the accuracy of this statement, suggesting that it is much lower than this. A test is proposed to take a random sample of 10 voters and ask each his reaction to the proposal.[2] Suppose that in the random sample of 10 voters, 4 say that they will vote against the mayor if he recommends a raise in fares and 6 say that they will not vote against him because of this. The problem is now one of statistical inference. We want to say something about total population of voters on the basis of this sample result. Specifically, we want to test whether or not the contention that the opposition to a subway fare increase is such that at least 8 out of 10 voters will vote against the mayor because of the proposal. In order to say whether the population parameter P is actually 0.8, we must construct a sampling distribution of all the possible outcomes of a sample of 10 we shall get if the actual proportion is 0.8. Table 5-1 gives the binomial distribution for $n = 10$ and $p = 0.8$. Figure 5-2 presents the frequency polygon for this sampling distribution.

[2] In reality a much larger sample would be drawn, and the question of the relationship between responses to a questionnaire and actual behavior in voting would have to be considered.

Looking at the sampling distribution tells us that, if the population P is actually 0.8, we shall observe a sample p of 0.4 or less only about 7 times out of 1,000. This figure of 7 times in 1,000 is the sum of the probabilities of each of the outcomes 0, 1, 2, 3, and 4. In other words, if the contention that the population parameter is 0.8 is true, the sample result is a rare event.

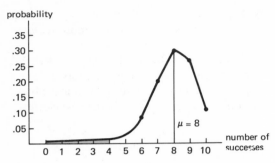

Figure 5-2. Frequency polygon.

We might therefore reject the notion that $P = 0.8$. However, if we do, we still run the risk of making an error. We shall be wrong less than 1 percent of the time if the same questions arise many times and we repeat the experiment many times. The sampling distribution, constructed on the basis of the binomial theorem, enables us to make the decision without taking more samples and also with a precise measure of our chances of being wrong.[3]

Certain facets of this example should be emphasized. For each trial (or individual drawn into the sample), the random variable X can take on only one of two values (vote or not vote). The observations are independent. This assumption implies that we are sampling from an infinite population. Flipping a coin, for example, can be considered an infinite process. Sampling from the population of all voters in a city can be considered an infinite process if we replace each person selected before the next one is selected. This raises the possibility of selecting the same person more than once and complicates the sampling problem.[4] In practice, we generally do not replace each observation before drawing the next. We therefore violate one of the principal assumptions (independence) underlying the binomial distribution. When this is so, the hypergeometric distribution should be used instead of the binomial. But when the sample is small relative to the population, that is, about 20 percent or less, the binomial and the hypergeometric distributions yield sufficiently similar results for most practical problems.

[3] This chance of being wrong is conventionally called error type 1. It is discussed in more detail in Chapter 7.
[4] See Chapter 15 for a further discussion of this point.

The hypergeometric distribution can be computed by means of the following:

$$P(X) = \frac{{}_R C_r {}_Q C_q}{{}_N C_n} \tag{5-1}$$

where N = total items in the population, n = items in the sample, R = number of successes in the population, r = number of successes in the sample, and Q and q = number of failures in the population and sample respectively. For example, if the population consists of 4 Democrats and 6 Republicans and a sample of 4 is drawn at random, what is the probability that the sample will have 3 Republicans?

$$P(3 \text{ Republicans}) = \frac{{}_6 C_3 {}_4 C_1}{{}_{10} C_4} = \frac{[6!/(3!3!)][4!/(1!3!)]}{10!/(4!6!)} \doteq 0.38$$

In this case, drawing a Republican is considered a success, and not drawing a Republican is considered a failure. The hypergeometric distribution is computed by taking the quotient of the product of all the ways 3 successes can be drawn from 6, times all the ways 1 failure can be drawn from 4, divided by the total number of samples of 4 taken from a population of 10. When this distribution is to be used because precision is required, there are tables available.[5]

What we did in the example concerning subway fares and votes was to construct a probability distribution in order to infer something about the population parameter. This was done by means of the binomial theorem. The resulting distribution is one of those in the general family called the binomial distribution. This is a family of distributions because it can take on different shapes depending on n and P. Figure 5-3 shows that the binomial distribution is positively skewed when P is small and negatively skewed

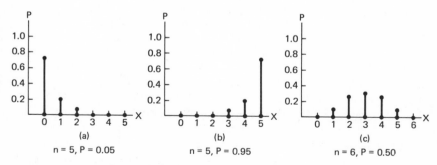

Figure 5-3. The binomial distribution.

[5] See G. J. Lieberman and D. B. Owen, *Tables of the Hypergeometric Distribution*, Stanford University Press, Stanford, Calif., 1960.

when P is large. As P approaches 0.5, the shape of the binomial becomes symmetrical.

The binomial distribution can be used in all cases where we are dealing with proportions and there are two possible outcomes. In most cases it is not necessary to compute the particular probability distribution for the binomial. If the sample is large and the value of P is about 0.5, the normal distribution can be used as a good approximation of the binomial. If the sample is small or P is either very small or very large, tables of the binomial distribution are available.[6] Tables of the binomial probability distribution have not been reproduced in this book because it gets relatively little use in most contemporary political science research.

A binomial probability distribution has a mean and a standard deviation. In this case we talk about expected rather than actual results. For example, if we toss a balanced coin 1,000 times, what is the number of heads we may expect? The answer is np, or the number of tosses times the probability of heads. For 1,000 tosses the expected number of heads is $1,000(1/2) = 500$.

The mean of a binomial distribution can therefore be defined as

$$\mu = np \tag{5-2}$$

The standard deviation is

$$\sigma = \sqrt{\sum(X_i - \mu)^2 P(X_i)} \tag{5-3}$$

For example, if we flip one coin two times, we can get 0, 1, or 2 heads, with respective probabilities of 1/4, 1/2, and 1/4. The mean number of heads, therefore, is $2(1/2) = 1$, and the standard deviation is

$$\sigma = \sqrt{(0 - 1)^2 1/4 + (1 - 1)^2 1/2 + (2 - 1)^2 1/4} = \sqrt{1/2}.$$

Rather than compute the standard deviation of the binomial distribution in this manner, we can derive the equivalent computing equation:

$$\sigma = \sqrt{nPQ} \tag{5-4}$$

If we suppose for the moment that about 68 percent of all the events are within plus or minus one standard deviation of the mean — the reason for this assumption is given below — we may say that in 1,000 flips of an actual coin we expect to observe 500 heads, but about 68 percent of the time the actual number will fluctuate between $500 \pm \sqrt{1,000(1/2)(1/2)}$, or from about 484 to 516. It is apparent, therefore, that we should not be too surprised if in 1,000 flips of a coin we observe 510 heads.

[6] See *Tables of the Binomial Probability Distribution*, U.S. Bureau of Standards, Applied Mathematics Series No. 6, 1949.

THE NORMAL DISTRIBUTION

The normal distribution is one of the most important in statistics because it has a large number of applications. When we discuss the normal distribution, it must first be emphasized that we are discussing a continuous rather than a discrete distribution. With continuous data, variables can take on any real-number value. Probabilities, therefore, are represented as areas under a curve rather than as particular points. As an illustration, suppose that we are able to construct a frequency distribution of the weight of all individuals in the United States after weighing each one on an extremely accurate scale. The probability of finding a person weighing a given number of pounds, if one person is selected at random from this population, is equal to the proportion of people who are this given weight. But if our weighing scale is extremely accurate, the probability of finding someone who weighs a specific amount is zero, because the area at the specific point is zero. For example, suppose that we started by cutting off the interval from 159.5 to 160.5 pounds. The probability of finding someone who weighs from 159.5 to 160.5 pounds is equal to the number of individuals who fall into this class interval divided by the total population. Suppose that there are 50,000 persons in the United States out of about 200 million who fall into this small interval. If we draw one member at random from the population, the probability that he will weigh from 159.5 to 160.5 pounds is 50,000/200,000,000. Suppose that we now make the interval smaller, from 159.9 to 160.1, and that there are 1,000 people who fall into this interval. Then the probability of a person's weighing between 159.9 and 160.1 is 1,000/200,000,000. As we make the interval infinitely smaller so as to approach a specific weight, such as 160 pounds, the probability of finding a person of this weight comes closer and closer to zero. With a continuous distribution we are dealing with variables that can take on an infinite number of values and a population that is infinitely large. Therefore, the actual probability is zero.

For this reason we speak of a probability density function when discussing continuous data. The normal curve is a probability density function that has constant properties given by the mathematical function rule that determines its specific shape when two parameters are known. This function rule is

$$f(X) = \frac{1}{\sigma\sqrt{2\pi}} e^{-x^2/2\sigma^2} \tag{5-5}$$

where π and e are the constants 3.1416 (the ratio of a circle's circumference to its diameter) and 2.7183 (the base of the Naperian logarithms) respectively. The unknown parameters are σ and x^2; the latter is $(X_i - \mu)^2$. Once σ and

x^2 are specified, the particular shape of the normal curve is known. To illustrate how this function rule is used to define the curve, consider the simple equation

$$f(X) = -3X^2$$

This function is graphed in Figure 5-4 for small integer values of X.

X	−3	−2	−1	0	1	2	3
f(X)	−27	−12	−3	0	−3	−12	−27

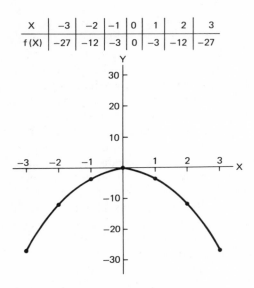

Figure 5-4.   Graph of $f(X) = -3X^2$.

It is much more laborious to graph a normal curve by the same method, but it can be done (actually, logarithms have to be used to compute exact values of Y for various values of x^2 and σ). The important thing to note about the function rule for the normal curve is that, if an ordinate is erected at the mean, the exponent $-x^2/2\sigma^2$ becomes zero. This ensures that the normal curve is unimodal and its highest point is at the mean. Because the exponent is a squared quantity, either side of the distribution has exactly the same density. This ensures that the normal curve is symmetrical. Because the exponent has a negative sign, the larger x^2, the smaller the density under the curve. The density decreases as x^2 becomes large in such a way as to ensure that given areas lie within one, two, and three standard deviations from the mean. The general properties of the normal curve given by the function rule, therefore, are: it is unimodal with its mean, median, and mode at the same point; it is symmetrical; and the areas of the curve within plus or minus one, two, and three standard deviations from the mean remain constant. There will always be

> 68.26 percent within ± 1 standard deviation
> 95.46 percent within ± 2 standard deviations
> 99.+ percent within ± 3 standard deviations

Because of these general properties it is not necessary to construct a normal curve for any particular distribution. For any particular normal distribution, such as that in Figure 5-5, in order to find the area within the limits marked by a and b, it is necessary to use the calculus to evaluate the definite integral of Equation 5-5, that is, $\int_a^b f(X)\, dX$. But because the areas

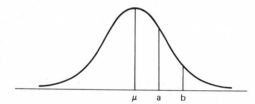

Figure 5-5.

that lie between any two points on a standard normal curve have been computed and tabled, we can simply refer to the standard normal curve to find areas once we convert the particular observation to standard form.

The standard normal curve is a special form of the normal distribution in which the mean is zero and standard deviations are in units of one, referred to as Z scores. One standard deviation is one Z. For example, if the $\bar{X}$ of a particular distribution is 80 and its standard deviation is 20, a score of 100 is exactly one standard deviation to the right of the mean. We can convert this to a Z score by means of the equation

$$Z = \frac{X_i - \bar{X}}{s}$$

$$= \frac{100 - 80}{20} \tag{5-6}$$

$$= 1$$

There is exactly one standard deviation, or one Z score, between 80 and 100. We said above that 34.13 percent of the normal curve lies within one standard deviation of the mean. Hence, if we want to know the proportion of a frequency distribution that lies between the mean of 80 and a score of 100, we convert to a standard (Z) score and then consult the table of areas under the standard normal curve in the Appendix. The area, as we said, is equivalent to probability. The use of the standard normal curve table is important enough to warrant some illustrations and discussion here.

The standard normal curve table in the Appendix shows the proportion of the total area that lies between the mean and any other point on the curve. Only one-half of the curve areas are given in the table, because the curve is perfectly symmetrical. To use the table, first find the Z score for any particular observation by Equation (5-6). The value for Z is sometimes called the *standard normal deviate*, and it represents, in standard form, the number of standard deviations the variable X_i is above or below its own mean.

To illustrate the use of the standard normal curve, consider the following. Suppose that we are told that the mean IQ of all people in the United States is 100 with a standard deviation of 10 and that the distribution of intelligence is normal. Now consider the following questions. What proportion of people have an IQ of between 80 and 100? We first compute the Z score:

$$Z = \frac{80 - 100}{10} = -2.0$$

We want to know the area of the curve that lies between the mean of 100 and the score 80. This is shown as a shaded area in Figure 5-6. The score

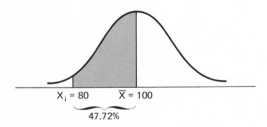

Figure 5-6. Area between the mean and two Z scores to the left.

80 lies exactly two standard deviations to the left of the mean (hence the negative sign). Table 1 in the Appendix tells us that this is about 47.72 percent of the total area lying to the left of the mean; that is, 47.72 percent of all people have an IQ of between 80 and 100 if our assumptions about intelligence being normally distributed and $\sigma = 10$ are correct.

What proportion of people have an IQ between 90 and 115? The number of standard deviations encompassed by these figures in standard normal deviates is

$$Z = \frac{90 - 100}{10} = -1.0$$

$$Z = \frac{115 - 100}{10} = 1.5$$

The observation 90 lies one Z score to the left, and the observation 115 lies 1.5 Z scores to the right of the mean. This area is shown in Figure 5-7.

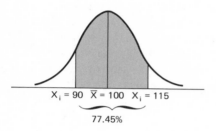

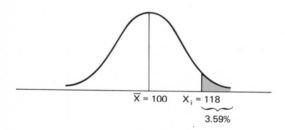

$X_i = 90$ $\overline{X} = 100$ $X_i = 115$

77.45%

Figure 5-7. Area of the standard normal curve between the scores 90 and 115.

Table 1 in the Appendix tells us that 34.13 percent of the total curve lies between the mean and one standard normal deviate to the left and that 43.32 percent of the curve lies between the mean and 1.5 standard normal deviates to the right. The total area encompassed by the scores 90 and 115 is therefore $34.13 + 43.32 = 77.45$ percent; or about 77 percent of all individuals have an IQ between 90 and 115.

What proportion of individuals have an IQ of more than 118? This area is shown in Figure 5-8. The score 118 is 1.8 Z scores to the right of the

$\overline{X} = 100$ $X_i = 118$

3.59%

Figure 5-8. Area to the right of the score 118.

mean of 100. The table of areas under the normal curve tells us that 46.41 percent of the curve lies between the mean and 1.8 Z scores. But we asked what proportion lies to the right of this figure. To obtain this, we must subtract 0.4641 from 0.5000. When we do so, we find that about 3.59 percent of the area lies to the right of the score 118. Note that, Table 1 in the Appendix gives only one-half of the total curve;[7] if we want to know how much of the total curve lies outside 1.8 standard normal deviates of the mean on

[7] Note that we subtract from 0.5000 rather than 1.0000, because the table gives only one-half of the total curve.

either side, we have to double the figure 0.4641 and subtract from 1.0000; or, conversely, double 3.59 to find 7.18 percent.

The normal curve is a probability distribution described in general by Equation (5-5). The normal curve is also an approximate description of certain empirical phenomena when they are cast in the form of a frequency distribution. This is true of such phenomena as the height of men, intelligence score, and measurement errors. Certain statistics, such as the mean $\bar{X}$, also form a normal distribution if the sampling process is random and is repeated an infinite number of times. This is an important fact for statistical inference. If we draw samples from a population continuously, compute the mean for each sample, and then cast them in a frequency distribution, we find that the frequency distribution of $\bar{X}$ forms a normal curve, with the majority of $\bar{X}$'s (68.26 percent) falling within plus or minus one standard deviation from the mean of this distribution. We consider, in Chapter 6, how these facts can be used in estimation problems.

THE POISSON DISTRIBUTION

Before turning to the question of statistical estimation, let us briefly consider one more probability distribution. The Poisson distribution has limited practical application in political science at present, but it can become more important as the simulation of certain relatively infrequent events becomes more important. The Poisson is like the binomial distribution except that it applies where P or Q is very small. This is so, for example, in events such as arrivals at airports, accidents, and defects in packages.

The Poisson probability distribution is defined as

$$p(r|m) = \frac{e^{-m}m^r}{r!} \qquad r = 0, 1, 2, \ldots \qquad (5\text{-}7)$$

The probability of r successes, given a mean number of successes of m, is given by Equation (5-7), where e is the constant 2.7183.

As an illustration, suppose that the mean number of bills referred back each day to a committee of the House of Representatives is one. The probability that three bills will be referred back to a committee on a particular day is

$$p(3|1) = \frac{2.7183^{-1} \times 1^3}{3!} = 0.061$$

Tables of the Poisson probability distribution are available that can be used to determine the probability of various r given different values of m.[8] They are not reproduced in this book because of their limited practical research application in political science at present.

[8] See, for example, E. C. Molina, *Poisson's Exponential Binomial Limit*, D. Van Nostrand Company, Inc., Princeton, N. J., 1949.

READINGS AND SELECTED REFERENCES

Feller, William. *An Introduction to Probability Theory and Its Applications*, 7th ed., John Wiley & Sons, Inc., New York, 1957, Chaps. 6–7.

Games, Paul, and George Klare. *Elementary Statistics*, McGraw-Hill Book Company, New York, Chaps. 7–8.

Hammond, Kenneth, and James Householder. *Introduction to the Statistical Method*, Alfred A. Knopf, Inc., New York, 1963, Chap. 8.

Mosteller, Frederick, Robert Rourke, and George Thomas. *Probability and Statistics*, Addison-Wesley Publishing Company, Inc., 1961, Reading, Mass., 1961.

Parl, Boris. *Basic Statistics*, Doubleday & Company, Inc., Garden City, N.Y., Chaps. 11–13.

Spurr, William A., and Charles P. Bonini. *Statistical Analysis for Business Decisions*, Richard D. Irwin, Inc., Homewood, Ill., 1967, Chap. 8.

Yamane, Taro. *Statistics: An Introductory Analysis*, Harper & Row, Publishers, New York, 1967, Chap. 7.

6

Statistical Estimation:
Means and Proportions

PROBABILITY AND NONPROBABILITY SAMPLES

Let us repeat some of the argument to this point and introduce the idea of probability sampling. Inferential statistics is concerned with making estimates of unknown population parameters on the basis of known sample statistics and with testing hypotheses. Sampling distributions of the statistics we are concerned with are used to do this. But sampling distributions constructed on the basis of the mathematics of probability cannot be used to make inferential statements unless the sample used is a probability sample. There are various kinds of techniques for drawing different probability samples that are described in Chapter 15. At this point it is necessary to introduce the idea of *simple random sampling*, because the kind of estimation considered in this and subsequent chapters assumes that the sample is drawn by this process. As a matter of fact, the majority of conventional statistical techniques assumes that simple random sampling is used.

Simple random sampling may be defined as the process of selecting observations in such a way as to ensure that each member of the universe from which the sample is drawn has an equal chance of being included in the sample. If the universe contains 100,000 individuals, simple random sampling ensures that, for a sample of one, each member of the universe has a chance equal to 1/100,000 of being included in the sample. If some members of the universe have a greater or lesser chance of being included, the sample cannot be considered a simple random sample. (This also means that there must be independence of selection. We consider this in Chapter 15.)

A simple random sample is not necessarily representative of the universe; that is, it is not necessarily the same as the universe. It is a chance phenomenon. It is possible that a particular random sample may be quite unrepresentative of the universe it is drawn from. There are several reasons why the sample may deviate from the true population figures. (*a*) *Error introduced by sampling*. This is usually called *sampling error*. It is strictly a chance phenomenon, because we select only part of the population and each time we select we can get a different result. This source of error is considered in this chapter. (*b*) *Error introduced by nonprobability sampling.* If the basis for drawing a sample is the judgment of individuals or some "hit or miss" procedure, errors will be introduced into the sample. Statistical inference has no way of controlling and measuring this kind of error. (*c*) *Errors of measurement.* The amount of error introduced when the measurement instruments are not valid and reliable can sometimes be larger than that introduced by sampling. Taking account of measurement error, as well as some other sources of error, is considered in more detail in Chapter 15.

We said that in simple random sampling each member of the population has an equal chance of being drawn into the sample. There are several physical operations that ensure this result. (*a*) The names of each individual (or unit) can be written on a piece of paper or card, the cards shuffled, and a sample member drawn. This process is continued until the desired sample size is reached. (*b*) The individuals in the universe can be listed and numbered serially, and a table of random numbers can be consulted. Each individual whose number appears in the table of random numbers is selected for the sample. (*c*) The individuals can be numbered serially, and every nth person can be selected after the first person is selected at random. The selection is made on the basis of the ratio desired in the sample. Basically, each of these methods approximates a simple random sample. There are several questions and problems associated with each method. They are considered in Chapter 15. In the present chapter we suppose that a simple random sample has been drawn and our task is to estimate, first, the mean μ of the population and, second, the proportion P in the population.

ESTIMATION OF MEANS   .

The best way to explain estimation is by means of an example. Suppose that a random sample of 400 is taken from a large population, such as all the people who voted for Goldwater in the 1964 presidential election. Suppose further that the problem is to estimate the mean μ income of all those who voted for Goldwater. Let us say that the sample mean income $\bar{X}$ is $10,000 and the sample standard deviation s is $5,000. We want to say something about the population mean μ on the basis of the sample result. As stated in Chapter 5, in order to do so, we must have a sampling distribution — a hypothetical probability distribution. Let us consider now how we may construct such a distribution and make a decision about μ on the basis of $\bar{X}$.

First, we know that, if we take another random sample of 400 from the same population and compute the mean income of all those who voted for Goldwater in the second sample, we shall not get a mean of exactly $10,000.

It may be $10,200, or $9,500, or even as much as $14,000. If we were able to repeat the process of drawing random samples of 400 from the same population an infinite number of times, each time computing the mean income of Goldwater voters, the sample means could be cast in the form of a frequency distribution. Such a frequency distribution is called a *sampling distribution of means*. In an actual situation, of course, we are not able to draw random samples continually in order to construct a sampling distribution to use as the basis for making a decision about μ. Usually we take just one sample. Then, through the mathematics of probability, we construct a sampling distribution of all of the means that we might get if we were able to repeat the sampling process an infinite number of times. Even better, we can use the standard normal curve as an approximation of the sampling distribution of means. This is possible because of two important theorems: the central limit theorem and the law of large numbers. Before considering these theorems, let us illustrate what we said in this paragraph by means of an example.

Table 6-1　Frequency Distribution of Marriages per 1,000 Population, Aged 15–44*

Stated class interval	*Frequency*
8–11	6
11–14	3
14–17	11
17–20	18
20–23	10
23–26	1
26–29	1
	50

*Adapted from: Bruce M. Russett *et al., World Handbook of Political and Social Indicators,* Yale University Press, New Haven, Conn., 1964. Stated class interval represents number of marriages per 1,000 population aged 15–44.

We shall construct a small sampling distribution of means. Recall that in statistical inference we deal with three frequency distributions: (*a*) the population distribution, which is unknown; (*b*) the distribution obtained by a simple random sample; and (*c*) the sampling distribution of the statistic we are using to estimate a population parameter, which in this case is the mean. The population distribution that will be used is the distribution of marriages per 1,000 population aged 15 to 44 for 50 countries.[1] The popu-

---

[1] The data are taken from Bruce M. Russett et al., *World Handbook of Political and Social Indicators*, Yale University Press, New Haven, 1964, pp. 229–230.

lation parameter μ is known, because we are using it as an illustration. It is 17.2. Usually we do not know the population parameter; we are trying to estimate it. The tabular frequency distribution and the curve for the population data are shown in Table 6-1 and Figure 6-1.

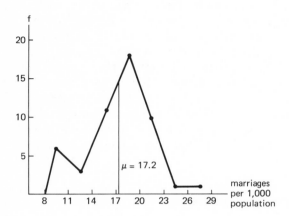

Figure 6-1. Frequency distribution of marriages per 1,000 population, aged 15–44. Adapted from Bruce M. Russett et al., *World Handbook of Political and Social Indicators*, Yale University Press, New Haven, Conn., 1964.

As a simulation of what a sampling distribution of means looks like, 50 samples of size $n = 5$ each were drawn by simple random sampling from this population with replacement. This was done by first numbering the 50 countries serially from 01 to 50 and then consulting the table of random numbers in the Appendix. A spot was chosen at random in this table, and successive two-digit numbers were perused. Each time a two-digit number between 01 and 50 occurred, the country having the number was selected for the sample. To give the reader a better understanding of these data, we might mention that the countries that are highest in marriages per 1,000 population aged 15 to 44 are the U.S.S.R. (26.6), Rumania (23.1), and West Germany (22.6). The United States is relatively high, with a rate of 21.5. The countries that are lowest are El Salvador (8.1), Honduras (8.7), and British Guiana (9.3).

The frequency distribution for the 50 samples of size 5 is shown in Table 6-2. The sample means range from a low of 14.30 to a high of 19.54. The means of the 50 means is 17.04, which is slightly less than the population mean of 17.20. The frequency polygon in Figure 6-2 comprises a sampling distribution of the means. If all possible samples of size $n = 5$ were drawn from the same population, the shape of the sampling distribution would be nearly normal. How many possible samples of size 5 (each one different) can be drawn from this population? The answer is $_{50}C_5 = 50!/[5!(45!)] =$ 2,118,760. It can readily be seen that, if we had to construct a sampling distribution of means of all possible samples of size 400 from a population of 10,000, which is $10,000!/[400!(9,600!)]$, our labor would be enormous!

Table 6-2 Frequency Distribution of Sample Means

$\overline{X}$	f
14–15	10
15–16	5
16–17	7
17–18	10
18–19	14
19–20	4
	50

Happily, we can use the general families of sampling distributions described in Chapter 5 rather than resort to this method. When we are dealing with means and our samples are larger than 30, we can use the standard normal distribution as an approximation of the sampling distribution of means. This is because of two important theorems: the law of large numbers and the central limit theorem. We state them in nonrigorous form here. The *law of large numbers* can be stated as follows: If we draw an infinite number

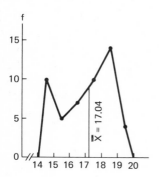

Figure 6-2. Sampling distribution of 50 means.

of samples from a population, compute the mean of each sample and return its elements before drawing the next sample, we can expect that the mean of these sample means, which we designate $\overline{\overline{X}}$, will be equal to the population mean; that is, $\overline{\overline{X}} = \mu$. The *central limit theorem* can also be stated nonrigorously for sample means as follows: If we draw an infinite number of samples

from a population, compute the mean of each sample and return its elements
before drawing the next sample, we can expect that the sampling distribution
of these sample means will approach the shape of the normal curve and that
the standard deviation of this sampling distribution of sample means will
be equal to $\sigma/\sqrt{n}$, where σ is the standard deviation of the population from
which the samples are drawn.

These two important theorems enable us to use the standard normal
curve as the sampling distribution of means that we should be likely to get
if we actually repeated the sampling process an infinite number of times.
It can be seen in Figure 6-2 that we get only a rough approximation of this
when 50 samples of size 5 are drawn from a population of 50. The mean of
the 50 sample means is 17.04 rather than 17.2, because we do not have enough
samples. But it can be seen that the approximation is close for as few as
50 samples. The standard deviation of the sampling distribution of means
in Table 6-2 is 1.69. The central limit theorem tells us to expect the standard
deviation of the sampling distribution of means to equal $\sigma/\sqrt{n}$. This is
called the *standard error of the mean*, and it is designated symbolically $\sigma\bar{x}$.
Because the standard deviation in the population from which the samples
were drawn (see Table 6-1) is 4.02, the standard deviation of the sampling
distribution computed by the central limit theorem is

$$\sigma\bar{x} = \frac{\sigma}{\sqrt{n}} = \frac{4.02}{5} = 1.79$$

The standard deviation of the 50 samples in Table 6-2 is less than this,
because there are too few samples and because the sample ($n = 5$) is more
than 5 percent of the population ($N = 50$). The correction factor to use
when the sample is more than 5 percent of the population is considered later
in this chapter. The standard deviation of the sampling distribution of means
in Table 6-2, as well as the mean, is only a close approximation of what the
central limit theorem and the law of large numbers tell us to expect, because
not all the assumptions of these theorems have been met. But when they
are, we can use the standard normal curve as the sampling distribution of
means for estimation problems. Let us now consider this question in detail.

POINT AND INTERVAL ESTIMATION

When we say something about an unknown population parameter on the
basis of a sample statistic, we can do either of two things: estimate the
interval in which the population parameter is likely to fall or estimate the
exact value of the population parameter and measure the amount of error

we are likely to be making. Consider again the problem of estimating the income of those who voted for Goldwater in 1964. Because we know that, if we drew a second sample from the same population, the mean income would not be exactly the same as the first mean, we should not be correct in saying that the mean income of those who voted for Goldwater is exactly $10,000 (the first sample result). Instead, we might say either of two things: (*a*) the mean income of those who voted for Goldwater is in the interval of $9,510 to $10,490, and we are 95 percent confident that this is correct; or (*b*) the mean income of those who voted for Goldwater is $10,000, and we are 95 percent confident that the error is not larger than $490. How do we measure the width of the interval for an interval estimate or the amount of error for a point estimate? What is meant by the statement that we are 95 percent confident that we are correct?

As the reader may know, we can use the normal curve to answer these two questions. Recall that for the normal curve we know how much of any area lies within standard deviations from the mean. We know that 95 percent of the total area of the normal curve is within plus or minus 1.96 standard deviations, or Z scores, from the mean; 98 percent is within plus or minus 2.33; and 99 percent is within plus or minus 2.58.[2] We can take the particular sample mean and construct an interval around it in which we can be fairly confident that the true population parameter lies. How confident depends on how large an interval is constructed around the sample mean. If the interval encompasses 1.96 standard deviations from the standard normal curve, we can be 95 percent confident that the true population μ is in this interval. What we are doing, in effect, is constructing an interval around the sample mean in which 95 percent of all sample means would fall if the true population mean were to equal the sample mean got by sampling.

In order to do this, the particular statistic we have got by random sampling must be converted to a standard score, because we do not want to construct a normal distribution for each particular sample; that is, in the case of a sample mean $\bar{X}$, we want to convert it to a Z, or standard form, so that it is possible to go directly to a table of areas under the standard normal curve. The method of converting the sample mean to a standard normal form is the same as that used for converting a particular observation to a standard score form. In this case, however, we are dealing with a sampling distribution of means, and the particular sample mean is treated as an individual observation. The mean of the sampling distribution, which the central limit theorem tells us is equal to μ, is the mean, therefore, that is subtracted from the particular sample mean. Furthermore, the standard deviation we divide by is that of the sampling distribution of means given above as $\sigma/\sqrt{n}$.

---

[2] If you are not convinced that these numbers are correct, you should turn to the table of areas under the standard normal curve and verify them.

Therefore, the equation for converting the particular sample mean to standard form is

$$Z = \frac{\bar{X} - \mu}{\sigma/\sqrt{n}} \tag{6-1}$$

In order to construct an interval around the standard form of $\bar{X}$ in which we can be 95 percent confident that the mean of the population lies, we want an interval in which it will be within plus or minus 1.96 standard deviations:

$$-1.96 > \frac{\bar{X} - \mu}{\sigma/\sqrt{n}} < 1.96$$

that is, the standard form of $\bar{X}$ will be in the interval that is within ± 1.96 standard deviations of the mean of the sampling distribution of $\bar{X}$, which is μ. There are two unknowns in this inequality: μ and σ. We can perform some simple algebra in order to eliminate all but the parameter in which we are interested, μ, from the middle term:

$$-1.96 \frac{\sigma}{\sqrt{n}} > \bar{X} - \mu < 1.96 \frac{\sigma}{\sqrt{n}} \qquad \text{multiplying by } \frac{\sigma}{\sqrt{n}}$$

$$-\bar{X} - 1.96 \frac{\sigma}{\sqrt{n}} > -\mu < 1.96 \frac{\sigma}{\sqrt{n}} - \bar{X} \quad \text{subtracting } \bar{X}$$

$$\bar{X} + 1.96 \frac{\sigma}{\sqrt{n}} > \mu < \bar{X} - 1.96 \frac{\sigma}{\sqrt{n}} \qquad \text{multiplying by } -1$$

Thus, a 95 percent confidence interval for μ is

$$\mu = \bar{X} \pm 1.96 \frac{\sigma}{\sqrt{n}} \tag{6-2}$$

We still have an unknown on the right side of this equation: σ. If we knew σ, we should not need to estimate μ, because we would know it. Therefore, we must estimate σ in order to estimate μ. If we have a large sample, we can use s as an estimate of σ. (In this case, a large sample is generally considered to be above 30.) Thus, the 95 percent confidence interval equation for μ for large samples[3] becomes

$$\mu = \bar{X} \pm 1.96 \frac{s}{\sqrt{n}} \tag{6-3}$$

[3] Note that Equation (3-13) is used to compute s. If s is computed by $s = \sqrt{\Sigma x^2/n}$, then Equation (6-3) should use $s/\sqrt{n-1}$ as the standard error of the mean. In the latter case, the symbol designation for the standard error of the mean becomes $s_{\bar{x}}$ instead of $\sigma_{\bar{x}}$.

Returning now to the example of estimating the income of people who voted for Goldwater on the basis of a sample of 400, we have

$$\mu = \$10,000 \pm 1.96 \frac{5,000}{\sqrt{400}}$$

$$= \$10,000 \pm \$490$$

$$= \$9,510 \text{ to } \$10,490$$

We can now say that we are 95 percent confident that the population mean is in the interval of \$9,510 to \$10,490. Another way of stating this is in terms of repeated sampling. If we repeated the process an infinite number of times, 95 percent of the time we should be correct in saying that the population mean is in this interval. Because of the chance that the sample mean might be extremely biased, 5 percent of the time the true population mean is not in this interval. It can be seen from the last statement that a confidence interval might be used to test a particular hypothesis. Although the question of hypothesis testing is considered in more detail in Chapter 7, at this point it will be instructive to consider the following example. Suppose that someone asserts that the mean income of Goldwater voters in the 1964 presidential election is \$6,000 and that a random sample of 400 produces a mean of \$5,400 with a standard deviation of \$4,000. The 95 percent confidence interval for μ is

$$\mu = \$5,400 \pm 1.96 \frac{4,000}{\sqrt{400}}$$

$$= \$5,008 \text{ to } \$5,792$$

We can therefore reject the figure \$6,000 with only a 5 percent chance that we are wrong; that is, only 5 percent of the time, if the process of sampling were repeated many times, would we get a figure as low as \$5,400 if, in fact, the actual population mean were \$6,000.

Point estimation follows the same logic except that we want to say what the population parameter is and measure how far off the estimate is likely to be. For point estimation, the sample mean is taken as the maximum likelihood estimator of the population parameter. (The term "maximum likelihood" is explained below.) The level of confidence, as in interval estimation, corresponds to the area under the standard normal curve. The amount of error that might be made is equal to the standard deviation of the sampling distribution, which is called the standard error of the mean:

$$\frac{\sigma}{\sqrt{n}} = \frac{s}{\sqrt{n}} \qquad \text{for samples} > 30$$

Thus, for the earlier example concerning the income of Goldwater voters in the 1964 presidential election, it can be said that the mean income of voters is \$10,000, and we can be 95 percent confident that the amount of error is not greater than

$$(1.96)\frac{5,000}{\sqrt{400}} = \$490$$

Note that, if we want to be more confident that the estimate is correct, the amount of possible error increases. For example, to be 99 percent confident, the amount of error is

$$(2.58)\frac{5,000}{\sqrt{400}} = \$645$$

Similarly, for a 99 percent confidence interval we have

$$\mu = \bar{X} \pm 2.58\frac{s}{\sqrt{n}} = 10,000 \pm 2.58\frac{5,000}{\sqrt{400}} = \$9,355 \text{ to } \$10,645$$

Hence, to have more confidence in our estimate, the range of the interval must be wider and the possible amount of error larger.

The Standard Error of the Mean

At this point, a few important characteristics of the standard error of the mean should be emphasized. First, fix firmly in your mind the idea that it is the standard deviation of the distribution of sample means when we use the standard normal curve as the sampling distribution of means. The standard error measures how far off any particular sample mean is likely to be as an estimate of μ. When the sample is small compared with the population (less than 5 percent), the standard error of the mean is

$$\sigma\bar{x} = \frac{\sigma}{\sqrt{n}}$$

where σ is the standard deviation of the population and n is the sample size. But when we do not know σ, we can estimate it with s if the sample is large (above 30), giving

$$s\bar{x} = \frac{s}{\sqrt{n}}$$

When the sample is a large percentage of the total population (> 5 percent), the f.p.c. factor has to be used. This is considered later in this chapter. At this point note simply that the standard error of the mean is defined by the equation above for the conditions specified.

Determining Sample Size

The same reasoning we used to make point and interval estimates can be used to determine what sample size is needed in order to say something about a population mean. Let us reverse the argument. Suppose that we want to estimate the income of all voters in the 1964 presidential election and that we are willing to be 95 percent confident that the error is not more than $100. What sample size do we need? As stated above, the amount of error that will be made for 95 percent confidence is 1.96 $(\sigma/\sqrt{n})$. We want this quantity to be equal to or less than $100; that is, 1.96 $(\sigma/\sqrt{n}) \leq \$100$. With a little algebra we can move the variable we want to determine to the right-hand side of the equation, leaving the rest on the left-hand side:

$$1.96 \frac{\sigma(\sqrt{n})}{\sqrt{n}} \leq \$100\sqrt{n} \qquad \text{multiplying both sides by } \sqrt{n}$$

$$1.96/100 \; \sigma \leq \sqrt{n} \qquad \text{simplifying and dividing by \$100}$$

or $\quad (1.96/100\sigma)^2 \leq n$

Again, we still have an unknown σ that needs to be determined. In this case, we cannot substitute s, because we have not drawn the sample yet. The only alternative, therefore, is to estimate σ. There are two ways to do this. We can rely on judgment based on past knowledge of the particular variable we are studying. For example, for income, the standard deviation of the sample of income earners in Chapter 2 is $7,435. We know that the income distribution of voters will be a little more homogeneous, because fewer low-income people vote. Therefore, we might take $6,000 as a rough estimate of the population standard deviation and compute the sample size on the basis of this:

$$[1.96/100(6,000)]^2 = n \doteq 13,830$$

that is, we need a sample of about 13,830 to be 95 percent confident that our error is not larger than $100.

The other alternative for estimating σ is to take 5/4 of the interquartile deviation.[4] About 25 percent of income earners make less than $4,000 a year, and about 25 percent make more than $10,000. The interquartile deviation is thus $6,000, and 5/4 of this is $7,500. This time we do not assume that the income of voters is more homogeneous than the population of income earners. The sample size needed when we use 5/4 of the interquartile deviation as an estimate of σ is

$$[1.96/100(7,500)]^2 = n \doteq 21,609$$

In both cases, the sample size needed is rather large. This brings us

[4] See the discussion in Chapter 3 for the rationale for this.

face to face with the fact that how accurate we can be is a function of how much money is available for sampling. It is rather costly to sample 11,000 to 21,000 people. Therefore, in estimating the mean income, if sufficient funds are not available for a sample of 11,000, the only alternatives are either to decrease the level of confidence or to increase the amount of error we are willing to tolerate. For example, if we maintain the maximum error of $100 but are satisfied to be 80 percent confident that our error is not larger than $100, the sample size required is

$$[1.28/100(7,500)]^2 = n \doteq 9,216$$

However, if the problem is important enough to require at least a 95 percent level of confidence, we might increase the amount of error to $500. Then the sample size needed is

$$[1.96/500(7,500)]^2 = n \doteq 864$$

Finally, note that, if we were estimating something that had a relatively small standard deviation, such as mean years of formal schooling, then the sample size needed to be 95 percent confident that the error is not greater than one year is

$$[1.96/1(3.8)]^2 = n = 56$$

Thus, it is apparent that the sample size needed depends on three factors: (*a*) the level of confidence desired, (*b*) the amount of error that can be tolerated, and (*c*) the amount of dispersion in the population we are generalizing about. It is also apparent that the most important of the three is the amount of dispersion in the population. If we are generalizing about a completely homogeneous population, a sample of one is sufficient.

Relationship to Population Size

Note that the size of the population we want to generalize about is not directly related to the sample size needed except insofar as it may affect the standard deviation. For example, the population of all income earners in the United States is approximately 78 million people. The population of undergraduate students at an average university is about 12,000. In the latter case, a smaller sample would be required to estimate the mean income, not because it is a smaller population, but because it is more homogeneous. The standard deviation of all income earners is about $7,435, but for college undergraduates it is more likely about $1,500.

Estimation of μ for Small Samples

If the sample size being used is under 30, the normal curve cannot be used as the sampling distribution of means. The distribution of sample means for small samples is more spread out than the normal curve, as shown in Figure 6-3. For small samples the distribution of sample means can be approximated better by the *t* distribution. The shape of the *t* distribution is known and tabled for various degrees of freedom, which are *n* — 1 in this case.[5] An intuitive understanding of the nature of this distribution is sufficient for the purposes of this book.

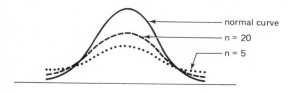

Figure 6-3.

In general, the smaller the sample, the more likely that the mean of the sample will be a less efficient estimation of the population mean. There will be a greater number of sample means farther from the true population mean than in the case of larger samples. The *t* distribution takes these facts into account.

The *t* distribution is symmetrical, but it is not normal, because a larger area falls outside of one, two, or three standard deviations than in the case of the normal distribution. The *t* distribution is thus platykurtic. The exact shape of the sampling distribution of *t* depends on the size of the sample. There is a *t* distribution for samples of *n* = 12, another for *n* = 13, and so on. As Figure 6-2 shows, the *t* distribution approaches the normal distribution as *n* increases.

We can use the same reasoning in making interval and point estimations of the population mean based on small samples as we did for large samples except that for small samples we convert the particular sample mean into a *t* rather than a *Z* score. For a 95 percent confidence interval estimate for a small sample, we replace the standard deviations of the normal curve in which 95 percent of all observations fall with the comparable areas of the *t* distribution. For example, for 95 percent confidence, we want the area under the *t* distribution in which 95 percent of the area lies, or ±*t*.025. The latter refers to the *t* value outside of which 5 percent of the total area lies, with 2 1/2 percent on either side. Note that the *t* distribution is set up in a manner opposite the normal distribution; that is, the numbers in the body of the table in the Appendix refer to standard deviations of the *t* distribution for different sample sizes (as will be explained later, degrees of free-

[5] The concept degrees of freedom is considered in Chapter 8. See Table 2 in the Appendix for areas under the *t* distribution.

dom are always $n - 1$ for the t distribution). The probabilities are at the top of the table, and they refer to the area of the t distribution that falls outside of the t value specified in the table.

The table of the t distribution in the Appendix shows that for a sample of 20, for example, the t value for $t.025$ is 2.093; that is, 2 1/2 percent of the area falls in each tail to the right and left of the interval $t = \pm 2.093$, and therefore 95 percent of the area falls within this interval. The t value that must be used for a 95 percent confidence interval for samples of $n = 20$ is thus 2.093.

As an example, suppose that we have a random sample of 20, the mean income of the sample is \$5,400, and the standard deviation is \$4,000. The 95 percent confidence interval will be

$$\mu = \bar{X} \pm t.025 \, \frac{s}{n - 1}$$

$$= \$5,400 \pm 2.093 \, \frac{4,000}{19}$$

$$= \$5,400 \pm \$441$$

$$= \$4,959 - \$5,841$$

When the sample is small and we use the t distribution, the confidence interval will be wider than that for a larger sample using the normal curve.

ESTIMATION OF PROPORTIONS

Just as the normal curve can be used as the sampling distribution of means when the sample size is above 30, the normal curve can also be used as the sampling distribution of proportions when the sample size is greater than 30 and P is not extremely small or large.[6] Suppose that we wanted to estimate the proportion of Americans who oppose a particular government policy and we take a sample of 400 to do so. Suppose that the proportion of individuals in this sample who oppose the policy is 0.40. If we drew a second sample and asked the same question, we should not come up with $p = 0.40$ but might get $p = 0.48$, or $p = 0.44$, or even $p = 0.30$. If we repeated the sampling process, computed p for each sample, and cast the resulting proportions in a frequency distribution, we should find that they would form a normal distribution. In order to construct a confidence interval for p, therefore, we want to convert the particular p we get from one sample into a normal or standard score, as was done for $\bar{X}$.

[6] When P is extremely small or large, the Poisson distribution should be used.

In Chapter 5 we said that for a binomial population, the mean and standard deviation are the expected number of events we should get in the long run, defined as

$$\mu = nP$$

$$\sigma = \sqrt{nPQ}$$

The second equation defines the number of successes we should expect in n trials. When estimating proportions, we are concerned with the sampling distribution of p rather than the frequency distribution of an actual experiment, and therefore we want the standard error of a proportion; that is, we are interested in the distribution of the proportion of successes in n trials. To get this, we use:

$$\sigma_p = \sqrt{\frac{PQ}{n}}$$

In contrast to the standard error of the mean the size of the standard error of a proportion depends on the size of the sample. For example, if $n = 10$ and $P = 0.5$,

$$\sigma_p = \sqrt{\frac{(0.5)(0.5)}{10}} = 0.1581$$

but, if $n = 100$ and $P = 0.5$,

$$\sigma_p = \sqrt{\frac{(0.5)(0.5)}{100}} = 0.0500$$

A little more formally stated, we can say that, if we take successive large samples from a population and compute p for each sample, the sampling distribution of p will approximate a normal distribution with a mean equal to $\mu = nP = P$ and a standard deviation equal to $\sigma_p = \sqrt{PQ/n}$, where N and P refer to population parameters and n is the sample size. We can convert the particular p we get for any sample into a Z or standard normal score by the following:

$$Z = \frac{p - P}{\sqrt{PQ/n}} \tag{6-4}$$

As with interval estimates for the mean, for a 95 percent confidence interval we want to construct an interval in which the standard form of p will be plus or minus 1.96 standard deviations from P, or

$$-1.96 > \frac{p - P}{\sqrt{PQ/n}} < 1.96$$

By algebra, we get

$$-1.96 \sqrt{\frac{PQ}{n}} > p - P < 1.96 \sqrt{\frac{PQ}{n}}$$

$$p + 1.96 \sqrt{\frac{PQ}{n}} > P < p - 1.96 \sqrt{\frac{PQ}{n}}$$

or

$$P = p \pm 1.96 \sqrt{\frac{PQ}{n}}$$

for the 95 percent confidence interval.

We said above that the uppercase P and Q refer to population parameters. Of course, we do not know them; our task is to estimate them. For large samples we can use the sample p and q as estimates of P and Q. Thus, the 95 percent confidence interval for large samples is given by

$$P = p \pm 1.96 \sqrt{\frac{pq}{n}} \qquad (6\text{-}5)$$

Returning to the example above, if the proportion of individuals in a random sample of 400 who oppose the government's policy is 0.40, the 95 percent confidence interval is

$$P = 0.40 \pm 1.96 \sqrt{\frac{(0.40)(0.60)}{400}}$$

$$= 0.40 \pm 0.025$$

$$= 0.375 \text{ to } 0.425$$

We conclude that we can be 95 percent confident that the population parameter is in this interval.

Point estimation for proportions is identical in reasoning. For the 95 percent level of confidence we can state that $p = P$ and be 95 percent confident that our error is not larger than $1.96\sqrt{pq/n}$. For the example above, this is $1.96\sqrt{(0.40)(0.60)/400} = 0.025$.

Determining Sample Size for Proportions

The amount of error we are likely to make in estimating proportions is given by $\sigma_p = \sqrt{PQ/n}$. If we want this to be less than a given amount, represented by E, and want to be 95 percent confident that the error is not larger than this amount, we set one side equal to the other and get

$$1.96 \sqrt{\frac{PQ}{n}} = E$$

Solving for n, we get

$$1.96 \sqrt{\frac{PQ}{n}} = E$$

$$1.96 \sqrt{PQ} = E(\sqrt{n})$$

$$\frac{1.96}{E} \sqrt{PQ} = \sqrt{n}$$

$$\left(\frac{1.96}{E}\right)^2 PQ = n \qquad\qquad (6\text{-}6)$$

We do not know PQ because they are the parameters we are trying to estimate and we are forced into the position of trying to find estimates for P and Q so as to determine how large a sample we need in order to get an accurate estimate of PQ. This is a little like having to guess what your opponent in chess is going to do so that you can take action that forces him to take action and you then know what he does. The difference is that in sampling, we have more solid information on which to base the guess than is usually so in a chess game. First of all, we know that P must lie in the interval between 0 and 1. If we let $P = 0.50$, then Q will also be 0.50, because $Q = 1 - P$; and the product of $PQ = 0.25$. Now let P be any value other than 0.50. For example, if it is 0.60, then $Q = 0.40$, and $PQ = (0.6)(0.4) = 0.24$. Note that any value of P other than 0.50 produces a product $PQ < 0.25$; that is, the largest number that PQ can be is 0.25. Therefore, we can always be on the safe side estimating $P = 0.5$, because the standard error of the proportion σ_{p} will be larger for any sample size for $P = 0.5$ than for any other value for P. Therefore, we can be conservative and substitute 0.25 or $1/4$ into Equation (6-6) in place of PQ and get

$$\left(\frac{1.96}{E}\right)^2 \left(\frac{1}{4}\right) = n \qquad\qquad (6\text{-}7)$$

For example, if we want to estimate the sample size needed to estimate the proportion of Americans who oppose a particular government policy and want to be 95 percent confident that our error is not more than 2 percent, we have

$$\left(\frac{1.96}{0.02}\right)^2 \left(\frac{1}{4}\right) = 2{,}401$$

If we are willing to tolerate a 5 percent error, n becomes

$$\left(\frac{1.96}{0.05}\right)^2 \left(\frac{1}{4}\right) = 384$$

In contrast to estimating means, the size of the sample needed to estimate proportions depends on how large an error we are willing to tolerate rather than on how homogeneous the population is to which we want to generalize, because the standard error of proportions cannot fluctuate so widely as for interval-scale variables. In general, therefore, smaller sample sizes are needed for estimating proportions than for means. Note that, as for means, the sample size needed for estimating proportions does not depend on the size of the population to which we want to generalize.

Estimating Proportions for Small Samples

For small samples the normal curve is not a good sampling distribution of proportions. In this case the binomial distribution must be used. Suppose that we had a sample of 20 and the sample $p = 0.40$. If tables of the binomial distribution are not available, a binomial distribution for $p = 0.40$ and $n = 20$ could be constructed using the binomial theorem. If we did so, we should find that 95 percent of all the cases fall within plus or minus 2.3 standard deviations of the mean. Therefore, the 95 percent confidence interval in this case is

$$P = 0.4 \pm 2.3 \sqrt{\frac{(0.4)(0.6)}{20}}$$

$$= 0.4 \pm 0.251$$

$$= 0.149 \text{ to } 0.651$$

However, a note of caution should be introduced here, because complex problems are involved in making inferences using the binomial distribution. Among other things, the confidence intervals are not symmetrical if $p \neq 0.5$. Small samples are difficult to deal with unless some prior information is available. We consider this question in Chapter 15.

CORRECTIONS FOR SAMPLING FROM A SMALL UNIVERSE

As we said earlier in this chapter, if the universe we are sampling from is so small that the sample comprises a large proportion of the universe (5 percent or more) or if we are sampling without replacement, then the standard error of the mean and the standard error of the proportion must be modified. For example, suppose that we want to determine the proportion of a congressional committee, consisting of 10 members, that has invested in communications stock. Suppose that we are able to speak with 9 of the 10 members and are able to learn that of the 9, there are 3 who invest and 6 who do not. Using the technique of the last section, for $n = 1$, $p = 3/9$, a 95 percent confidence interval runs from 0.07 to 0.69. But it is obvious that this is not meaningful at all. The tenth member of the committee either does or does not invest in communications stock. If he does, $P = 0.4$; and if he does not, $P = 0.3$. Thus, saying that we are 95 percent confident that the population parameter is between 0.07 and 0.69 for this case is not very good. We know very well that it is either 0.4 or 0.3. In this case, a confidence interval is obviously not called for.

Consider a more realistic example. Suppose that we take a sample of $n = 43$ of the House of Representatives. The sample size is about 10 percent of the population size. Suppose that we use this sample to estimate two parameters: the mean income of all representatives and the proportion of those who perceive their role to be a delegate rather than a trustee. To construct confidence intervals or make point estimates, the standard error of the proportion and the standard error of the mean would be multiplied by $\sqrt{1 - n/N}$, where N is the population size and n is the sample size. This is called the *finite population correction factor* and is considered in more detail in Chapter 15. Given the correction factor, the standard error of the mean becomes

$$\sigma_{\bar{x}} = \frac{\sigma}{\sqrt{n}}\sqrt{1 - \frac{n}{N}} \tag{6-8}$$

or $\qquad\qquad s_{\bar{x}} = \dfrac{s}{\sqrt{n}}\sqrt{1 - \dfrac{n}{N}} \qquad$ for samples > 30

and the standard error of a proportion becomes

$$\sigma_p = \sqrt{\frac{PQ}{n}}\sqrt{1 - \frac{n}{N}} \tag{6-9}$$

or $\qquad\qquad s_p = \sqrt{\dfrac{pq}{n}}\sqrt{1 - \dfrac{n}{N}} \qquad$ for samples > 30

MAXIMUM LIKELIHOOD ESTIMATION

So far in this chapter we have described how to make point and interval estimates, but we have not said anything about the method of estimation itself. There are various methods of estimation. In this book we consider only two: maximum likelihood and least squares. Least squares is taken up in Chapter 9. What we have been describing in the present chapter is maximum-likelihood estimation. The principle of maximum likelihood provides estimators that are efficient, consistent, and sufficient. An *unbiased estimator* is one that produces an expected value of a statistic equal to the population parameter to be estimated. For example, the expected number of heads in 1,000 flips of a coin is 500, and the expected proportion is 0.50. We also know that the mean of a sampling distribution of proportions if n is large enough is equal to the population proportion. Similarly, the mean of a sampling distribution of all possible means is expected to be equal to the population parameter μ. Also, the most likely sample mean $\bar{X}$ is the mean of all possible sample means. Therefore, $E(\bar{X}) = \mu$, and we can say that $\bar{X}$ is an unbiased estimator of μ. In simpler terms, a biased estimator is one that consistently under- or overestimates the parameter. Thus, the sample median would not be an unbiased estimator of μ except under one condition (the reader should be able to identify this condition).

A *consistent estimator* is one that approaches the population parameter as n increases. Suppose that $\mu = 155$. If we have a random sample of 20 and $\bar{X} = 150$, then as n is increased, $\bar{X} \rightarrow 155$. For example, for a sample of 10,000, we might expect $\bar{X} = 154.9$. Thus, $\bar{X}$ is a consistent estimator of μ.

An *efficient estimator* is one whose dispersion is smallest. For example, $\bar{X}$ and Md are both unbiased estimators of μ when we have a normal distribution. But the dispersion of Md is greater than that of $\bar{X}$.[7] Therefore, although both may be used as estimators of μ when we have a normal distribution, $\bar{X}$ is a more efficient estimator.

A *sufficient estimator* is one that uses all the information a sample contains about the population parameter. $\bar{X}$ is a sufficient estimator of μ because no other sample statistic, such as s, can add further information about μ.

Maximum-likelihood estimation provides a method of obtaining efficient, consistent, and sufficient estimators. In the example used at the beginning of this chapter, the sample mean was $10,000. In terms of the maximum likelihood, we say that the population parameter μ is most likely $10,000. Why is this so? The population parameter is either $10,000 or it is not. By converting to a Z score, we saw that the sample was not likely to have been drawn from a population with $\mu = $14,000$, or $12,000, or $11,500, and so on. It is only if we assume that $\mu = $10,000$ that the sample result is most likely. Thus, the most likely value of μ is $10,000.

The mathematics of maximum-likelihood methods are beyond the scope of this book. Suffice it to say that $\bar{X}$ and p are maximum-likelihood estimators of μ and P respectively. The intuitive understanding of maximum-likelihood estimation presented above is sufficient for the purposes of this book.

We have described in this chapter how sampling distributions are used as an aid in making estimates of the population parameters μ and P. This, of course, does not exhaust the topic of estimation. We could now discuss estimating other population parameters, such as σ, or, as we shall see later in the book, parameters such as ρ (correlation coefficient) and β (regression

---

[7] The latter was given above as $\sigma/\sqrt{n}$, and the former is $\pi\sigma/2\sqrt{n}$.

coefficient). However, at this point it is better to turn to the other major facet of statistical inference — hypothesis testing. We shall return later to additional estimation questions but give them much less emphasis, because they are secondary to the questions of testing hypotheses and measuring relations between variables.

READINGS AND SELECTED REFERENCES

Blalock, Hubert, Jr. *Social Statistics*, McGraw-Hill Book Company, New York, 1960, Chap. 12.

Diamond, Solomon. *Information and Error*, Basic Books, Inc., Publishers, New York, 1959, Chap. 6.

Edwards, Allen. *Statistical Analysis*, Holt, Rinehart and Winston, Inc., New York, 1965, Chaps. 8–10.

Freund, John. *Modern Elementary Statistics*, Prentice-Hall, Inc., Englewood Cliffs, N. J., 1960, Chap. 10.

Games, Paul, and George Klare. *Elementary Statistics*, McGraw-Hill Book Company, New York, 1967, Chap. 9.

Spurr, William A., and Charles P. Bonini. *Statistical Analysis for Business Decisions*, Richard D. Irwin, Inc., Homewood, Ill., 1967, Chaps. 11, 13.

Yamane, Taro. *Statistics: An Introductory Analysis*, Harper & Row, Publishers, New York, 1967, Chap. 10.

7
Hypothesis Testing: Difference of Means and Proportions

Estimating population parameters, such as μ and P, is an important aspect of statistical inference. In political science, their applications are usually in survey research and opinion sampling. Another important area of statistical inference concerns hypothesis testing of the relationship among variables. In conventional terminology, we are usually interested in the relationship between a *dependent* variable, or the phenomenon to be explained, and one or more *independent* variables, or the things that are to do the explaining. In this chapter we first describe the procedure of hypothesis testing, usually called the classical or conventional method (contemporary developments are considered in Chapter 14). It is called the classical method because it is that aspect of statistical inference which was first developed and widely used. We then consider two methods of testing hypotheses about the relationship between two variables: difference of means and difference of proportions.

In conventional statistical decision making, or hypothesis testing, the hypothesis that is tested is called the *null hypothesis*. It is usually symbolized as H_0. The research hypotheses are symbolized as $H_1, H_2, \ldots, H_n$. We first illustrate the method by means of examples and then consider it more generally.

Suppose that we are doing research concerning the relationship between the political milieu of a community and the role perception of legislators.[1] Specifically, suppose that the research hypothesis is that, as the political milieu becomes more competitive, legislators tend to adhere more to a delegate perception of their role. Let the symbols Y_i and X_i represent the dependent and independent variables respectively. The research hypothesis in this case would be stated as follows:

$$H_1: \text{As } X_i \text{ increases, } Y_i \text{ increases}$$

(or X_i and Y_i are positively related).

[1] This example was suggested by the work of Heinz Eulau and his colleagues concerning the role perception of state legislators. See Heinz Eulau, John C. Wahlke, William Buchanan, and LeRoy Ferguson, *The Legislative System*, John Wiley & Sons, Inc., New York, 1962.

The null hypothesis can take on any one of several different forms, depending on the research design being used. We shall have more to say about research design in a moment. For the present we note that, if the research design aims to determine if there is a correlation between Y_i and X_i, the null hypothesis would be

$$H_0: \rho = 0$$

where ρ represents the correlation in the population.

If, instead, the question is to be answered by a comparison of different groups, the null hypothesis would be

$$H_0: \mu_1 = \mu_2 = \cdots = \mu_n$$

where μ represents the mean score for role perception for each of the different groups.

In the former case, the null hypothesis states that there is no correlation between the variables in the population. In the latter case, the null hypothesis states that the mean value of role perception is the same for all samples of legislators, regardless of the political milieu they are drawn from. In the latter case, the research would have been designed so that we could select a sample of legislators from competitive political milieus and another from noncompetitive political milieus, and compute the mean value of role perception for each group.

These examples illustrate that the specific form of the null hypothesis depends on the particular statistical test being used. (And, as we shall see in a moment, the kind of statistical test that is appropriate depends on the kind of research design being used.) The different forms that the null hypothesis may take will become clearer as the various tests are considered. At this point, a general discussion of why the null hypothesis is stated in the form of no difference is necessary.

The null hypothesis uses the rule of negative inference (*modus tolens*) in logic. We try to prove a hypothesis by disproving its contrary, because it is simpler to disprove a specific negative hypothesis than an open-ended positive one. The reasoning is as follows. We hypothesize that the contrary H_0 of the research hypothesis H_1 is true. We deduce what consequences C_j are likely to occur if this contrary is true. We observe empirical results to see if C_j occurs. If C_j has occurred, we affirm H_0. If C_j has not occurred, we reject H_0. In the form of the classical syllogism, we have

1. If H_0, then C_j (the deductive premise).
2. C_j is true (the empirical observation).
3. Therefore, H_0 is true (the conclusion).

Note that, although it is stated in absolute form here, the conclusion in statistics is probably true or false and not absolutely so. As we said above, probability theory helps measure the chances of making an error in accepting the conclusion. In contrast to the form of the classical syllogism, in statistical hypothesis testing, the deductive question is usually in the form "How likely is it that we shall observe C_j if H_0 is true?" The empirical observation is the sample result, such as a particular $\bar{X}$. The conclusion is generally in the form "If H_0 is true, we shall observe a sample value of $\bar{X}$ only about 1 percent of the time. Given this, we may reject H_0, but always with a risk of making an error."

Two kinds of error are distinguished in conventional statistical hypothesis testing: error type 1 and error type 2, symbolized as α and β respectively. Let us consider them in detail.

TYPE 1 AND TYPE 2 ERROR

Suppose that we believe that 60 percent of a particular population, such as all voters in the United States, favor a hard line against communism. We do not know the actual population parameter P. The hypothesis $P = 0.60$ is therefore the null hypothesis.[2] Suppose now that we take a random sample of 400 voters and find that the proportion in the sample who favor a hard line against communism is $p = 0.55$. If the population parameter is $P = 0.60$, how likely is it that we shall observe a sample $p = 0.55$? To answer this we set up a sampling distribution of all the p's we may obtain under the assumption of the null hypothesis that $P = 0.60$. Of course, because we have a large sample here, we can use the normal curve as the sampling distribution of p. Therefore, we want to convert the particular p we have observed into a Z score in accordance with the equation given in Chapter 6:

$$Z = \frac{0.55 - 0.60}{\sqrt{[(0.55)(0.45)]/400}}$$
$$= -2.01$$

The table of the normal curve areas (see the Appendix) tells us that a Z score of -2.01 has a probability of 2.22 percent, that is, we shall rarely observe a sample $p = 0.55$ if we are sampling from a population whose parameter is $P = 0.60$. We may therefore reject H_0: $P = 0.60$, but in so doing we are not absolutely sure that we are right. We run the risk of rejecting a true hypothesis.

This is error type 1. It is the risk of rejecting the null hypothesis when

[2] We shall see below that an open-ended null hypothesis, such as $P > 0.60$, cannot be tested statistically, because there is no way to obtain a sampling distribution. The hypothesis $P = 0.60$ must have a theoretical rationale. We do not simply pull such a statement out of the air. Hence, a theory is essential to research because it tells us where to begin. In the present case, the figure was pulled out of the air to use as an illustration. It has no theoretical rationale.

it is true, and it is measured by the sampling distribution of the statistic we are testing. It also is called the *significance level*. "Significance" here refers only to statistical significance, not research significance, and is really nothing more than the probability of observing the sample result if the null hypothesis is true. In the example above, we may reject the null hypothesis that $P = 0.60$, because the risk of making error type 1 is $\alpha < 0.05$.

What level of risk is acceptable in hypothesis testing? That is, what risk of error type 1 is considered too great? The levels of statistical significance that we generally see reported in research are 0.01 and 0.05, or the 1 percent and 5 percent levels. These are the conventionally acceptable levels of risk of error type 1. If the risk is less than 5 percent, or 1 percent, we usually reject the null hypothesis. But there is nothing sacred about these levels. The risk we are willing to take of making error type 1 depends entirely on the kind of research we are doing. For example, if we were testing a particular medicine in which the consequence of rejecting a true hypothesis would be the loss of human life, we might want to set the risk of α error at 0.00001 rather than 0.01.

What are the factors that should be taken into account in setting the critical levels of α? There are a number of them. One is the cost of accepting the alternative hypothesis. For example, suppose that the hypothesis $P = 0.60$ refers to a standard for the mixture of metal in a certain product where 60 percent is the desired proportion of a certain component. The sample result, $p = 0.55$, tells us that the production process is probably not achieving this standard. But to accept this conclusion might mean that all the machines in the company would have to be reset at great cost, which would have to be taken into consideration before a decision could be made. Certainly it would be foolish to make a decision based on the sample evidence alone that the true proportion is not likely to be 0.60.

We shall have more to say about decision criteria, such as cost, later in this book. For political scientists, the consequence of wrongly rejecting a true hypothesis is that we might mislead our colleagues and students. However, given the healthy skepticism that exists in the discipline, this is not too likely. Can we set an acceptable level of risk for this kind of consequence? The answer seems apparent. It is partially a subjective matter. This is also true in measuring the consequences of a wrong decision in other areas. What level of probability did the Manhattan Project scientists require for the hypothesis that an uncontrollable chain reaction would not occur after the first atomic pile was activated beyond a critical level?[3] Is there any objective answer to this question?

The moral of all of this is plain. By far the better practice is to report the probability of error type 1 rather than arbitrarily selecting a critical value for α. The decision to accept or reject a null hypothesis is the responsi-

[3] See R. Rudner, "Remarks on Value Judgments in Scientific Validation," *Scientific Monthly*, vol. 79, pp. 151–153, 1954.

bility of the researcher, and cannot easily be dispensed with by appeals to an arbitrary value of probability.

Thus far, we have been discussing the question of rejecting the null hypothesis when it may be true. Suppose that we do not reject it. Then we obviously cannot make error type 1. However, we are still not free of error, because if we do not reject the null hypothesis, we accept it, and when we do, we are confronted with another type of error, that of accepting a false hypothesis. The null hypothesis may be false although we accept it. This is error type 2, or β error. In conventional statistical decision making, there are two kinds of error that we can make and two ways of being right. They are shown in Table 7-1. At the head of the table are two possible real-world

Table 7-1   Alternative Outcomes in Testing Statistical Hypotheses

		The null hypothesis is:	
		True	*False*
The decision is to:	Accept it	Correct	Error type 2
	Reject it	Error type 1	Correct

situations, or *states of nature*, as they are sometimes called. In the rows at the left are the two decision alternatives. In the cells are the consequences. For example, if the real-world situation is that the null hypothesis is true and we accept it, we have made a correct decision. If the null hypothesis is false and we accept it, we have made error type 2. If the null hypothesis is true and we reject it, we have made error type 1. The desire, of course, is to use decision procedures that always lead to correct decisions. But it is one of the fundamental premises of statistical decision making that a world completely free of error is not possible. The most we may do is to try to reduce the probability of making errors.

Errors type 1 and type 2 are inversely related. For a given sample size, if we decrease the chances of making error type 1, by making the level of α very small, we increase the chances of making error type 2. The lower the α value we set, the fewer the hypotheses we shall reject. But at the same time the chances are then increased of accepting more hypotheses that are false. Under these conditions it would seem impossible to reduce the probability of making errors. This is not so, for it is possible to reduce the combined probability of making errors type 1 and 2. Let us consider how.

The probability of making error type 2 depends on how close the sample value is to the true population parameter. If the sample value is very close, the chances of making error type 2 are very great. This is another way of saying that the chances of making error type 2 depend on how false the null hypothesis is. If it is very wrong, the chances of making error type 2 are very small.

To illustrate the probability of making error type 2, consider the following example. Suppose that the true population mean in a particular situation is μ = \$6,000 (and, of course, we do not know this). Suppose that we hypothesize that the population mean is H_0: μ = \$5,950. Finally, suppose that a random sample of 400 yields a mean of $\bar{X}$ = \$5,900 and a standard deviation of s = \$2,000. If we were testing a hypothesis in this case, we should ask, "How likely is it that we shall obtain $\bar{X}$ = \$5,900 if the population parameter μ is \$5,950?" Of course, it is very likely. Exactly how likely can be determined by converting the sample mean $\bar{X}$ to standard form, as follows:

$$Z = \frac{5,900 - 5,950}{2,000/\sqrt{400}} = -0.5$$

This tells us that the area of the standard normal curve between a mean of \$5,950 and an observation of \$5,900 is 0.1915. Because we did not predict the direction in this case, we want to know how much of the standard normal curve is within ± 0.5 Z scores of the hypothesized mean of \$5,950. This is 2×0.1915, or 0.3830. The area outside of this is therefore $1 - 0.3830 = 0.6170$. Thus the chance of observing a sample mean \$50 away from the true population mean of \$5,950 is very great. The chance of making error type 1 would be 61.7 percent if we rejected the null hypothesis. Thus we cannot reject the null hypothesis μ = \$5,950 in this case. But, of course, we have accepted a false hypothesis, because we said above that μ = \$6,000.

Before the reader concludes that we have done a little sleight of hand here, because we set up the assumption that μ = \$6,000, let us pursue the example a little further. Suppose that the true population mean is \$7,000. Now consider what a sample is likely to produce if the population value actually is \$7,000. We know that 95 percent of the time we are likely to obtain a sample mean within $\pm 1.96\sigma_{\bar{x}}$ of \$7,000 if μ = \$7,000. About 5 percent of the time the sample $\bar{X}$ will be outside of this range. If σ = \$2,000, the 95 percent confidence interval for a population whose mean is \$7,000 and for samples of size 400 is

$$= \$7,000 \pm 1.96 \, \frac{2,000}{\sqrt{400}}$$

$$= \$7,000 \pm \$196$$

$$= \$6,804 \text{ to } \$7,196$$

that is, a random sample is likely to produce a mean $\bar{X}$ within this range 95 percent of the time.

Now, suppose that the null hypothesis is H_0: μ = \$5,950. How likely is it that we shall accept this hypothesis if the true population mean μ = \$7,000?

Very unlikely, because most samples will yield a mean between $6,804 and $7,196, and we shall reject H_0: μ = $5,950 under these conditions.

In simpler terms, this example says that, as the null hypothesis gets closer and closer to the true population mean, the chances of making error type 2 increase. When there is very little difference between the null hypothesis and the population parameter, we are almost certain to make error type 2, because a sample is not likely to detect the difference. For example, if the population parameter, which we do not know, is $7,000 and we hypothesize that it is $7,010 (the null hypothesis), we are almost certain to make error type 2, because most samples will support the null hypothesis. However, the consequences of error in this case are very small, because the error is very small.

The measurement of error type 2 depends on the sample size, the level of α, and what the null hypothesis and true population parameters are. Let us consider a hypothetical case. Suppose that H_0: μ = $6,000, σ = $5,000, and n = 100. What value of $\bar{X}$ do we need to reject H_0 at the 0.05 level of significance? This is

$$1.96 = \pm \frac{\bar{X} - 6,000}{5,000/\sqrt{100}} = \pm 980$$

Hence, a sample mean less than $5,020 or more than $6,980 would lead us to reject H_0 with a 5 percent risk of error type 1.

Now, suppose that the true mean is $5,000. What is the probability of accepting the hypothesis μ = $6,000? Given the decision to reject H_0 for values of $\bar{X}$ less than $5,020, we can compute β by finding out how much of the area of the sampling distribution of $\bar{X}$ for μ = $5,000 is to the right of $5,020. This is

$$Z = \frac{5,020 - 5,000}{5,000/\sqrt{100}} = 0.04$$

The table of normal curve areas (see the Appendix) tells us that 1.6 percent of the total area lies between the mean of $5,000 and an observation that is 0.04 standard deviations to the right of this. This is shown in Figure 7-1a.

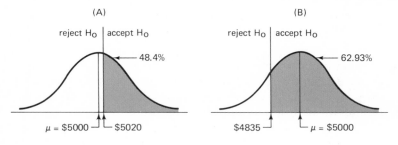

(A)

reject H_0 | accept H_0

48.4%

μ = $5000 — $5020

(B)

reject H_0 | accept H_0

62.93%

$4835 — μ = $5000

Figure 7-1. Probability of β error.

Remember that we are testing the hypothesis H_0: μ = $6,000. Given α = 0.05, we shall reject H_0 if the sample mean is less than $5,020. But we shall accept it for all sample means larger than $5,020, and hence the shaded area (48.4 percent) is the probability of making error type 2 under these conditions.

To demonstrate that α and β are inversely related, suppose that we set α = 0.01. Then we shall reject H_0: μ = $6,000 for values of $\bar{X}$ less than $4,835 (and more than $7,135 — the reader should verify this). β in this case is increased to 62.93 percent, as shown in the shaded area in Figure 7-1b.

Finally, to show that β increases as $\mu \rightarrow H_0$, the reader should compute β for increasing values of μ. He should find that for μ = $5,500, β = 0.8315; for μ = $5,700, β = 0.9131; and for μ = $5,900, β = 0.9608.

We may show some of this in the form of an *operating characteristics curve*. This is done in Figure 7-2. As the true mean μ moves farther and

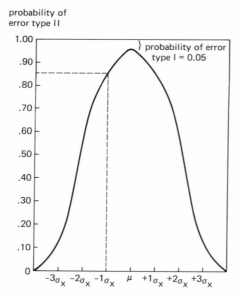

Figure 7-2. Operating characteristics curve for H_0: μ = $6,000, β = $2,000, n = 400, and α = 0.05.

farther away from the hypothesized mean, the chances of making error type 2 diminish. For given values of α, the exact probability depends on how many standard errors the true mean is from the hypothesized mean. If it is one standard error above or below the hypothesized mean, the probability of error type 2 is about 0.85, as shown by the dotted line in Figure 7-2.

Another closely related concept is the *power* of a statistical test. The power of a test is the probability that a correct decision will be made. A more powerful test enables us to make the correct decision with a smaller sample than a less powerful test. The power of a test is

$$1 - \beta = 1 - p \text{ (accepting } H_0 | H_0 \text{ is false)}$$

The power of a test thus depends on the probability of error type 2, which we said before depends on how wrong H_0 is. If H_1: $\mu = \$6,000$ and we hypothesize that $\mu = \$5,000$, a sample will enable us to decide which is correct. But if $\mu = \$6,000$ and we hypothesize that $\mu = \$5,950$, it will be difficult to decide which is correct on the basis of one sample. Where the alternative hypotheses are very close, β will be large, and $1 - \beta$ will be small; that is, we should need a very powerful test to make the correct decision in this case.

As we said before, the exact values of β and $1 - \beta$ depend on the specific conditions of the test. But we may note in general that β can be reduced by increasing the size of the sample, because $\sigma_{\bar{x}}$ decreases as n increases. As $n \rightarrow N$, $\bar{X} \rightarrow \mu$, $\sigma_{\bar{x}} \rightarrow 0$, and $\beta \rightarrow 0$.

ONE- AND TWO-TAIL ALTERNATIVES

A statistical test may be one-tailed or two-tailed, depending on whether or not we have predicted direction. If we predict the direction of the difference, a one-tailed test is appropriate. If we are unable to predict direction, a two-tailed test is necessary. The following illustration should clarify this distinction.

Suppose that the research and null hypotheses are

$$H_1: P < 0.50$$

$$H_0: P = 0.50$$

In this case we have predicted the direction of the difference between the hypothesized population parameter and the research hypothesis. To make this meaningful, consider research situations where this may be so. Suppose that, because the student body in a particular college is liberal, someone argues that 50 percent of the students believe that the conscription should be replaced by a voluntary army. Alternatively, a protagonist argues that the actual proportion is much less. Consider another example. We may argue that, given American mores in general, 70 percent of all elected representatives in the United States adhere to a delegate rather than a trustee perception of their role. An alternative hypothesis, based on the notion that the spirit of independence is predominant, is that much less than 70 percent adhere to a delegate perception. There are a number of cases of this kind in which a directional research hypothesis may be used. But in all cases there has to be a theoretical rationale for specifying a particular value for the null hypothesis.

The sampling distribution we construct is based on the null hypothesis, which, in this example, is H_0: $P = 0.50$; that is, assuming that the population proportion is 0.50, what is the sampling distribution of p for samples of a given size? For large samples the standard normal curve can be used as an approximation of the sampling distribution of p. This is shown in Figure 7-3. The shaded area in the left tail of this sampling distribution is the one-

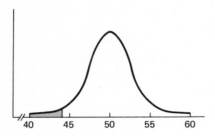

Figure 7-3. Sampling distribution of p and rejection region for H_0: $P = 0.50$ $\alpha = 0.05, n = 200$.

tailed area of rejection for $\alpha = 0.05$. If the sample p for a sample of 200 falls in this region, we shall reject H_0: $P = 0.50$, with a risk of 5 percent of making error type 1.

How do we arrive at the p values for this critical region? First, the table of areas under the standard normal curve (see the Appendix) tells us that the point to the left of which 5 percent of the total area of the curve lies is the point that is -1.65 Z scores to the left of the mean. The proportion at this point, under the assumption that $P = 0.50$, is

$$-1.65 = \frac{p - 0.50}{\sqrt{(0.50)(0.50)/200}}$$

$$(0.035)(-1.65) = p - 0.50$$

$$p = 0.442$$

that is, if $P = 0.50$, the observation $p = 0.442$ is the point below which 5 percent of all observations fall in a standard normal distribution. This is the shaded area in Figure 7-3. Therefore, if a random sample is drawn and the proportion in this sample is $p = 0.43$, that is, 43 percent favor a voluntary army, we shall reject the null hypothesis that $P = 0.50$, running a 5 percent risk of rejecting a true hypothesis.

Let us now suppose that there is no theoretical reason for predicting that P is likely to be either less than or greater than 0.50 but only that it is different from 0.50. The null hypothesis is still that the true population proportion is 50 percent, and we suppose that there is a theoretical reason for this. In this case a two-tailed test of the sample result is required. The competing hypotheses are

$$H_1: P \neq 0.50$$

$$H_0: P = 0.50$$

Because we have not predicted direction, the rejection regions for $P = 0.50$ lie in both ends of the sampling distribution: 2 1/2 percent in the lower tail and 2 1/2 percent in the upper tail. This is shown in Figure 7-4.

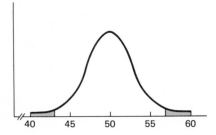

Figure 7-4. Sampling distribution of p and rejection region for H_0: $P = 0.50$, $\alpha = 0.50$, $n = 200$.

How do we arrive at the proportions for this rejection region? We know that exactly 2 1/2 percent of the area of the standard normal curve lies above 1.96 Z scores to the right of the mean and 2 1/2 percent of the area lies outside of the point that is 1.96 Z scores to the left. Hence, we want the p values for $\pm 1.96\ Z$ scores, or

$$1.96 = \frac{p - 0.50}{\sqrt{(0.50)(0.50)/200}}$$

$$0.431 \geq p \leq 0.569$$

Therefore, if a particular random sample yields a $p = 0.42$, we shall reject H_0: $P = 0.50$, with a 5 percent risk of rejecting a true hypothesis. Similarly, if a sample yields a $p = 0.57$, we shall reject H_0: $P = 0.50$. Note that with a two-tailed alternative we need a more extreme value for p in order to reject the null hypothesis than is so with a one-tailed alternative.

SUMMARY OF CONVENTIONAL STATISTICAL HYPOTHESIS TESTING

What we have been discussing thus far is the classical or conventional method of making decisions about hypotheses. The hypothesis we make the decision about is the null hypothesis. The steps involved in this procedure may be outlined as follows:

1. *Formulate the null hypothesis.* If it is a test of a relationship between two or more variables, the null hypothesis will usually state that no relationship exists. If it is a test of a single parameter value, the null hypothesis will specify it.

Note that in the cases we have discussed thus far, the alternative, or research, hypothesis has been open-ended. However, the null hypothesis must be explicit, because the sampling distribution is that which we shall get if the null hypothesis is true, and we must have a value for it in order to construct it. The alternative, or research, hypothesis may be open-ended or a specific value. For example, the null hypothesis may state that $P = 0.50$, and the alternative hypothesis that $P = 0.30$. However, it is seldom possible to be so explicit about the alternative hypothesis in political science research, because the theory we are working with must be very well developed before we have a good reason for postulating a specific value for a population parameter. If it is possible to do so, we can minimize not only the probability of making error type 1 but also that of error type 2. This is not considered in detail here, because it is seldom the case in political science research that a specific parameter value can be predicted.

2. *Select α and β levels.* Conventionally α levels of 0.01 and 0.05 are selected but, as we said above, there is nothing magical about these levels. It is important to select an α level before the sample result is computed in order to avoid allowing the sample results to enter into the decision of what will be an acceptable criterion for the rejection of H_0. For example, suppose that we do not predict the direction of difference beforehand and the distribution in Figure 7-4 is the appropriate one. Suppose also that the sample result is $p = 0.44$. We shall not be able to reject the null hypothesis at the 5 percent level of α, nor say then that we suspected that $P < 0.50$ all along and we can now, therefore, apply a one-tailed test, in which case we can now reject H_0: $P = 0.50$ at the 0.05 level of significance.

We usually specify the level of α, but not the level of β, because we cannot usually give a specific value to the alternative hypothesis. We can compute β only if it is possible to specify a value for H_1. However, because α and β are inversely related and if we cannot specify a value for H_1, we may allow the decision about α to be based on which kind of error is more serious: rejecting a true hypothesis or accepting a false one. If we think that the consequences will be worse if we reject a true hypothesis, we set α at a very small level, say, $\alpha = 0.0001$. We are less likely to reject a true hypothesis in this case. Note that from the perspective of the research hypothesis, we would accept fewer when α is small. But, of course, we are more likely to accept a false hypothesis. However, if accepting a false hypothesis is a more serious error, we set the significance level α relatively high, say, $\alpha = 0.10$, and we thereby reduce the level of β. Recall also that, if it is important enough (subjectively measured) and there are sufficient funds (objectively measured), we can avoid this dilemma somewhat by simply increasing the sample size.

3. *Select the appropriate sampling distribution.* The sampling distribution we use to make a decision depends on the null hypothesis, the particular parameter, and the sample size. The following are some of the sampling distributions appropriate to different parameters and sample sizes (they are all discussed in this book): (*a*) the standard normal curve for difference of means or proportions, $n \geq 30$; (*b*) the t distribution for difference of means, $n < 30$; (*c*) the binomial distribution, or the Poisson distribution if p or q is very small, for difference of proportions, $n < 30$; (*d*) the F distribution for difference among several means; (*e*) the F distribution for correlation of two or more variables; and (*f*) the chi-square distribution for relationships between two or more proportions.

4. *Draw probability sample and measure sample results.* As we said earlier, most often simple random sampling is used. However, other forms of sampling are possible, and are discussed in Chapter 15.

5. *Make decision about the null hypothesis.* The decision is dichotomous: to accept or reject H_0. Frequently in research, the sample results may be on the borderline for the selected level of significance, but, as suggested in item 2 above, the level should be set beforehand, and the decision accepted on this basis; not after the sample results have been studied.

EXPERIMENTAL AND NONEXPERIMENTAL RESEARCH DESIGN

One further distinction is necessary before turning to the difference of means and proportions tests. This is the question of research design. We cannot go into all the complicated questions involved in research design here. All we can do is to suppose that the reader has some familiarization with the question. However, it is necessary to make some elementary distinctions here, because the kind of statistical test that is appropriate to a particular case depends on the kind of research design being used.

Although it is somewhat of an oversimplification, it is possible to distinguish two principle kinds of research design (there are, of course, many variations of these two): experimental and nonexperimental. The basic logic of each is quite simple. In classical experimental research design, the researcher creates an artificial testing situation in order to control all variables other than the one(s) he is interested in. He manipulates this variable physically and observes the effects on the dependent variable. The manipulation of one variable leading to a corresponding change in the other is what philosophers of science call a *causal* relationship. It is relatively easy to infer cause in this case, because the results of the manipulation are immediately apparent in a controlled situation.

In nonexperimental research design the investigator observes events as they occur in the real world and studies the relationship between variables. If he wants to study the effects of a particular variable, he controls the effects of all others by mathematical manipulation of the value of the variables or by the method of selecting samples. In this case, it is not so easy to infer causal relationships, because it is never clear whether all the relevant variables other than the one being studied have been controlled. We say more about the problem of making causal inferences in Chapter 9. At this point we illustrate the distinction just made with some examples.

The classical laboratory experiment can best be described in terms of an example taken from biology. Suppose that we are interested in whether a particular drug affects learning ability in animals. A group of animals is selected at random. Random sampling is used as a device for controlling the other variables that may be related to the experimental outcome. The supposition is that, by random sampling, other variables will be distributed in the group of experimental animals as in the population at large. The animals are given a group of tasks to perform, and the time it takes to learn to do them is measured and recorded. Then the animals are given the drug (the independent variable) and another (similar) group of tasks to perform. The time it takes to learn the second group of tasks is also measured and recorded. The mean time it takes to perform the tasks before and after the administration of the drug is compared. If the difference is too large to be attributable to sampling error, we may infer that the independent variable has produced ("caused") the difference.

This is generally called a *before-after* or *within-subject* research design. It has defects, such as the fact that it does not control for the possible transfer of learning from the before situation to the after. However, it is important to note that the independent variable has been physically manipulated in this case in order to understand whether it has an effect and the measures have been taken on the same individuals and paired.

To try to handle the problem of transfer of learning in the within-subject design, a *between-subjects*, or control and experimental group, design may be set up. In this case, the measurements are taken on different groups rather than on the same group so as to control the effect of the experiment itself. Two groups are selected at random. Each group is given the same tasks to perform. One group, called the *control group*, is not administered the drug before it is given the tasks, and the second group, called the *experimental group*, is. The mean time it takes each group to learn the tasks is recorded and compared. The difference may be ascribed to sampling variation, which can be measured, and to the effect of the independent variable. Note that, in this case, measures of different individuals are compared.

Either one of these designs can be applied in a laboratory or in the real world. In the latter case, we select samples from existing populations rather than creating populations. It is more appropriate to refer to observations in the world as they occur as *quasi-experimental* design in order to distinguish them from the classical experiment. In within-subject design, each observation is paired. For example, we may observe a sample of individuals before a particular event, such as an election, and compute a mean score of these individuals. We may then observe the same individuals after the event, that is, the election, compute a mean score, and compare the before and after scores. This form, which we might call *matched pairs*, is considered in an example below. In between-subjects design, we draw two independent random samples, compute scores for each, and compare the mean of each group. For example, we may compare legislators from two states on a variable, such as conservatism, using this design. An example of this kind is considered below.

Another way of understanding the difference between the two kinds of research design is as follows. In the within-subject each score is paired for the same individual. They are therefore dependent observations, because an individual who scores high on an observation before the exposure to the independent variable is also likely to do so after; his after score will be a function of his before score. In the between-subjects design, each observation made in one sample is independent of those made in the other.

The distinction is important because the procedure for testing a difference between means for each kind of research design is different. In the next section we discuss the procedure for testing a difference between means for independent observations. In a later section, entitled "Matched Pairs," we discuss the procedure for testing a difference between means for paired observations.

In general the distinction is as follows. For dependent, or paired, observations, we obtain two measures of each subject in the sample: one before the introduction of the independent variable and one after. We may do so in a laboratory, where we are able to manipulate the independent variable physically, or in the real world, where the independent variable is "introduced" in the normal course of events, such as an election's occurring.

For independent observations we draw two samples at random, expose one to the test situation but not the other. Only one measure of each subject is obtained, and the results of the two groups are compared. This may be done in a laboratory, where the independent variable is physically controlled, or in the real world, where one group is exposed to a certain kind of experience in the normal course of events but the other is not.

The statistical test that is appropriate to determining whether there is a difference between means depends on whether the design is for independent observations or matched pairs. In the next two sections we describe the difference of means test appropriate to independent observations. Then the test appropriate to matched pairs (the before-after design) is described.

DIFFERENCE OF MEANS FOR INDEPENDENT OBSERVATIONS

Suppose that the theory we are working with tells us to expect that economic liberalism is related to the size of a community. To determine whether or not this is true based on experimental design, we take a group of communities, measure their economic liberalism, increase their size, and see whether or not their economic liberalism also changes. It is apparent that the question in this case cannot be answered on the basis of experimental design, because the manipulation just described is not possible. Instead, we may sample from existing populations of different size and determine if there is a difference in economic liberalism in the size groups.

Suppose that we are able to classify a large number of communities into two categories: large and small. We select random samples of n_1 and n_2 from each category and measure the mean score of economic liberalism of each, perhaps by taking something like the per capita expenditures for welfare. The mean of each group is compared. Suppose that a difference is found. Is it statistically significant?

Recall now that, if we sample from the same two groups again, compute the mean for the new sample, and take the difference, we shall not find the same difference as for the first pair of samples. In other words, there will be some sampling fluctuation in the differences we obtain each time. In order to determine whether the difference we find for samples of size n_1 and n_2 is attributable to chance, we set up a sampling distribution of the difference between means for all possible samples we may select from these two populations. However, it is not necessary actually to compute this sampling distribution, because we can use either the normal distribution, if the sample size ≥ 30, or the t distribution, if the sample size < 30, as approximations of the sampling distributions we should get if the sampling process is repeated an infinite number of times.

The central limit theorem is the rationale for this. It may be stated as follows in this situation: If we take a large number of independent random samples of size n_1 and n_2 from the same population and compute $\bar{X}_1 - \bar{X}_2$ for each pair of samples, this quantity will be distributed normally with a mean equal to $\mu_1 - \mu_2$ and a standard error of $\sigma_{\bar{X}_1 - \bar{X}_2} = \sqrt{\sigma_{\bar{X}_1}^2 + \sigma_{\bar{X}_2}^2}$.

For example, suppose that the difference between the population means is $\mu_1 - \mu_2 = 30 - 20 = 10$. The central limit theorem tells us to expect that the mean of the difference of means of a very large number of samples will approach 10. To illustrate this, suppose that five independent random samples are drawn from each population, the mean $\bar{X}$ score for economic liberalism is computed for each community each time, and one mean is subtracted from the other. Suppose that the following are the results:

Sample number	$\bar{X}_1 - \bar{X}_2 = d_i$
1	$32 - 20 = 12$
2	$30 - 22 = 8$
3	$31 - 20 = 11$
4	$28 - 22 = 6$
5	$28 - 19 = 9$
	$\overline{46}$

$$\bar{X}_d = 46/5 = 9.2$$

In the first sample the mean $\bar{X}$ for community 1 is 32, that for community 2 is 20, and so on. In this hypothetical case the mean of the differences for the five samples is not 10, and we should not expect it to be so with such a small number of samples. But, in the long run, we should expect it to approach 10 as a limit, and the frequency distribution of the differences will form a normal curve. The frequency distribution of the differences between the means of each pair of samples is the sampling distribution of the difference of two means. As in the sampling distribution of a single mean, it is not necessary to construct a sampling distribution of the difference of means, because we can use the normal distribution as an approximation of the sampling distribution.

Consider now the standard deviation of the sampling distribution:

$$\sigma_{\bar{X}_1 - \bar{X}_2} = \sqrt{\sigma_{\bar{X}_1}^2 + \sigma_{\bar{X}_2}^2}$$

The standard error of the differences of the means is the sum of the standard errors of each mean rather than the difference of the separate standard errors, because there are now two sources of error, one from each of the samples, and therefore the individual standard errors must be added. As with any standard deviation, the standard error of the difference of means tells us how much deviation from the population difference of means to expect by chance, or sampling error. Let us consider how to measure this. The standard error of the mean is $\sigma_{\bar{X}} = \sigma/\sqrt{n}$. The square of the standard error of the mean is $\sigma_{\bar{X}}^2 = \sigma^2/n$. When we combine the standard error of each sample and extract the square root, the result is

$$\sigma_{\bar{X}_1 - \bar{X}_2} = \sqrt{\frac{\sigma_1^2}{n_1} + \frac{\sigma_2^2}{n_2}} \tag{7-1}$$

The standard error of the difference between means is thus

$$\sigma_{\bar{X}_1 - \bar{X}_2} = \sqrt{\sigma_{\bar{X}_1}^2 + \sigma_{\bar{X}_2}^2} = \sqrt{\frac{\sigma_1^2}{n_1} + \frac{\sigma_2^2}{n_2}} = \frac{\sigma_1}{\sqrt{n_1}} + \frac{\sigma_2}{\sqrt{n_2}}$$

If the standard normal curve can be used as the sampling distribution of the difference of means when $n > 30$, in order to test whether any particular observed difference may be expected to be produced by chance, we convert the particular observed difference found in two samples to a standardized or Z score. In this case, $\bar{X}_1 - \bar{X}_2$ is treated as an individual observation, and we subtract from it the mean of the sampling distribution, which is equal to $\mu_1 - \mu_2$, and, as with an individual observation, divide by the standard deviation:

$$Z = \frac{(\bar{X}_1 - \bar{X}_2) - (\mu_1 - \mu_2)}{\sqrt{\sigma_1^2/n_1 + \sigma_2^2/n_2}} \tag{7-2}$$

However, because we hypothesize that $\mu_1 - \mu_2 = 0$, Equation (7-2) reduces to

$$Z = \frac{\bar{X}_1 - \bar{X}_2}{\sqrt{\sigma_1^2/n_1 + \sigma_2^2/n_2}} \tag{7-3}$$

Before we can use this to find a Z value, however, we have to solve for the one unknown that still remains, σ^2. In the case of samples ≥ 30, we can use s^2 as an estimate of σ^2, giving

$$Z = \frac{\bar{X}_1 - \bar{X}_2}{\sqrt{s_1^2/n_1 + s_2^2/n_2}} \tag{7-4}$$

To test whether a particular observed difference between means may have been produced by chance, we simply use Equation (7-4) and then consult the table of areas under the normal curve.

Note that Equation (7-4) is nothing more than the difference between the two sample means, divided by the standard deviation of the sampling distribution of the difference between means. The latter is the amount of error we expect to observe by chance. If the distribution of the difference of means is normal, about 68 percent of the time the error will be within ± one standard error of the mean of the sampling distribution. Given the equation for the standard error, it is apparent that the size of the standard error depends on the size of the samples and the variance in the populations we are sampling from. The larger the sample sizes relative to the population variances, the smaller the standard error. With a small standard error and a relatively large difference between sample means, the resulting quotient is large. Under these conditions we tend to reject the null hypothesis, which, in the difference of means test, is H_0: $\mu_1 = \mu_2$, or there is no difference between means in the populations.

As an illustration of the application of the difference of means test we take data from a study by Herbert McClosky concerning the relationship between personality and attitudes. The particular data that will be used is the difference in isolationism between Democratic and Republican leaders who are "high articulates."[4] Isolationism was measured in this study by a nine-item attitude scale. The score for any one person, therefore, could be from 1 to 9, with a higher score representing more isolationism. The mean scores for a sample of Democratic and Republican leaders who are high articulates are shown in Table 7-2. The difference in means of the two groups

Table 7-2   Mean Isolationism Scores for Democratic and Republican Leaders*

	Democratic Leaders	*Republican Leaders*
$\bar{X}$	1.94	3.04
n	638	384

*Adapted from: Herbert McClosky, "Personality and Attitude Correlates of Foreign Policy Orientation," in James N. Rosenau (ed.), *Domestic Sources of Foreign Policy*, The Free Press of Glencoe, New York, 1967, pp. 51–109.

is $3.04 - 1.94 = 1.1$. Note that the mean scores for both groups are relatively low. However, the Republican leaders appear to be higher in isolationism than the Democratic leaders. Could the observed difference in means of 1.1 have been produced by sampling error? To answer this, we use the normal curve as the sampling distribution of the difference of means we should obtain if we repeatedly sampled from the same populations. We therefore convert the particular difference to standard form by means of Equation (7-4). To do so requires that we know s_1 and s_2. These were not reported in the source from which the data in Table 7-2 were taken. However, for purposes of illustration, we can estimate what s_1 and s_2 are likely to be. We can estimate the standard deviation of a frequency distribution to be about 1/5 of the range. Because the range of isolationism in this case must be 1 to 9 (and

[4] Herbert McClosky, "Personality and Attitude Correlates of Foreign Policy Orientation," in James N. Rosenau (ed.), *Domestic Sources of Foreign Policy*, The Free Press of Glencoe, New York, 1967, pp. 51–109. "High articulates" are defined by McClosky as college graduates who also score high on an intellectualism scale. Strictly speaking, the isolationism scale is an ordinal rather than an interval scale, and a mean should not be computed. However, we follow McClosky here and treat the numerical scores as if they did constitute an interval scale.

this is true of both samples), we have $1/5(8) = 1.6$ as the estimation of both s_1 and s_2. Note that Equation (7-4) applies even if the standard deviations of each sample are different.[5]

The Z score for the difference between means of the Democratic and Republican leaders is

$$Z = \frac{3.04 - 1.94}{\sqrt{2.56/384 + 2.56/638}}$$

$$= 11.0$$

Is a one-tailed or a two-tailed test appropriate here? If we predict that the Republican leaders are higher in isolationism than the Democratic leaders, a one-tail test is appropriate. We ask, 'What is the probability of observing a mean $\bar{X}_1$ as large or larger than 3.04 under the assumption that there is no difference between $\bar{X}_1$ and $\bar{X}_2$?'' The table of areas under the normal curve (see the Appendix) tells us that much less than 1 percent of the curve lies outside of the area bounded by the mean and 11 standard scores above it. For a one-tailed test we conclude that the chances of observing a difference of means as large as this in one direction is much less than 1 percent if the population difference is zero. Therefore, we reject H_0: $\mu_1 = \mu_2$, with less than 1 percent chance of making error type 1.

If we do not predict the possible direction of the difference, we apply a two-tailed test. In this case, the research hypothesis is $\mu_1 - \mu_2 \neq 0$, and the null hypothesis is the same as before. To test the null hypothesis, we find the area that lies both to the right and to the left of 11 standard errors of the mean. This is also much less than 1 percent of the total area. Once again we conclude that the null hypothesis is wrong, and the chance of making error type 1 is less than 1 percent.

Do we now accept the conclusion that there is a difference in isolationism between Democrats and Republicans in the populations from which these samples were drawn? Yes. Remember that we can be wrong in making this decision, but the chances are very small. Also note that the significance we have considered here is statistical significance, not research significance. For example, it could be the case that a difference in scores of 1.1 on this particular scale would be unimportant in predicting how Republican and Democratic leaders would vote on various legislation concerning foreign policy. This is an entirely different question, with which the test we just made is not concerned. It is worth emphasizing that statistical significance does not mean research significance. A difference of 0.5 could be statistically significant in the sense that it is the actual difference in the population, but it might not be theoretically significant. The question of strength of relationship is the subject of later chapters, and we return to this question then.

[5] When the population variances are equal, it is more efficient to make a pooled estimate of the common variance based on each sample. However, this is more important when the sample is small (<30), and thus it is considered in the next section.

Confidence Intervals for Difference of Means

We can use the standard error of the difference between means to estimate the difference between means in the population. To estimate $\mu_1 - \mu_2$, we use $\bar{X}_1 - \bar{X}_2$. Because we know that the sampling distribution of $\bar{X}_1 - \bar{X}_2$ forms a normal curve, we also know that about 68 percent of such differences are within $\pm \sigma_{\bar{X}_1 - \bar{X}_2}$, and that 95 percent of such differences are within $1.96 \sigma_{\bar{X}_1 - \bar{X}_2}$, and so on. Hence, for the example, to construct an interval around $\bar{X}_1 - \bar{X}_2$ within which we are 95 percent confident that $\mu_1 - \mu_2$ lies, we have

$$\mu_1 - \mu_2 = 1.1 \pm 1.96 \sqrt{2.56/384 + 2.56/638}$$
$$= 1.1 \pm .196$$

We can therefore estimate that the difference in the population means is between 0.804 and 1.296, and be 95 percent confident that this estimate is correct.

DIFFERENCE OF MEANS FOR INDEPENDENT OBSERVATIONS (n < 30)

If the two independent random samples are small, the standard normal curve is not a good approximation of the sampling distribution of the difference of means. In this case, the t distribution must be used. However, we must now consider whether or not $\sigma_1^2 = \sigma_2^2$. If the first population variance does equal the second, it is more efficient to make a pooled estimate of the common variance σ^2 rather than two independent estimates. Of course, because we are sampling from each population, we do not know whether or not $\sigma_1^2 = \sigma_2^2$. However, we can test whether or not this is likely to be true, using the sample variances and the F distribution. This is considered in Chapter 11 and therefore is not considered here. In this chapter we simply consider each situation separately; that is, where $\sigma_1^2 = \sigma_2^2$ and where $\sigma_1^2 \neq \sigma_2^2$.

When either one or both of the sample sizes are less than 30, we test whether a particular observed difference between means may have been produced by chance by converting them to a t score. The same procedure as that in the previous section is used. We subtract from the particular observed difference of means, the mean of the sampling distribution of means, and divide by the standard error or standard deviation of the sampling distribution. But because we hypothesize that $\mu_1 - \mu_2 = 0$, the conversion equation becomes

$$t = \frac{\bar{X}_1 - \bar{X}_2}{\sqrt{\sigma_1^2/n_1 + \sigma_2^2/n_2}} \tag{7-5}$$

We do not know $\sigma_1{}^2$ and $\sigma_2{}^2$. But if we know that $\sigma_1{}^2 \neq \sigma_2{}^2$, we make two separate estimates, one for each variance, using $s_1{}^2$ and $s_2{}^2$ respectively. In this case we divide by $n_1 - 1$ and $n_2 - 1$ rather than n_1 and n_2, because we have lost one degree of freedom in each case. (The concept of degrees of freedom is discussed in Chapter 8). The equation then becomes

$$t = \frac{\bar{X}_1 - \bar{X}_2}{\sqrt{s_1{}^2/(n_1 - 1) + s_2{}^2/(n_2 - 1)}} \tag{7-6}$$

Because this is essentially the same as the procedure for converting to a Z score when sample sizes are > 30, it is not illustrated by an example.

When $\sigma_1{}^2 = \sigma_2{}^2$ it is more efficient to make a pooled estimate rather than two separate estimates of σ^2. A pooled estimate of σ^2 simply means that we make one estimate that is a weighted average of the two sample variances:

$$\frac{\sum (X_1 - \bar{X}_1)^2 + \sum (X_2 - \bar{X}_2)^2}{n_1 + n_2 - 2} \tag{7-7}$$

If we symbolize this as s^2, the estimated standard error of the difference of means becomes

$$\sigma_{\bar{X}_1 - \bar{X}_2} = \sqrt{\frac{s^2}{n_1} + \frac{s^2}{n_2}}$$

where $s^2 = [\sum (X_1 - \bar{X}_1) + \sum (X_2 - \bar{X}_2)^2/](n_1 + n_2 - 2)$. Substituting this into Equation (7-6), we get

$$t = \frac{\bar{X}_1 - \bar{X}_2}{\sqrt{s^2/n_1 + s^2/n_2}} \tag{7-9}$$

Another, slightly shorter method of computing $\sigma_{\bar{X}_1 - \bar{X}_2}$ can be obtained by first simplifying the expression

$$\sigma_{\bar{X}_1 - \bar{X}_2} = \sqrt{\frac{\sigma^2}{n_1} + \frac{\sigma^2}{n_2}} = \sigma \sqrt{\frac{1}{n_1} + \frac{1}{n_2}}$$

which is true if $\sigma_1{}^2 = \sigma_2{}^2$. Now the common standard deviation σ can be estimated by weighting each sample s by its respective n and dividing by the appropriate degrees of freedom:

$$\sigma = \sqrt{\frac{n_1 s_1{}^2 + n_2 s_2{}^2}{n_1 + n_2 - 2}} \tag{7-10}$$

We can now bring back the quantity $\sqrt{1/n_1 + 1/n_2}$ and obtain

$$\sigma_{\bar{X}_1 - \bar{X}_2} = \sqrt{\frac{n_1 s_1{}^2 + n_2 s_2{}^2}{n_1 + n_2 - 2}} \sqrt{\frac{1}{n_1} + \frac{1}{n_2}} \tag{7-11}$$

The t formula then becomes

$$t = \frac{\bar{X}_1 - \bar{X}_2}{\sqrt{(n_1 s_1^2 + n_2 s_2^2)/(n_1 + n_2 - 2)} \ \sqrt{1/n_1 + 1/n_2}} \qquad (7\text{-}12)$$

Equations (7-9) and (7-12) are mathematically equivalent. When the population variances are the same, instead of converting to a t score by means of Equation (7-6), we do so by Equation (7-9) or (7-12). These are equivalent and produce the same result. The last is simply a short method for arriving at the same thing. In both cases, we use the t distribution as the sampling distribution of the difference between means and obtain a pooled estimate of the variance rather than adding the two separate standard errors.

We illustrate the difference of means test for small samples where the population variances are not equal with data taken from Guetzkow and Simon. The study concerns the relationship between communications networks and problem-solving ability.[6] The experimenters set up three groups of students selected at random. Each group was given the same tasks to perform, but the mode of communication permitted among people in a group was different for each group. One group was to communicate in an "all channel" pattern, the second in a "wheel" pattern, and the third in a "circle" pattern.[7] The time it took each group to solve problems was measured and compared. Because each group was selected at random, we may assume that the difference in mean time for each group can be attributed to sampling fluctuations and to the independent variable (the communications networks).

The data for two of the three groups are shown in Table 7-3. (The third

Table 7-3 Time of Task Trials for Two Communications Nets (Minutes)*

	$\bar{X}$	s	n
Wheel	19.12	3.09	15
Circle	29.45	5.08	21

*Adapted from: Harold Guetzkow and Herbert Simon, "The Impact of Certain Communications Nets upon Organization and Performance in Task-oriented Groups," in Albert H. Rubenstein and Chadwick Haberstroh, *Some Theories of Organization,* Richard D. Irwin, Inc., Homewood, Ill., and the Dorsey Press, 1966, p. 435.

[6] Harold Guetzkow and Herbert Simon, "The Impact of Certain Communications Nets upon Organization and Performance in Task-oriented Groups," in Albert H. Rubenstein and Chadwick Haberstroh, *Some Theories of Organization,* The Dorsey Press, Homewood, Ill., 1966, pp. 425–443.

[7] The three groups listed correspond roughly to the most open to the most restricted communications patterns; the all channel being the most open. The conclusion is that the intermediate group with some restrictions did best.

group, the "all channel" group, was intermediate between these two with a mean time of 24.38 minutes). The general findings of the study was that the "wheel" group was faster than the other two. For reasons that will become clearer in Chapter 11, we test here only the difference in means of the wheel and circle groups.[8]

The difference in mean time that it took each group to perform the tasks seems quite large: 10.33 minutes. Could this observed difference in means have been produced by sampling error? We use a two-tailed test for this, because direction was not predicted.

The null hypothesis is the same as before: H_0: $\mu_1 = \mu_2$. The null hypothesis will be rejected at $\alpha \leq 0.01$. To determine the probability of observing a difference as large or larger than 10.33, if the actual difference between means is zero, we convert to a t score, using a pooled estimate of σ^2. Equation (7-12) is used:

$$t = \frac{29.45 - 19.12}{\sqrt{[21(5.08)^2 + 15(3.09)^2]/(21 + 15 - 2)} \sqrt{1/21 + 1/15}}$$

$$= 6.8$$

The table of areas under the t distribution in the Appendix tells us that for $n_1 + n_2 - 2$ degrees of freedom, which are always the degrees of freedom in this test, a t of 6.8 has a two-tailed probability of <0.0001. We may therefore reject the null hypothesis of H_0: $\mu_1 = \mu_2$.

Note that the population in this case can be considered in two ways. The samples are from the finite population of all students in the particular university where this experiment was conducted, or they are from the infinite (analytical) population of all individuals performing similar tasks under similar conditions.

DIFFERENCE OF MEANS FOR MATCHED PAIRS

When the research design is of the before-after type, the tests just described do not apply. When the observations are made on the same subjects before and after an event, the difference of means of the two sets of scores does not form a normal or a t distribution. However, the mean difference of the individual scores does.

As an illustration of the appropriate test in a before-after situation, we use data taken from Charles Jones. One of the things Jones was interested

---

[8] The data in Table 7-3 show that $s_1^2 \neq s_2^2$, and if an F test showed that this is also likely to be true of the population variances, the appropriate equation for testing a difference between means in this case would be (7-6). However, for purposes of this illustration we assume that $\sigma_1^2 = \sigma_2^2$.

in was whether congressional elections had an effect on the voting behavior of congressmen. The dependent variable is scores given by the *Congressional Quarterly* to congressmen. For the Republicans Jones used the Economic Support Scores, and for Democrats the Federal Role Support Scores.[9] We use only the sample of Republicans here, because the results of both groups were essentially the same. Economic Support Scores for 129 incumbent Republicans who were reelected in 1958 were available for both before and after the election. The data are shown in Table 7-4.

Table 7-4   Economic Support Scores for 129 Republican Congressmen before and after the 1958 Election*

	1 $\bar{X}$ ESS before 1958 election	2 $\bar{X}$ ESS after 1958 election	3 $\bar{X}_2 - \bar{X}_1$
All 129	50.5	59.7	9.2
Group *A* (48)	52.6	63.1	10.5
Group *B* (60)	51.6	58.3	6.7
Group *C* (21)	43.0	56.2	13.2

*Adapted from: Charles O. Jones, "The Role of the Campaign in Congressional Politics," in M. Kent Jennings and L. Harmon Zeigler, *The Electoral Process,* Prentice-Hall, Inc., Englewood Cliffs, N.J., 1966, pp. 21–41.

Jones divided the total 129 Congressmen into three groups. Group *A* is 48 congressmen who lost more than 5 percent of their margin of victory since their previous election. Group *B* is 60 congressmen who lost less than 5 percent of their margin of victory. Group *C* is 21 congressmen who increased their margin of victory. Part of the question Jones was concerned with was whether those whose election was safer (group *C*) would change less as a result of the election. In the example to be considered here, we are concerned only with one question: "Was the behavior of the total group of 129 congressmen in regard to the amount of economic support they gave to the administration any different after the 1958 election than before it?"

In order to test this question, we have to know the ESS scores for each of the 129 congressmen both before and after the election. These scores should be paired, a difference between each obtained, and the mean of this difference computed, as shown in Table 7-5. The scores used in this table

[9] Charles O. Jones, "The Role of the Campaign in Congressional Politcs," in M. Kent Jennings and L. Harmon Zeigler, *The Electoral Process*, Prentice-Hall, Inc., Englewood Cliffs, N. J., 1966, pp. 21–41. The Economic and Federal Role Support Scores refer to how much support the Congress is giving to the administration's proposals. A higher score means a greater amount of support. Strictly speaking, a *Z* or *t* test is not appropriate unless the sample is random. The 129 congressmen that Jones used are not a random sample, but, for purposes of illustration, assume that they are.

are hypothetical for illustrative purposes. The statistic to be tested here is the mean difference between the paired scores. In an actual case, this would be obtained by taking the mean of column 4 in Table 7-5. Note that we do not actually have to compute it this way, because the mean of column 4 is

Table 7-5 Matched Pairs of ESS Scores of Congressmen before and after the 1958 Election

1 *Congressman*	2 *ESS before 1958 election*	3 *ESS after 1958 election*	4 $D =$ *(col. 3 − col. 2)*
Mr. *A*	50.5	58.1	7.6
Mr. *B*	40.0	54.3	14.3
Mr. *C*	55.6	65.8	10.2
.	.	.	.
.	.	.	.
.	.	.	.

the same as the difference between the means of columns 2 and 3 (this is because of rule 1 of summation). Hence, we know that the mean of the difference of the scores of the individual congressmen is 9.2, as given in row 1 in Table 7-4.

We assume in this example that the 129 congressmen were selected by random sampling. If we continually sampled from the same population, each time computing the mean of the difference in ESS scores before and after the election, we should get some fluctuation around the true population mean. In this case we hypothesize that there is no difference in the true population mean (which, in effect, is a hypothesis that the election has no effect), or $H_0: \mu_D = 0$. If the mean difference of this sample (Table 7-4) is too different from 0 to be attributed to sampling error, we shall reject the hypothesis. To test this, we need a sampling distribution of all the mean differences we should obtain if we repeated the sampling process an infinite number of times. It can be shown that the mean difference for matched pairs will be distributed in a normal distribution for samples ≥ 30 (and in a t distribution for samples of less than 30). Therefore, to test whether the observed difference in Table 7-4 may have been the result of sampling error, we convert it to a Z score:

$$Z = \frac{\bar{X}_D - \mu_D}{\sigma_D/\sqrt{n}} \qquad (7\text{-}13)$$

Now, we are faced with the familiar problem of having one unknown: σ_D.

As usual, we can use s_D as an estimate of σ_D. The s_D in this case refers to the standard deviation of the differences of individual scores.

The sample standard deviation was not reported in the Jones data, and so we must estimate it in order to complete the illustration. This time we estimate it to be the average of the standard deviations of the three separate groups, which was reported in the study. This average is 12.33.

Note that in Equation (7-13) we subtract, from the actual mean difference of matched scores, the mean difference of the actual population, and divide by the standard error of the sampling distribution. Because the sampling distribution of the mean difference is a distribution of sample means, the standard error in this case is exactly the same as the standard error of the mean. Also note that the null hypothesis is $H_0: \mu_D = 0$. Therefore, Equation (7-13) becomes

$$Z = \frac{\bar{X}_D}{s_D/\sqrt{n}} \tag{7-14}$$

For the data in Table 7-5, we have

$$Z = \frac{9.2}{12.23/\sqrt{129}} = 8.5$$

The table in the Appendix tells us that a Z of 8.5 has a probability of much less than 0.01, and we can therefore reject the null hypothesis that $\mu_D = 0$.

For samples of less than 30, we convert to a t distribution by

$$t = \frac{\bar{X}_D}{s_D/\sqrt{n-1}} \tag{7-15}$$

Several different equations for testing a mean difference can be derived from

$$Z = \frac{\bar{X}_D}{s_D/\sqrt{n}} \tag{7-14}$$

$$= \frac{\bar{X}_D}{\sqrt{s_D^2/n}} \tag{7-16}$$

$$= \frac{\bar{X}_D}{\sqrt{\dfrac{\sum d^2}{n(n-1)}}} \tag{7-17}$$

where $d^2 = (X_i - Y_i)^2$. It can also be shown that

$$s_D^2 = s_X^2 + s_Y^2 - 2r_{XY}s_X s_Y$$

that is, the variance of the difference between paired scores is equal to the sum of the variances of the X and Y scores minus the product of two times the correlation between X and Y and the standard deviations of X and Y (where X and Y represent before and after measures of the same individuals). If we substitute this in Equation (7.16), we get

$$Z = \frac{\bar{X}_1 - \bar{X}_2}{\sqrt{(s_X{}^2 + s_Y{}^2 - 2r_{XY}s_Xs_Y)/n}} \tag{7-18}$$

Note that this is almost identical to Equation (7-4) except that in Equation (7-18) we subtract the correlation between the two sets of observations from the sum of the individual variances.

DIFFERENCE OF PROPORTIONS

The mathematical model for differences of proportions for large samples (>30) is the same as that for differences of means. If a large number of independent random samples of size n_1 and n_2 are drawn from the same population, the sampling distribution of $p_1 - p_2$ will be normal with a standard error of

$$\sigma_{p_1 - p_2} = \sqrt{\frac{P_1 Q_1}{n_1} + \frac{P_2 Q_2}{n_2}} \tag{7-19}$$

The null hypothesis to be tested is that $P_1 = P_2$. In the differences of proportions test, a pooled estimate of P_1 and P_2 can be made, and the standard error becomes

$$\sigma_{p_1 - p_2} = \sqrt{PQ\left(\frac{1}{n_1} + \frac{1}{n_2}\right)} \tag{7-20}$$

where $P = (n_1 p_1 + n_2 p_2)/(n_1 + n_2)$.

We can test the significance of an observed difference of proportions by converting to a Z score in the usual manner:

$$Z = \frac{(p_1 - p_2) - (P_1 - P_2)}{\sqrt{PQ(1/n_1 + 1/n_2)}}$$

$$= \frac{p_1 - p_2}{\sqrt{PQ(1/n_1 + 1/n_2)}}$$

since H_0: $P_1 - P_2 = 0$. To illustrate the difference of proportions test, we use data from Matthews and Prothro's sample of Southern voters.[10] One of the things they found was that 31 percent of the 655 white voters in their sample and 36 percent of the 480 Negro voters were strong Democrats.

[10] See Donald R. Matthews and James W. Prothro, "The Concept of Party Image and Its Importance for the Southern Electorate," in M. Kent Jennings and L. Harmon Ziegler, *The Electoral Process*, Prentice-Hall, Inc., Englewood Cliffs, N. J., 1966, pp. 139–174.

Suppose that we are interested in knowing whether Negroes in the South are stronger Democrats than whites, hypothesizing that we expect this to be so. Because we have a directional research hypothesis, a one-tailed test is possible. Before converting to a Z score, we must first make a pooled estimate of P, as follows:

$$P = \frac{n_1 p_1 + n_2 p_2}{n_1 + n_2}$$
$$= \frac{480(0.36) + 655(0.31)}{480 + 655}$$
$$= 0.33$$

Therefore

$$Z = \frac{0.36 - 0.31}{\sqrt{(0.33)(0.67)(1/480 + 1/655)}}$$
$$= 1.8$$

The probability of observing a Z of 1.8 or more under the null hypothesis for a one-tailed test is less than 0.05. We can therefore reject the null hypothesis $H_0: P_1 = P_2$, with a risk of making error type 1 of less than 0.05. Our conclusion is that Negroes are stronger Democrats than whites.

We have now discussed the two major aspects of inferential statistics: making estimates and testing hypotheses. These aspects of statistics have their widest application in survey research and in research where experimental design is possible. In the next several chapters we consider hypothesis testing for nonexperimental design and introduce an important new aspect of statistics: measures of strength of relationship.

READINGS AND SELECTED REFERENCES

Blalock, Hubert, Jr. *Social Statistics*, McGraw-Hill Book Company, New York, 1960, Chaps. 11, 13.

Freund, John. *Modern Elementary Statistics*, Prentice-Hall, Inc., Englewood Cliffs, N. J., 1960, Chaps. 7, 11.

Games, Paul, and George Klare. *Elementary Statistics*, McGraw-Hill Book Company, New York, 1967, Chaps. X–XII, XV.

Hammond, Kenneth, and James Householder. *Introduction to the Statistical Method*, Alfred A. Knopf, Inc., New York, 1963, Chap. 9.

Hays, William L. *Statistics for Psychologists*, Holt, Rinehart and Winston, Inc., New York, 1963, Chap. 9.

Parl, Boris. *Basic Statistics*, Doubleday & Company, Inc., Garden City, N.Y., 1967, Chaps. 17–20.

Spurr, William A., and Charles P. Bonini. *Statistical Analysis for Business Decisions*, Richard D. Irwin, Inc., Homewood, Ill., 1967, Chaps. 11–13.

Wyatt, Woodrow, and Charles Bridges. *Statistics for the Behavioral Sciences*, D. C. Heath and Company, Boston, 1967, Chaps. 6, 7, 9.

Yamane, Taro. *Statistics: An Introductory Analysis*, Harper & Row, Publishers, New York, 1967, Chaps. 8, 9.

8

Tests and Measures for Nominal and Ordinal-Scale Data

Thus far we have discussed the foundations of statistical inference and methods of describing various aspects of frequency distributions and of making estimates and testing hypotheses. We come now to that part of statistics which is more significant in terms of contemporary research in political science, that part which deals not only with inferring population characteristics but also with measuring how strongly variables are related. In this and the next several chapters we tie more closely together the questions of level of measurement and the statistical methods appropriate to each. In this chapter we consider the methods of analyzing relationships among variables measured on nominal and ordinal scale. In Chapters 9 and 10 we consider methods appropriate to interval- (or ratio-) scale data.

Ideally the researcher strives to measure variables as precisely as possible so that he can use the more powerful statistical techniques described in Chapters 9 and 10 to analyze data. Practically, however, it is often difficult to do so. Ideally, the researcher uses random samples drawn from specific populations so that he can use statistics to make inferences about populations rather than simply about data at hand. Practically, however, he cannot always do so.

In these cases, the researcher must use what are sometimes called nonparametric statistics (considered in the first section of this chapter) or order statistics (considered in the second section of this chapter). Statistics which do not use parameter values, such as σ, or which do not make assumptions about the population distribution are called nonparametric or distribution-free statistics.

As we shall see, parametric statistics suppose, among other things, that the population from which the samples are drawn is normally distributed. Nonparametric statistics do not. In general, nonparametric statistics make fewer assumptions about the population than parametric statistics. Because they do, they are less powerful. The power of a statistical test is measured in terms of error type 2 $(1 - \beta)$. The more powerful a test, the more likely we are to make a correct decision for the same sample size; that is, the more likely we are to accept a true hypothesis. Because the power of a test is measured in terms of error type 2, it is necessary to measure the probability of error type 2 for each null hypothesis in order to get a measure of the power of the various tests to be described in this chapter. But because this is seldom an important factor in research, we do not consider the matter (the interested reader should consult Siegel). In general, however, it can be noted that the tests to be described in this chapter require a larger sample than those in Chapters 9 and 10 in order to have the same power.

TESTS AND MEASURES FOR NOMINAL-SCALE DATA

In a great deal of current political science research most variables are measured only in terms of categories. Variables such as party affiliation, religion, social class, region, blocs, and alliances, are nominal-scale variables. In analyzing relationships among variables of this kind, we generally cast them in what are called *contingency tables*, or matrices. A matrix, or contingency table, is any table of rows and columns, such as Table 8-3. There are several methods that can be used to extract information from tables like this. One of the most widely used is the chi-square distribution.

Significance Tests

Chi-square Distribution.   The chi-square distribution is appropriate when we have one or more variables measured on a nominal scale and want to determine if the frequencies or proportions we observe in a sample also hold in the population. It is an inferential method for testing the null hypothesis that the sample was drawn from a population in which the relative frequencies in each group are the same. Let us first consider the test for one variable, then describe the mathematical model, and finally consider a test for two variables.

We said above that nonparametric statistics do not require as many assumptions about a population as parametric statistics. However, there are important assumptions. The chi-square test assumes independence of observations, random sampling, nominal-scale measurement, and a large enough sample (≥ 50) to apply the general chi-square distribution. The role of these assumptions will become clearer as we proceed. Let us now consider an example of one variable.

Suppose that we are interested in determining whether the position of a candidate on a ballot influences his chances of being elected. For simplicity we postulate an election in which there are 10 candidates for the same office and in which the position of all the candidates on the ballot in previous elections was determined by chance. Let us say that data are available for the past 1,000 elections for this office. The figures in row O in Table 8-1 represent the observed frequencies of winners for each of the 10 positions on the ballot. If we take as the null hypothesis the idea that position on the ballot makes no difference in the outcome of the election, we shall expect the number of win-

Table 8-1   Position on Ballot and Number of Election Wins

	Position on Ballot										
	1	2	3	4	5	6	7	8	9	10	
E	100	100	100	100	100	100	100	100	100	100	1,000
O	150	100	120	130	80	90	90	70	80	90	1,000

ners for each position to be as shown in row E. Some difference between the observed and expected frequencies will occur by chance. The chi-square distribution can be used to tell us the probabilities of observing differences between observed and expected frequencies. Thus we must first measure the magnitude of the difference.

We do so by subtracting the expected from the observed frequencies and squaring before summing in order to avoid a zero result. Because the sum of the squared differences between observed and expected frequencies is affected by the number of observations as well as the magnitude of the difference, we may correct for this and get a measure of relative difference by dividing each squared difference by the expected frequencies:

$$\chi^2 = \sum_{i=1}^{k} \frac{(O_i - E_i)^2}{E_i} \tag{8-1}$$

For the data in Table 8-1 we have

$$\chi^2 = \frac{(150 - 100)^2}{100} + \frac{(100 - 100)^2}{100} + \cdots + \frac{(90 - 100)^2}{100}$$

$$= 58$$

The distribution we test the null hypothesis against is a positively skewed distribution whose shape depends on the number of degrees of freedom or the number of categories in the sample (the term "degrees of freedom" is explained below). Figure 8-1 shows the general shape of the chi-square distribution for three different amounts of degrees of freedom. Note that, as the number of degrees of freedom, or categories in this case, becomes larger, the area under the curve for larger values of chi square increases. Thus, the larger the number of categories, the larger the chi-square value needed in order to reject the null hypothesis of equal frequencies.

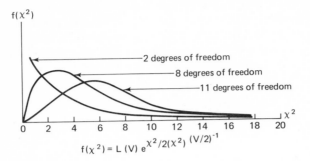

$$f(\chi^2) = L\,(V)\,e^{\chi^2/2(\chi^2)}\,{}^{(V/2)^{-1}}$$

Figure 8-1. The chi-square distribution for different degrees of freedom.

Let us now consider the mathematical model for the chi-square distribution. Suppose that we have three categories in which we are to place three things at random, and the probability of placing one of them in any of the three categories is the same, that is, $1/3$. There are 10 different ways in which these three things can be placed in the three categories, as shown in Table 8-2.

Table 8-2 Possible Outcomes for Placing Three Items in Three Categories

Outcome Number	Categories			P	χ^2
	1	2	3		
1	1	1	1	6/27	0
2	0	1	2	3/27	2
3	0	2	1	3/27	2
4	1	0	2	3/27	2
5	2	0	1	3/27	2
6	1	2	0	3/27	2
7	2	1	0	3/27	2
8	3	0	0	1/27	6
9	0	3	0	1/27	6
10	0	0	3	1/27	6

Thus, one possible outcome is getting one item in each category. Another is getting none in the first category, one in the second, two in the third, and so on. The probability of getting any one of these can be computed directly by the multinomial probability rule:

$$p(U_1, U_2, U_3) = \frac{N!}{U_1!\,U_2!\cdots U_k!}\,P_1^{U_1} P_2^{U_2} \cdots P_k^{U_k} \qquad (8\text{-}2)$$

For example, the probability of getting one in each category is

$$p(1,1,1) = \frac{3!}{1!1!1!} (1/3)(1/3)(1/3) = 6/27$$

and, for the other frequencies,

$$P(0,1,2) = \frac{3!}{0!1!2!} (1/3)^0(1/3)^1(1/3)^2 = 3/27$$

$$P(0,2,1) = \frac{3!}{0!2!1!} (1/3)^0(1/3)^2(1/3)^1 = 3/27$$

$$\begin{matrix} \cdot & & \cdot \\ \cdot & & \cdot \\ \cdot & & \cdot \end{matrix}$$

$$P(0,0,3) = \frac{3!}{0!0!3!} (1/3)^0(1/3)^0(1/3)^3 = 1/27$$

Only 1 of the 10 different outcomes has a chi-square value of 0; that is, the outcome $(1,1,1)$. The chi-square value is computed by Equation (8-1) for each of the possible outcomes. Thus, the chi-square value for the outcome $(1,1,1)$ is

$$\chi^2 = \frac{(1-1)^2}{1} + \frac{(1-1)^2}{1} + \frac{(1-1)^2}{1} = 0$$

When we place these results in a table as follows, we have a chi-square distribution:

χ^2	p
0	6/27
2	18/27
6	3/27

The probability of a particular chi-square value is the sum of the probabilities of all the ways the chi-square value may occur. Thus, the probability of observing a chi-square value of 6 is

$$p(\chi^2 = 6) = p(3,0,0) + p(0,3,0) + p(0,0,3)$$
$$= 1/27 + 1/27 + 1/27$$
$$= 3/27$$

Now suppose that for a sample size of 3 we observe the distribution of $(0,0,3)$. The chi-square value for this observation, as we said, is 6. The probability of observing a chi-square value of 6 under the null hypothesis that

$f_1 = f_2 = f_3$ is 3/27, or about 0.11, that is, if the null hypothesis is true, we should expect to find a distribution of (0,0,3) about 11 percent of the time if we repeated this experiment many times. This probability of 0.11 is referred to as the significance level of this particular observation.

In the example of 1,000 elections it is apparent that computing chi-square values by this method would be extremely time-consuming. Fortunately, tables providing good approximations are available. In the example about placing three things in three categories, a chi-square value of 6 for two degrees of freedom is shown in Table 3 in the Appendix to have a probability of about 0.05. In this case, the tabled approximations of the chi-square distribution are not very good because of the small size of the sample. However, as the sample size increases, the approximations provided by the chi-square distribution become better and better. For a sample size as small as 6 the approximation is close. If n is sufficiently large, the sampling distribution of chi-square as defined by the equation in Figure 8-1, is a close approximation of the exact chi-square distribution computed by the multinomial rule.

In general, sample sizes of about 50 or more are required for very close approximations by the chi-square distribution. When the sample is smaller than 50 or there are less than five cases in each category, different tests are required. As we shall see, there are several alternative tests that are better under different conditions. To return to the example concerning position on a ballot, the number of categories is 10, and the number of degrees of freedom is $k - 1 = 10 - 1 = 9$. The chi-square table shows that a chi-square value of 58 for nine degrees of freedom has a probability of less than 0.001. We therefore reject the null hypothesis and conclude that position on a ballot does influence the chances of winning an election. The direction of the influence must be ascertained from the data. The hypothetical data in Table 8-1 were set up to show that being listed at the top of a ballot is an advantage.

Table 8-3   Political Milieu and Interest in Elections

Interest in elections	Political milieu			Total
	Democratic	*Competitive*	*Republican*	
Much	A 9 (8.98)	B 21 (18.65)	C 8 (10.36)	38
Some	D 13 (11.82)	E 23 (24.55)	F 14 (13.64)	50
Little	G 4 (5.20)	H 10 (10.80)	I 8 (6.00)	22
Total	26	54	30	110

Let us now consider a test for two or more variables. The mathematical model and sampling distribution for two or more variables is the same as for one. As an example let us take data from the work of Eulau and his colleagues concerning the relationship between political milieu and the amount of interest in elections. The data are shown in Table 8-3.[1]

To understand the rationale of the chi-square test, consider the following. If we draw one city at random from this sample of 110 cities, what is the probability that it will be classified as Democratic? Because there are 26 out of the total of 110 classified as Democratic, the probability is 26/110. The probability that it will be a city classified as Democratic and in which there is perceived to be as much interest in elections is 9/110, and so on. The chi-square test, as we said above, is based on the null hypothesis of equal frequencies, or, in this case, equal proportions in each cell: $H_0: p_1 = p_2 = \cdots = p_k$. Thus, if there is independence between the two variables of interest in elections and political milieu, we shall expect to get in the cross classification of Democratic and much interest in elections $38/110 \times 26 = 8.98$ observations. This is the expected frequency for this cell under the null hypothesis. The expected frequencies for each of the other cells under the null hypothesis are placed in parentheses. The reasoning for computing the expected frequency for the cell of Democratic milieu and much interest in elections may be stated as follows: If the two variables are independent, we shall expect to find in this cell that proportion of those cities classified as Democratic which those classified as having much interest in elections comprise of the total sample. The same reasoning can be applied for every cell. In short form, the method for computing the expected frequencies for each cell is

For cell A:

$$\frac{(A + D + G)(A + B + C)}{n} = 8.98$$

For cell B:

$$\frac{(B + E + H)(A + B + C)}{n} = 18.65$$

$$\cdot$$
$$\cdot$$
$$\cdot$$

For cell I:

$$\frac{(C + F + I)(G + H + I)}{n} = 6.00$$

or the product of the respective marginal totals for the cell divided by the total number in the sample.

---

[1] The numbers in the cells outside of the parentheses represent cities that fall into each of the cross classifications. The basis of the milieu classification is as follows: cities in which more than 60 percent of the registered voters are Democrats are classified as Democratic milieu, those with less than 40 percent are classified as Republican, and those between 40 and 60 percent are classified as competitive. The interest in elections classification is based on responses of city councilmen to questions concerning the amount of interest they perceive to exist in elections. See Heinz Eulau, Betty Zisk, and Kenneth Prewitt, "Latent Partisanship in Nonpartisan Elections: Effects of Political Milieu and Mobilization," in M. Kent Jennings and L. Harmon Zeigler, *The Electoral Process*, Prentice-Hall, Inc., Englewood Cliffs, N. J., 1966, pp. 208–238. The table is adapted from Table 9. Note that a chi-square test is appropriate only if random sampling has been used.

Let us consider this in more general terms.[2] Consider the matrix labeled as follows:

	C_1	C_2	
R_1	$C_1 \cap R_1$	$C_2 \cap R_1$	$\sum R_1$
R_2	$C_1 \cap R_2$	$C_2 \cap R_2$	$\sum R_2$
	$\sum C_1$	$\sum C_2$	n

The null hypothesis for the chi-square test is based on the definition of statistical independence presented in Chapter 4. It can therefore be stated as follows:

$$p(R_1|C_1) = p(R_1|C_2) = p(R_1)$$

and
$$p(R_2|C_1) = p(R_2|C_2) = p(R_2)$$

Now, because $p(C_1) = \sum C_1/n$ and $p(R_1) = \sum R_1/n$, then $p(C_1 \cap R_1) = \sum C_1/n \cdot \sum R_1/n = \sum C_1 \cdot \sum R_1/n^2$ and the expected frequency in cell one is

$$E(C_1 \cap R_1) = p(C_1 \cap R_1) \cdot n$$
$$= \frac{\sum C_1 \cdot \sum R_1}{n^2} \cdot n$$
$$= \frac{\sum C_1 \cdot \sum R_1}{n}$$

The reasoning for the other cells is the same.

The equation for computing chi-square is the same as that for one variable described above:

$$\chi^2 = \sum \frac{(f_0 - f_e)^2}{f_e}$$

The chi-square value can be computed directly with this equation, as follows:

$$\chi^2 = \frac{(9 - 8.98)^2}{8.98} + \frac{(21 - 18.65)^2}{18.65} + \cdots + \frac{(8 - 6)^2}{6} = 2.06$$

However, there is a shorter computing equation that can be derived by simple algebra:

$$\chi^2 = \sum \frac{(f_0 - f_e)^2}{f_e}$$
$$= \sum \frac{(f_0^2 - 2f_0 f_e + f_e^2)}{f_e}$$
$$= \sum \frac{f_0^2}{f_e} - n$$

[2] I am indebted to Benjamin Walter for this explanation.

since $\sum f_0 = \sum f_e = n$. For the data in Table 8-3 we have

	f_0^2	f_0^2/f_e
a	81	9.0200
b	441	23.6461
c	64	6.1776
d	169	14.2978
e	529	21.5478
f	196	14.3695
g	16	3.0796
h	100	9.2592
i	64	10.6666
		112.0615

$$\chi^2 = 112.06 - 110 = 2.06$$

To determine whether or not a chi-square of 2.06 may have occurred by chance under the null hypothesis, we need to look up this value in the table of chi-square distribution in the Appendix under the appropriate degrees of freedom. Let us first consider the very important concept of degrees of freedom.

In general, in mathematics, when we have a set of equations, known as simultaneous equations, containing unknowns, we cannot solve for these unknowns unless there are no more unknowns than there are equations. As a simple example of this point, consider the following two equations:

$$A = b + cD$$
$$AD = b + cD^2$$

If only two of the terms in these equations are unknown, we can solve for them. If $A = 2$ and $D = 4$, the only unknowns are b and c. Hence

$$2 = b + c(4)$$
$$2 \cdot 4 = b + c(4^2)$$

and

$$2 = b + 4c$$
$$8 = b + 16c$$

We can subtract the first from the second to solve for c:

$$8 = b + 16c$$
$$\underline{-2 = b + 4c}$$
$$6 = 12c$$
$$c = 1/2$$

Because we now know c, we can substitute it in the first equation to solve for b:

$$2 = b + 1/2(4)$$
$$b = 0$$

Therefore

$$2 = 0 + 1/2(4)$$

$$2 = 2$$

If three of the terms in these two equations had been unknown, we should not have been able to solve for b and c without making an assumption about the third unknown. For example, we might assume it to be equal to zero or some other arbitrary number. In this case, we lose one degree of freedom, because all the other unknowns depend on this assumption. This can also be illustrated in terms of a matrix, as follows:

$a_{(5)}$	b	10
c	d	14
6	18	24

Once one of the values in this matrix is given (the value in cell a, for example), then all the other cell values are determined. Cell b must be 5, because the sum of row values is 10; cell c must be 1; and cell d must be 13. In this case, we have lost one degree of freedom in solving for the three cell values. In general, in statistics, the degrees of freedom are $n - 1$, because we usually make one assumption in computing unknown population parameters. For example, when we use s as an estimate of σ, we lose one degree of freedom, and hence we divide by $n - 1$ rather than by n in computing s. In the case of the chi-square test, however, the degrees of freedom are $(r - 1)(c - 1)$ — rows minus one times columns minus one. In the example in Table 8-3, there are four degrees of freedom. The chi-square distribution in the Appendix shows that a value of 2.06 for four degrees has a probability of between 0.5 and 0.75. In this case, we cannot reject the null hypothesis, because the observed frequency can occur by chance very often under the null hypothesis. We conclude, therefore, that political milieu does not appear to be related to interest in elections in the population from which this sample is (assumed to have been) drawn.

Fisher's Exact Test. If the sample size is under 50, or if there are several cells in the matrix that have less than five in a cell, the chi-square distribution will yield false results. The Fisher exact test is appropriate under these conditions (see Figure 8-2). The question we want to answer is the same as for the chi-square test; that is, assuming the independence of 1 and 2, what is the probability of observing a distribution as extreme or more extreme than that in Figure 8-2? To determine this, we must measure the exact probability of getting the observed distribution and then the exact probabilities of getting more extreme distributions than this.

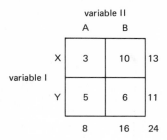

Figure 8-2.

With fixed marginals, the more extreme distributions can be found easily by decreasing the observations in cells a and d by one and increasing those in b and c by one until the number in cell a becomes 0. If the smallest number of observations is not in cell a, this will not occur. Therefore, the table should be set up so that the cell with the smallest number of observations is a (see Figure 8-3). The probabilities of getting each of these can be computed by

$$p(0_i) = \frac{(a+b)!(c+d)!(a+c)!(b+d)!}{n!a!b!c!d!}$$

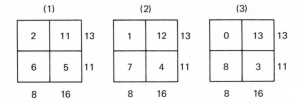

Figure 8-3.

For example, the probability of getting the set of observations in Figure 8-2 is

$$p(0_i) = \frac{13!11!16!8!}{24!3!10!5!6!} = 0.179$$

The probabilities for the more extreme distributions are computed in the same manner, and the probabilities added together. The null hypothesis is the same as for the chi-square test: H_0: $p_1 = p_2 = \cdots = p_k$. In this example, we are not able to reject H_0 at $\alpha \leq 0.05$.

It is obvious that the computational work for this test is extensive. Tables that can be used instead are available (see Siegel, cited at the end of this chapter).

The chi-square test is one of the more useful tests available for nominal-scale data. There are others that can be used. We discuss the binomial and runs tests briefly.

The Binomial Test.   The binomial model described in Chapter 5 can be used to test questions concerning proportions when the event has only two outcomes. For example, the probability of getting three heads in five flips of a fair coin, where we assume the probability of a head to be one-half, is $p(3,5;1/2) = {_3}C_5(1/2)^3(1/2)^2 = 10/32$. In other words, we have observed an event that can occur relatively frequently by chance if what we assume to be true about the probability of heads is true. In testing the null hypothesis, however, we test the probability of getting the observed result and results more extreme rather than simply the probability of the exact observation. In this example, we have $p(3) + p(4) + p(5)$ heads $= 10/32 + 5/32 + 1/32 = 1/2$. The null hypothesis in this case is $H_0: p = 1/2$. We do not reject it, because the chance of making error type 1 if we do is $\alpha = 0.5$.

This test has more interesting applications than answering questions about the fairness of a coin. As an example we consider the chance theory of Deutsch and Madow.[3] The theory suggests that individuals who are in positions of leadership do not necessarily make a higher proportion of correct decisions than anyone else. They only appear to. For example, suppose that individuals have no better than an equal chance of making correct decisions. Suppose that we observe a person in a leadership position and the 12 decisions he makes in a period of time. Suppose further that out of these 12 decisions, 9 are correct and 3 are incorrect. Could he have made as many as 9 correct decisions if his ability were not better than the ordinary person's, that is, no better than fifty-fifty? The binomial probability distribution shows that

$$p(9,12;1/2)\ = 0.0537$$
$$p(10,12;1/2) = 0.0161$$
$$p(11,12;1/2) = 0.0029$$
$$p(12,12;1/2) = \underline{0.0002}$$
$$0.0729$$

The answer is that 7 times in 100 the result can be obtained by chance. Thus, we may conclude that this person's abilities are no better than fifty-fifty. Of course, we may not be willing to accept the chance theory on the basis of observing one person. But note that, in order to reject the H_0: $p = 1/2$, at $\alpha \leq 0.05$, the person must make at least 10 out of 12 decisions correctly. Assuming that people in the higher levels of a bureaucracy make

[3] See Karl W. Deutsch and William G. Madow, "A Note on the Appearance of Wisdom in Large Bureaucratic Organizations," *Behavioral Science*, pp. 72–78, January, 1961.

relatively few decisions they are identifiably responsible for, it becomes apparent that it is difficult to judge when a person's ability to make correct decisions is better than $p = 0.5$.

The Runs Test. The runs test is a test of the randomness of a sequence of events. For example, if we were interested in determining if there is a tendency for the United States Supreme Court to consider certain kinds of cases in sequence, we might use the symbol $+$ to represent consideration of this kind of case and the symbol $-$ to represent consideration of a different kind of case. Suppose that in a series of 14 cases we observed the following sequence:

$$\underset{1}{+\ +}\ \ \underset{2}{-\ -\ -}\ \ \underset{3}{+}\ \ \underset{4}{-\ -\ -\ -}\ \ \underset{5}{+\ +}\ \ \underset{6}{-}\ \ \underset{7}{+}$$

There are 7 *runs* in this series; that is, the number of sequences of the same kind of event is 7. If there is no tendency for particular cases to be considered in a series, we shall expect to observe a random fluctuation back and forth. If there were nothing random at all in the process, we might expect to observe something like the following:

$$+\ +\ +\ +\ +\ +\ \ \ -\ -\ -\ -\ -\ -\ -\ -$$

or only 2 runs. However, the highest number of possible runs with these observations is 12:

$$+\ -\ +\ -\ +\ -\ +\ -\ +\ -\ +\ -\ -\ -$$

It is apparent that neither extreme, 2 runs or 12, is likely to be a random series. We are interested in determining how far we may depart from these two extremes and still consider the process nonrandom.

The distribution we are concerned with can be symbolized R_u. The sampling distribution of R_u is known. In this example, runs of 4 to 8 are considered random. In larger samples (>50) the normal curve can be used as an approximation of the runs distribution by converting to a Z score, as follows:

$$Z = \frac{R_u - \mu_{R_u}}{\sigma_{R_u}} = \frac{R_u - [2n_1 n_2/(n_1 + n_2) + 1]}{\sqrt{[2n_1 n_2 (2n_1 n_2 - n_1 - n_2)]/[(n_1 + n_2)^2(n_1 + n_2 - 1)]}}$$

Strength of Relationship

Thus far we have been concerned primarily with statistical description and significance. Statistical significance is not the same as practical or research significance. Consider again the example concerning the relationship of communications structure and problem solving. Suppose that the time it takes two population groups to solve similar problems in two different communications structures is 25 minutes and 10 seconds and 25 minutes and

30 seconds respectively, where the first group is highly restricted in its mode of communication and the second is not at all. A large enough sample will uncover the population difference, thus yielding a statistically significant finding. But the practical and theoretical consequences of a difference of 20 seconds may be nothing at all. The latter question concerns the strength of relationship between two variables, and we now consider how to measure it for nominal-scale data.

Measures Based on Chi Square. The size of the chi square we compute is a direct function of the size of the sample. For example, the data shown in Table 8-4 have the same proportion of cases in each example. But the chi-square value for case *B* is much larger than that for case *A*, because the sample is larger. This suggests that we may use chi square as a measure of strength of relationship by taking account of the sample size. We can do so by dividing by *n*, and we get a result called phi square: $\phi^2 = \chi^2/n$. We can take the square root of this to transform the data back into units it originally measured: $\phi = \sqrt{\chi^2/n}$.

Table 8-4   Illustration of Circumstance in which the Proportion of Observation in Each Cell is the Same but the χ^2 Value is Different

	A			B	
2	10		20	100	
3	50		20	500	

In a two-by-two table, the upper limit of this coefficient is one, and we may say that the relationship between the two variables is perfect. The lower limit is zero, and we may say that there is no relationship. However, when a table has more than two rows and columns, the upper limit is no longer one, and the coefficient becomes difficult to interpret (the reader should verify this himself). There are some alternatives to try to correct for this, but they all have only limited application (see Blalock for a discussion of this). Hence, a new set of measures has been developed in recent years that does not use chi square as a base. Let us consider these measures in detail.

Measures Based on Probability Reasoning. As an example, consider the hypothetical data in Table 8-5 concerning the relationship between social class and party affiliation. Suppose that the name of each person in this sample is written on a card, placed in a container, and thoroughly mixed. One of the names is picked at random. What is the probability that the person picked is a Democrat? Obviously, 60/110. Suppose that we are asked to predict whether the person picked at random is a Democrat or a Republican, and each time the slips are replaced before the next name is picked. It would be to our advantage to predict Democrat each time, because we make a smaller long-run error, or an error of $50/110 = 45$ percent in the long run. Suppose next that, when each slip is picked, we are told first whether the person is upper or lower class before we predict his party affiliation. If the person picked is upper class, we should predict a Republican, and if the person picked is lower class, we should predict Democrat. The long-run error would be $(20 + 10)/110 = 27$ percent. Thus, knowing whether each person is upper or lower class improves prediction by 18 percent. Symbolically stated, we have

$$p(\text{error} \mid A_j) = p(\text{error} \mid A_1)p(A_1) + p(\text{error} \mid A_2)p(A_2)$$

$$= \left(\frac{20}{60}\right)\left(\frac{60}{110}\right) + \left(\frac{10}{50}\right)\left(\frac{50}{110}\right)$$

$$= \frac{20}{110} + \frac{10}{110} = \frac{30}{110} = 0.27$$

that is, the probability of error knowing A_j (the column headings) is the probability of error knowing the first column heading times the probability of this kind being selected, plus the same for every subsequent column.

Table 8-5 Hypothetical Data Concerning the Relationship between Social Class and Party Affiliation

Party affiliation	Social class		
	Upper	*Lower*	*Total*
Republican	40	10	50
Democrat	20	40	60
Total	60	50	110

When this is subtracted from 0.45 we have tau beta, or τ_b. It tells how strongly the two variables are related, because if there is no relationship between them, the error in predicting B when A is known will be the same as that in predicting B when A is unknown. When the two variables are perfectly related, the error in predicting B when A is known is zero.

Tau beta measures the absolute reduction in error. A better measure considers the relative reduction. For example, if the error in prediction when A is not known is 45 percent and if it is reduced to zero when A is known, the relative error is reduced by 100 percent, although the absolute reduction is only 45 percent. Goodman and Kruskal have developed a measure of the relative reduction in error in prediction that is called lambda b, symbolized by λ_b. Given the discussion above, its definition should be obvious:

$$\lambda_b = \frac{p(\text{error}|A_j \text{ unknown}) - p(\text{error}|A_j \text{ known})}{p(\text{error}|A_j \text{ unknown})}$$

Thus, if the error is reduced by 18 percent knowing A and the total error not knowing A is 45 percent, we find out what proportion the first is of the second. In this hypothetical example, we have $\lambda_b = (0.45 - 0.27)/0.45 = 0.40$.

The beauty of this measure is that it varies between zero and one. We may interpret the resulting coefficient as follows: When the relationship is perfect, λ_b is one. When there is no relationship, λ_b is zero. Numbers in between indicate degrees of strength.

In order to ease the computational work, rather than computing the percentage error in the manner above, we can get an approximation by working directly with the frequencies in the cells by

$$\lambda_b = \frac{\sum \max f_{ij} - \max f_{\cdot j}}{n - \max f_{\cdot j}}$$

where $\sum \max f_{ij}$ is the sum of the maximum frequencies in the cells in each of the rows and $\max f_{\cdot j}$ is the maximum frequency in the row marginals. For the data in Table 8-5 we have $\lambda_b = (40 + 40 - 60)/(110 - 60) = 0.40$. Note that we obtain the same thing if we work simply with the minimum of the row marginals and the cell frequencies. Thus

$$\lambda_b = \frac{p(\text{error}|A_j \text{ unknown}) - p(\text{error}|A_j \text{ known})}{p(\text{error } A_j \text{ unknown})}$$

$$= \frac{50/110 - (20 + 10)/110}{50/110}$$

$$= \frac{50 - (20 + 10)}{50}$$

$$= 0.40$$

As another example, consider data taken from a study by Peabody concerning the relationship between the percentage of Negroes in a congressman's district and his vote for or against expansion of the rules committee.[4] In doing so, we note that λ_b is an asymmetrical measure, so that it is important

[4] Robert L. Peabody, "The Enlarged Rules Committee," in Robert Peabody and Nelson Polsby (eds.), *New Perspective on the House of Representatives*, Rand McNally & Company, Chicago, 1963.
    Note that λ_b is indeterminate when the population lies in one column or class. In this case it will be zero and cannot be interpreted. See Leo A. Goodman and William Kruskal, "Measures of Association for Cross Classifications," JOURNAL OF THE AMERICAN STATISTICAL ASSOCIATION, 49, 1954, pp. 732–764.

to put the correct variable along the rows of the table. In this example, we take the vote on the rules committee to be the dependent variable, and we want to see how strongly it is related to the percentage of Negroes in a district. The dependent variable is put in the rows of the table, as in Table 8-6.

Table 8-6 Percentage of Negroes in District and Vote for Expansion of the Rules Committee*

	0–9.9	10–19.9	20–29.9	30–39.9	40–49.9	≥50	Total
Number voted against	8	3	22	14	12	3	62
Number voted for	9	12	6	5	4	0	36
Total	17	15	28	19	16	3	98

*Adapted from: Robert L. Peabody, "The Enlarged Rules Committee," in Robert Peabody and Nelson Polsby (eds.), *New Perspective on the House of Representatives,* Rand-McNally & Co., Chicago, 1963.

The strength of the relationship between the two variables as given by lambda *b* is

$$\lambda_b = \frac{(9 + 12 + 22 + 14 + 12 + 3) - 62}{98 - 62}$$

$$= \frac{36 - (8 + 3 + 6 + 5 + 4 + 0)}{36} \doteq 0.28$$

It appears, therefore, that the percentage of Negroes in a Southern congressman's district is positively related to a vote against expansion, so that we may conclude that as the percentage of Negroes in a district goes up, the vote against expansion of the rules committee does also. However, the relationship is not very strong. Note that, if the table were turned around so that the percentage of Negroes were put in the rows of the table, the lambda *b* measure would be only 0.08. Of course, it does not make much sense to treat the percentage of Negroes in a district as the dependent variable in this case.

Point-biserial Correlation. When the problem involves an interval-scale variable combined with a nominal-scale variable, the point-biserial correlation coefficient can be used as a measure of strength of relationship. The point-biserial coefficient is an unbiased estimate of the product-moment correlation coefficient, which we describe in Chapter 9. For the moment, let us consider the point-biserial coefficient.

There are two ways to do so. We may assign arbitrary values to the dichotomized variable, such as zero and one. In the example in the preceding section, a one might be assigned to each vote against expansion of the rules committee and a zero to each vote for it. The equation for product-moment correlation (see Chapter 9) can then be used, and the result treated as any product-moment correlation. But a shorter method can be used by working directly with the means of each nominal-scale variable with the equation:

$$r_{pb} = \frac{\bar{Y}_1 - \bar{Y}_2}{s_Y} \sqrt{\frac{n_1 n_2}{n(n-1)}}$$

For the data in Table 8-6 we have

$$r_{pb} = \frac{29.5 - 20.3}{12.8} \sqrt{\frac{(62)(36)}{(98)(97)}} = 0.23$$

29.5 is the mean percentage of Negroes in the districts of the congressmen who voted against expansion of the rules committee, and 20.3 is the mean percentage of Negroes in the districts of the congressmen who voted for it.

The point-biserial coefficient is very close to that computed by λ_b, and because it is an unbiased estimate of the product-moment correlation coefficient, it is a better measure of strength of relationship. This is because we can use the square of r_{pb} in interpreting its meaning. This is considered in more detail in Chapter 9. Here we note that, when we square 0.23, we get 0.0529, and we can say that 5.29 percent of the variation in voting on expansion of the rules committee can be explained by the percentage of Negroes in a congressman's district. This also means that almost 95 percent has to be explained by other factors and that the percentage of Negroes in a district is therefore a minor factor.

The point-biserial correlation coefficient gives a measure of strength of relationship between a nominal-scale and an interval-scale variable. We can test whether the correlation found in a sample also exists in the population. To do so, we need a sampling distribution of r_{pb}. It is known that the t distribution can be used as an approximation of this when we convert to a t score by means of the following:

$$t = \frac{r_{pb}(n-2)}{1 - r_{pb}^2}$$

which for the data in the example is

$$t = \frac{0.23(96)}{1 - 0.053} = 23.32$$

A t of 23.32 with 96 degrees of freedom is significant beyond the 0.001 level, and we therefore reject the null hypothesis that the correlation in the population is zero.

An important point must now be emphasized. When we use a sampling distribution, such as the t distribution, to make a test about a population parameter, such as the correlation in the population, certain assumptions have to be met. In this case, we are concerned with statistical significance rather than strength of relationship. A test of statistical significance is never warranted unless the sample data are obtained by random or some other form of probability sampling. All the tests considered in this book suppose that the sample data were obtained by random sampling. Corrections for other forms of probability sampling are considered in Chapter 15. When sampling distributions are used to make inferences about population parameters, other assumptions are also made. As we said at the beginning of this chapter, with nonparametric statistics the assumptions are less restrictive than with parametric statistics. But unless the assumptions are met, the results obtained by a sampling distribution will be false. For the point-biserial correlation, in addition to the assumption of random sampling, two other assumptions are made: (*a*) the populations from which the samples were drawn are normally distributed (in this case, the populations are the percentages of Negroes in districts of congressmen who voted for and against expansion), and (*b*) the variances in both populations are the same. The rationale for these assumptions is considered in Chapter 9. Here we note that, unless these three assumptions are met, a test of statistical significance is not warranted, and the findings concerning strength of relationship, that is, $r_{pb} = 0.23$, hold only for the sample. In the present example, because the data were not obtained by random sampling, the conclusions hold only for the 98 congressmen in the sample.

TESTS AND MEASURES FOR ORDINAL-SCALE DATA

If we are able to measure the data on an ordinal rather than simply a nominal scale, we gain considerably in our ability to extract information from the data. Most attitudinal scales, such as the authoritarian scale, are ordinal-scale variables, although the numbers obtained by adding responses of individuals to the several questions or items in the scales themselves form a ratio scale. Strictly speaking, it is better to rank these scores and use the statistical tests to be described in this section rather than to treat the scores as if they formed a ratio scale and use the measures to be described in Chapter 9.

In this section we first consider the two principle measures of strength of relationship for ordinal-scale data: the Spearman rank-order correlation coefficient and the Kendal tau. We then consider several measures of statistical significance for ordinal-scale data. These different measures of statistical significance all have the same objective: to determine if the sample results also hold in the populations from which the samples were drawn (at random). In contrast to the measures for nominal-scale data, there are several tests of significance available for ordinal-scale data. The Kolmogorov-Smirnov test for a single variable; the runs test for two variables; the Mann-Whitney U test for two groups, which is comparable with the difference of means test for interval-scale data described above; and finally the Kruskal-Wallis test for j independent groups, that is, for more than two groups (the test is comparable with the analysis of variance test for interval-scale data to be considered in Chapter 11).

Strength of Relationship

Rank-order Correlation. There are two principal rank-order correlation measures: the Spearman correlation coefficient, symbolized r_s; and the Kendal correlation coefficient, generally symbolized τ. The first is simple and appropriate if there are not many ties in the data. The second is appropriate if there are.

The logic of the Spearman correlation coefficient is beautifully simple. Suppose that there are two variables we are interested in, anomie and conservatism, and we want to measure how strongly they are related.[5] A sample of 12 individuals is measured on each variable and given scores in which a higher score represents a greater amount of each attribute possessed by each individual. We suppose that the scores obtained on the 12 individuals are as shown in Table 8-7. We suppose that the measure we are using forms an ordinal rather than an interval scale. Hence, we must first rank the scores before proceeding. We may do so by ranking them first on one variable and then putting next to each person's rank on one variable his respective rank on the other, as in Table 8-8. If there is perfect agreement between the two variables, the person who ranks first on the anomie scale will also rank first on the conservatism scale, the one who ranks second on the anomie scale will also rank second on the conservatism scale, and the one who ranks nth on the anomie scale will also rank nth on the conservatism scale.

It can be seen that this is not true of the hypothetical data, because the person who ranks first on the anomie scale (person 4) ranks first on conservatism, but the person who ranks second on the anomie scale ranks third on the conservatism scale, and so on. Thus, $d_i = \sum (X_i - Y_i)^2$ seems to be a logical measure of the amount of disagreement, because as the agreement between the two sets of ranks becomes poorer, the sum of the squared difference in ranks also grows larger. However, this measure is affected by the size of the sample, and we must correct for this. Furthermore, it would be more

[5] This example is based on the work of McClosky and Schaar. See Herbert McClosky and John H. Schaar, "Psychological Dimensions of Anomy," *American Sociological Review*, vol. 30, pp. 14–20, February, 1965. Also Herbert McClosky "Conservatism and Personality," in S. Sidney Ulmer, *Introductory Readings in Political Behavior*, Rand McNally & Company, Chicago, 1961, pp. 33–45.

Table 8-7 Anomie and Conservatism Scores

	Scores	
Person	Anomie	Conservatism
1	82	42
2	98	46
3	87	39
4	40	37
5	116	65
6	113	88
7	111	86
8	83	56
9	85	62
10	126	92
11	106	54
12	117	81

Table 8-8 Ranks on Two Variables

	Rank on		
Person	Anomie	Conservatism	d_i^2
4	1	1	0
1	2	3	1
8	3	6	9
9	4	7	9
3	5	2	9
2	6	4	4
11	7	5	4
7	8	10	4
6	9	11	4
5	10	8	4
12	11	9	4
10	12	12	0
Total	. . .	. . .	52

convenient to have a measure that varies within the same limits no matter how small or large the sample. The Spearman equation takes care of these questions:

$$r_s = 1 - \frac{6 \sum d_i^2}{n(n^2 - 1)}$$

For the data above we have

$$r_s = 1 - \frac{6(52)}{12(12^2 - 1)} = 0.82$$

The conclusion is that the relationship between the two variables is very strong.[6]

If there are many ties, the Spearman coefficient will be a biased estimate of the population coefficient. The Kendal tau corrects for ties. Let us illustrate this by the use of simple data. Consider the following ranks of four observations on variables A and B:

Respondent	Variable A	Variable B
a	1	2
b	2	3
c	3	1
d	4	4

Instead of subtracting differences, we count the number of times the rank on variable B is out of order vis-a-vis variable A. For each time the order is correct, we give a plus one, and for each time the order is incorrect, we give a minus one. For example, the ranks of respondents a and b on variable B are in the same order as on variable A; that is, 2 and 3 move in the same direction as 1 and 2, and we give a plus one. The ranks of a and c on variable B (2,1) are in the opposite direction to the ranks on variable A (1,3), and we give a minus one. Doing the same for all possible pairs of respondents, we have

$$
\begin{array}{ll}
(2,3) \; +1 & (3,1) \; -1 \\
(2,1) \; -1 & (3,4) \; +1 \\
(2,4) \; +1 & (1,4) \; +1
\end{array}
$$

The algebraic sum of the plus and minus ones is two. The measure of strength of relationship is given by $\tau = s/[1/2(n)(n-1)]$, where s is the sum of the plus and minus ones. For the simple data in the illustration, we have $\tau = 2/[1/2(4)(3)] = 2/6 = .33$.

Now let us go to the question of ties, because this is the reason for preferring the more laborious Kendal tau procedure to the Spearman. Let us note first, however, that ties are scored as the average score of the numbers that would be given to each observation (respondent) if there had not been any ties. The tied scores and the scores that would be used are shown in Table 8-9. The correlation for ties is

$$
\tau = \frac{s}{\sqrt{1/2N(N-1) - t} \; \sqrt{1/2N(N-1) - v}}
$$

where $t = 1/2 \sum t_i(t_i - 1) = $ the number of ties in each set of ties in A and $v = 1/2 \sum v_i(v_i - 1) = $ the number of ties in each set of ties in B. For example, consider the data in Table 8-10. There is one group of ties of two and one group of ties of three for variable A. Thus

$$
t = 1/2[2(1) + 3(2)] = 4
$$

---

[6] Although we have reserved until later in the chapter the question of statistical significance for ordinal-scale variables, we should note here that r_s can be tested for significance by converting to the t distribution for samples >30 with the following:
$$
t = r_s (n-2)/(1 - r_s^2), \text{ with } n - 2 \text{ degrees of freedom.}
$$

And there is one group of ties of two for variable B. Thus, $v = 1/2[1(2)] = 1$. The s for these data is 24:

(4, 3) −1	(3, 1) −1	(1, 2) +1	(2, 5) +1	(5, 6) +1	(6,7.5) +1	(7.5,7.5) +1
(4, 1) −1	(3, 2) −1	(1, 5) +1	(2, 6) +1	(5,7.5) +1	(6,7.5) +1	(7.5, 9) +1
(4, 2) −1	(3, 5) +1	(1, 6) +1	(2,7.5) +1	(5,7.5) +1	(6, 9) +1	(7.5, 9) +1
(4, 5) +1	(3, 6) +1	(1,7.5) +1	(2,7.5) +1	(5, 9) +1		
(4, 6) +1	(3,7.5) +1	(1,7.5) +1	(2, 9) +1			
(4,7.5) +1	(3,7.5) +1	(1, 9) +1				
(4, 9) +1	(3, 9) +1					

The tau measure of strength of relationship is thus

$$\tau = \frac{24}{\sqrt{1/2(9)(8) - 4}\sqrt{1/2(9)(8) - 1}} = .73$$

Table 8-9   Illustration of Method for Computing Scores for Ties

Respondent	Rank	Score to be used
a	1	1
b	2	2.5 $\frac{2+3}{2} = 2.5$
c	2	2.5
d	3	4
e	4	5
f	5	7
g	5	7 $\frac{6+7+8}{3} = 7$
h	5	7
i	6	9
.	.	.
.	.	.
.	.	.

Table 8-10   Hypothetical Data for Computing Kendal Tau

| Respondent | Variables | |
	A	B
a	1	4
b	2.5	3
c	2.5	1
d	4	2
e	5	5
f	7	6
g	7	7.5
h	7	7.5
i	9	9

Significance Tests

Kolmogorov-Smirnov Test. The Kolmogorov-Smirnov test is similar in logic to the chi-square test, because it is concerned with the difference between an observed frequency distribution and a hypothetical one. But it is appropriate when variables are measured on an ordinal scale. It is concerned with the statistic $D = \max E_c - O_c$, where E_c represents the expected cumulative frequency and O_c the observed cumulative frequency. The sampling distribution of the statistic D is known, and it is therefore unnecessary to construct a sampling distribution for this test.

As an example, suppose that we are interested in determining if voters have a preference for the way candidates are portrayed in campaign pictures. A series of five photographs is taken of a candidate in which each pose is

Table 8-11 Hypothetical Data for Kolmogorov–Smirnov Test

	Rank of photograph				
	1	*2*	*3*	*4*	*5*
f	0	1	0	5	4
E_c	1/5	2/5	3/5	4/5	5/5
O_c	0/10	1/10	1/10	6/10	10/10
$E_c - O_c$	2/10	3/10	5/10	2/10	0/10

slightly different, ranging from serious (given a value of 1) to smiling (given a value of 5). The photographs are presented to a random sample of 10 people who are asked to rank them from the pose they like best to that which they like least. If a person has no preference, photographs of each rank of seriousness should be selected equally often. The null hypothesis, therefore, is the same as for the chi-square test: $H_0: f_1 = f_2 = f_3 = f_4 = f_5$. Suppose that the actual selection of ranks is as shown in Table 8-11; that is, no one prefers the most serious pose (photograph 1), one person prefers the next most serious pose, and so on. The maximum difference between the observed cumulative frequency and the expected cumulative frequency is 5/10, or 0.50. The table of the D distribution shows that the probability of observing a D this large or larger under the null hypothesis of no difference in frequencies is less than 0.01 (see Siegel for this table). We conclude, therefore, that there is a difference in preference; the data were set up to demonstrate a preference for a smiling pose.

Wald-Wolfowitz Runs Test. Next we consider an extension of the runs test to two samples. The example is a question about the difference in attitude of two congressional committees toward government intervention in the economy. Suppose that this is measured on an ordinal scale in which a higher score means a more favorable attitude. The null hypothesis is that there is no difference in attitude between the two committees. The data in Table 8-12

Table 8-12 Attitude Scores Concerning Government Intervention in the Economy

Committee A		Committee B	
86	118	55	7
69	45	40	9
72	141	22	76
65	104	58	26
113	41	16	36
65	50	15	20

are the hypothetical figures we use here. A visual inspection of the scores tends to show that the members of committee *A* are much more favorable toward government intervention in the economy. In order to test whether the difference may have been produced by chance with the Wald-Wolfowitz test, we must set it up so as to count the number of runs. The data are first cast in order of increasing scores for each group:

A	41	45	50	65	65	69	72	86	104	113	118	141
B	7	9	15	16	20	22	26	36	40	55	58	76

and then in order of sequence of all scores, and the number of runs identified as the sequences of scores of individuals from the same group:

7 9 15 16 20 22 26 36 40 41 45 50 55 58 65 65 69 72 76 86 104 113 118 141

B B B B B B B B B A A A B B A A A A B A A A A A

        1   2   3   4   5   6

The sampling distribution of the runs statistic shows that we need a number of runs of eight or less for significance at the 0.05 level (see Siegel for this table). We have six, and we therefore reject the null hypothesis that there is no difference between the two committees.

In general, if the null hypothesis is true, the scores of the two groups will be mixed, and the number of runs, therefore, will be large. Depending on the size of the samples, when R_u is small, the null hypothesis will be rejected. The sampling distribution of the R_u statistic is relatively uncomplicated. For large n, the sampling distribution of R_u is approximately normal. It has a mean of

$$\mu_{R_u} = \frac{2n_1n_2}{n_1 + n_2} + 1$$

and a variance of $$\sigma_{R_u}{}^2 = \frac{2n_1n_2(2n_1n_2 - n_1 - n_2)}{(n_1 + n_2)^2(n_1 + n_2 - 1)}$$

Hence, we can test a large sample R_u by

$$Z = \frac{R_u - \mu_{R_u}}{\sigma_{R_u}}$$

Note also that, if we measured these attitudes on an interval scale, the difference of means test and the t distribution could be used. The reader should compute t on the basis of the data in Table 8-12 to see if it produces the same result.

Mann-Whitney Test.   The Mann-Whitney test is similar to the runs test but more powerful. The null hypothesis is that the two populations from which the samples were drawn are the same. Suppose that the populations are the Republican and Democratic delegates to the presidential nominating conventions. Suppose that we want to test whether there is a difference between the two in preference for a candidate who has an elite origin. A random sample is drawn from each of the populations, and a test (series of questions) is administered to each, yielding the following (hypothetical) scores:

Sample 1 (Democrats): 35   31   26   26   20   14   12   10   4
Sample 2 (Republicans): 34   31   28   26   25   24   21   20   20   20

The scores for each sample have to be combined and ranked on a single scale, as in Table 8-13. First we obtain the sum of ranks for both samples and label them SR_1 and SR_2 respectively. Then we compute a statistic designated U by the following equation:

$$U_1 = n_1n_2 + \frac{n_1(n_1 + 1)}{2} - SR_1$$

For the data in Table 8-13 this is

$$U_1 = (9)(10) + \frac{9(9 + 1)}{2} - 102$$

$$= 33$$

**Table 8-13 Hypothetical Scores of Preference for a
Candidate of Elite Origin**

Sample 1 (n = 9)		Sample 2 (n = 10)	
Score	*Rank*	*Score*	*Rank*
35	1	34	2
31	3.5	31	3.5
26	7	28	5
26	7	26	7
20	13.5	25	9
14	16	24	10
12	17	21	11
10	18	20	13.5
4	19	20	13.5
		20	13.5
$SR_1 = 102.0$		$SR_2 = 88.0$	

We can also compute U_2 for the second sample, but it is unnecessary because the same results are obtained either way. Under the null hypothesis that the samples came from a population having the same preference for a candidate of elite origin, the sampling distribution of U has a mean of

$$\mu_U = \frac{n_1 n_2}{2}$$

and a variance of

$$\sigma_U{}^2 = \frac{n_1 n_2 (n_1 + n_2 + 1)}{12}$$

If each sample is larger than eight, the normal table can be used as the sampling distribution of U; otherwise the tables for critical values of U have to be used (these have not been reproduced in this book but can be found in Siegel). We can convert the sample U to a Z score by the following:

$$Z = \frac{U_1 - \mu_U}{\sigma_U}$$

For the hypothetical data in Table 8-13 we have

$$Z = \frac{33 - 45}{12.25} = -0.98$$

which is not significant at the 0.05 level. We cannot therefore reject the null hypothesis, and we conclude that the two populations are the same in their preference for a candidate of elite origin. Note that, if we tested the null hypothesis using U_2, the same result would obtain except that the Z score would be positive. U_2 would be

$$U_2 = n_1 n_2 + \frac{n_2(n_2 + 1)}{2} - SR_2$$

$$= (9)(10) + \frac{(10)(11)}{2} - 88 = 57$$

and hence Z would be 0.98.

Let us consider another use of the U statistic. In their research concerning the relationship between perception and action in the 1914 World War I crisis, Holsti, North, and Brody tested a hypothesis concerning the relationship between policy response and input action (perception of threat from the

Table 8-14   Reaction Scores for Dual Alliance and Triple Entente*

Dual Alliance (Sample A)	0.13; 1.38; −0.25; −2.76; −1.09; −2.19; −1.21; 0.04; −0.03; −1.70; −0.38; −0.18.
Triple Entente (Sample B)	−0.13; 1.67; 0.38; −1.45; 1.70; 0.42; 1.19; 0.99; 0.04; −0.40; 0.40; −0.15.

*Adapted from: Ole R. Holsti, Robert C. North, and Richard A. Brody, "Perception and Action in the 1914 Crisis," in J. David Singer, *Quantitative International Politics: Insights and Evidence,* The Free Press of Glencoe, New York, 1968, pp. 123–159.

other side) for the nations of the Triple Entente and those of the Dual Alliance.[7] They hypothesized that the nations of the Triple Entente would underreact, in terms of violence, to their perception of threat from the other side, whereas those of the Dual Alliance would overreact. To test this hypothesis, each group of nations was given reaction scores based on the difference between levels of violence in the actions of the opposing coalitions and the level of violence in the resulting action response. A sample of 12 time periods was taken, from June 27–July 2 to August 3–4, 1914. If the difference is negative, it means that the alliance was overreacting; whereas if the difference is positive, the alliance was underreacting. The scores for each alliance are shown in Table 8-14. The ranks for the two samples are therefore

Sample: *B B A B B B B B A   B   A   A B B A A A B A A A A A A*
Rank: 1 2 3 4 5 6 7 8 9 10.5 10.5 12 13 14 15 16 17 18 18 20 21 22 23 24

The sum of the ranks for each sample is

$$SR(A) = 190.5$$
$$SR(B) = 109.5$$

[7] See Ole R. Holsti, Robert C. North, and Richard A. Brody, "Perception and Action in the 1914 Crisis," in J. David Singer, *Quantitative International Politics: Insights and Evidence*, The Free Press of Glencoe, New York, 1968, pp. 123–159.

Computing U with $SR(B)$ and labeling it U_1, we have

$$U_1 = 12(12) + \frac{12(13)}{2} - 109.5 = 12.5$$

And we convert to a Z score:

$$Z = \frac{12.5 - 72}{17.32} = -2.33$$

Exactly 1.98 percent of the normal curve lies outside of the area that is ± 2.33 standard errors from the mean. We can therefore reject the null hypothesis that the populations are the same with less than a 5 percent chance of making error type 1.

Kruskal-Wallis Test. A direct extension of the Mann-Whitney test for j independent groups is the Kruskal-Wallis test. The data for the j groups are arranged in the same way as in the Mann-Whitney test, and the rank sum for each group is found (see Table 8-15). We may denote this by RS_j. The sum of these sums may be denoted by RS. We then test the null hypothesis by computing.

$$H = \frac{12}{n(n + 1)} \left(\sum \frac{RS_j^2}{n_j} \right) - 3(n + 1)$$

which is distributed approximately as a χ^2 distribution with $j - 1$ degrees of freedom. (If there are many ties, H should be divided by

$$C = 1 - \left(\frac{\sum_i^G (t_i^3 - t_i)}{n^3 - n} \right)$$

where t_i is the number of ties in each G set of ties.)

For example, suppose that we wanted to test the relationship between authoritarianism and power aspirations. Suppose that we have a random sample of elected officials each of whom has been measured on each scale and that the authoritarianism scale is assumed to be an ordinal scale, and the power-aspiration scale nominal. We assign individuals at random to each of the three levels of power aspiration (low, medium, and high), and the authoritarianism scores and ranks (in parentheses) for each of the groups is as indicated in Table 8-15.

The value of H for these data is

$$H = \frac{12}{15(16)} \left[\frac{(25.5)^2 + (39)^2 + (55.5)^2}{5} \right] - 3(16) = 4.5$$

Table 8-15   Hypothetical Data for the Kruskall-Wallis Test

	Power Aspirations	
Low	Medium	High
15 (8)	5 (1)	30 (14)
8 (3)	10 (4)	35 (15)
14 (7)	16 (9)	17 (10)
12 (5.5)	25 (13)	19 (11)
6 (2)	20 (12)	12 (5.5)
$RS_j =$ 25.5	39	55.5
RS = 120		

The χ^2 distribution for $j - 1 = 2$ degrees of freedom shows that this is not significant at ≤ 0.05 level. Hence, we cannot reject the null hypothesis that the population distributions are identical.

If there were many ties, we should use the correction for ties given above. However, this correction will not matter much unless there are many ties or n is small.

The concerns of this book more or less come to a focal point in the next two chapters. We have considered several important aspects of statistics thus far: methods for (*a*) describing aspects of frequency distributions, (*b*) making point and interval estimates, (*c*) testing hypotheses about a single variable and about two variables, and (*d*) measuring the extent of association between two variables for nominal and ordinal scale data. Now we turn to a powerful method that enables us to do several of these things at the same time, providing the variables are measured on an interval or ratio scale.

READINGS AND SELECTED REFERENCES

Blalock, Hubert, Jr. *Social Statistics*, McGraw-Hill Book Company, New York, 1960, Chaps. 14, 15.

Diamond, Solomon. *Information and Error*, Basic Books, Inc., Publishers, New York, 1959, Chaps. 9, 13.

Freund, John. *Modern Elementary Statistics*, Prentice-Hall, Inc., Englewood Cliffs, N. J., 1960, Chaps. 13, 16.

Hays, William L. *Statistics for Psychologists*, Holt, Rinehart and Winston, Inc., New York, 1963, Chaps. 17, 18.

McCarthy, Philip J. *Introduction to Statistical Reasoning*, McGraw-Hill Book Company, New York, 1957, Chap. 11.

Siegel, S. *Nonparametric Statistics for the Behavioral Sciences*, McGraw-Hill Book Company, New York, 1956.

9

Simple Linear Correlation and Regression

REGRESSION ANALYSIS

One idea of the goal of science is that the ultimate objective is to reach predictive statements. Perfect prediction means perfect knowledge, which, of course, is impossible. In most real situations the best that can be achieved is more or less rough approximations. The physical sciences have achieved relatively exact predictions. But it seems fairly clear that the social sciences are not likely to achieve a similar degree of exactness. They have relied more heavily, therefore, on statistics and probability theory than the physical sciences. Regression and correlation analysis were developed especially for use in situations where exact predictions are not possible but where the variables have been measured with a high degree of precision, that is, on an interval or ratio scale. In this chapter we consider simple linear regression first and then correlation.

As an example of an exact law in physical science, consider the law concerning the relationship between volume and temperature for various permanent gases. Various gases, kept at constant pressure, increase in volume by about 1/273 of their initial volume at 0°C for each degree of rise in temperature. With this law we can predict the volume of a gas knowing its temperature and initial volume. This is a direct linear relationship.

In a direct linear relationship, a change in one variable is accompanied by a constant change in another. For example, if we have two variables X and Y whose relationship is the following:

Y	0	2	4	6	8	10
X	0	4	8	12	16	20

the rate of change in Y for changes in X is

$$\frac{Y_2 - Y_1}{X_2 - X_1} = \frac{\Delta Y}{\Delta X} = \frac{1}{2}$$

where ΔY (delta Y) represents the amount of change in Y and ΔX the corresponding amount of change in X. This rate of change is the same no matter which pair of Y and X values we select. Such relationships are called linear because, when the coordinate points of the two variables are plotted, all the pairs of X and Y values fall on a straight line (see Figure 9-1).[1]

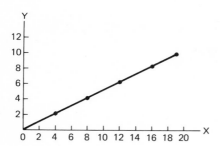

Figure 9-1.   Graph of $Y = \frac{1}{2}X$.

The slope of the line in Figure 9-1 is the rate of change in Y for changes in X, which we said above is $1/2$. The line in the figure may be described by an algebraic equation of the form $Y = A + B(X)$, where A is the point at which the line crosses the Y axis, usually called the Y intercept, and B is the slope or rate of change in Y for changes in X. Because, in the example above, the line crosses the Y axis at the origin (zero point), the equation in this case reduces to $Y = B(X) = 1/2(X)$.

Now, because there is a perfect linear relationship between the two variables, and we know the slope of the line, we can predict values of Y for any value of X. Hence, when $X = 8$, $Y = 1/2(8) = 4$; and when $X = 6$, $Y = 1/2(6) = 3$; and so on.

But even if two variables do not stand in such a perfect, straight-line relationship, it may be possible to make relatively accurate predictions about the Y variable based on knowledge of the X variable by postulating that the relationship approximates a straight line and measuring how far off from a postulated straight-line relationship the actual data falls. This form of approximation is called *linear regression*, whose simplest form, expressed as an algebraic equation, is one of the form $Y' = a + b(X)$. In this equation Y' (prime) is used to indicate that we are talking about predicted values of Y rather than actual values, and the lower case letters a and b to indicate the constants of the Y intercept and the slope when we have a sample (as we shall see, the Greek letters α and β are used to designate the same thing for the population).

[1] Coordinate points are plotted by finding the point at which *vectors*, or lines drawn out from each of the corresponding X and Y values, intersect. Thus if we draw a line up from point 1 on the X axis and a line out from the right of the Y axis at point 6, the two lines intersect at the coordinate points 1,6. We connect the points with a straight line to depict the relationship between the variables.

This equation is called a regression equation because of the work of its originator, Sir Francis Galton. In his work on heredity, Galton noted that the sons of very tall men tended to be tall, but not quite so tall as their fathers. And the sons of short fathers tended to be short, but not quite so short as their fathers. He reasoned, therefore, that, if we want to predict the height of an individual, knowing only the height of his father, we predict his height as falling between the height of his father and the average height of the population of his generation (the son's). He called this tendency of individuals to be not quite so tall or short as their parents the tendency of *regression toward mediocrity*.[2]

Let us now consider how to find numerical values for the linear regression equation $Y' = a + b(X)$ that somehow best describe the coordinate points when they do not fall on a straight line. We use an example concerning the relationship between a congressman's voting behavior and his constituents' attitudes. This example, derived from the work of Miller and Stokes, is a test of the part of democratic theory that states that a representative should vote the way his constituents want.[3]

Let us suppose that we measure a congressman's voting behavior on bills pertaining to civil rights on a scale ranging from 0 to 100, where 0 represents complete opposition to civil rights and 100 represents complete support. This is the dependent variable and is designated by Y_i. Let us suppose also that we measure the attitude of the congressmen's constituents on a scale ranging from 0 to 10, where 0 represents complete opposition

Table 9-1 Constituents' Attitudes and Representative's Voting Behavior

Congressman	*Votes on civil rights,* Y	*Constituents' attitudes toward civil rights,* X	X^2	XY
A	8	2	4	6
B	20	4	16	80
C	30	5	25	150
D	50	7	49	350
E	50	9	81	450
F	90	10	100	900
G	100	10	100	1,000
Total	348	47	375	2,946

[2] If we concluded from this that eventually everybody will be the same height, we should commit what is known as the *regression fallacy*. All this really says is that, if an individual is much taller than the mean of his generation, he is not likely to have a son much taller than the mean of his (the son's) generation.

[3] Warren E. Miller and Donald E. Stokes, "Constituency Influence in Congress," *American Political Science Review*, vol. 59, pp. 45–56, March, 1963. We use hypothetical data in Table 9-1 but the actual correlations in Chapter 10.

to civil rights and 10 complete support. This is the independent variable and is designated by X_i. For ease of illustration suppose that we have a random sample of seven congressmen. A much larger sample would be needed in an actual case. The hypothetical scores of the congressmen and their constituents are given in Table 9-1. When these are plotted on a graph, it can be seen that, although a line drawn through the coordinate points does not form a straight line, most of the points fall very close to a straight line. The dotted line in Figure 9-2 has been drawn freehand so as to keep it as close to the coordinates as possible. The freehand method of drawing such a line is considered in Chapter 13. Here we describe the *least-squares* method.

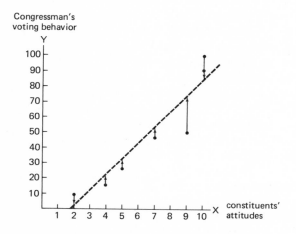

Figure 9-2. Scattergram for data in Table 9-1.

There is some scatter in the points of Figure 9-2, but there is also a systematic pattern running from the lower left-hand side to the upper right-hand side of the scattergram. It appears, therefore, that the relationship approximates a straight line. If we are able to find an equation for the line that best fits these points, we may be able to make fairly accurate predictions about Y_i based on knowledge of X_i.

How do we fit the line that best fits the points and find the values for the constants a and b in the regression equation $Y' = a + b(X)$? If we wanted a method that would give us an exactly correct prediction of Y the greatest number of times, we should predict the mode of Y no matter what the value of X. In this case, we should be correct two out of seven times, or $2/7 = 0.28$. However, if we want the estimate of Y to be as close to Y as possible over the long run, we use the method of least squares. The least-squares line is that line at which the vertical distances of the actual Y values from the predicted Y' values are at a minimum. In symbolic terms, we want to fit a line so that

$$\sum_{i=1}^{n} (Y_i - Y'_i)^2 = \text{minimum}$$

The distances from each Y to its predicted counterpart on the line are indicated by arrows in Figure 9-2. The predicted Y' values, of course, lie on a straight line and thus do not exactly correspond to actual Y values for given X values unless an actual Y value happens to fall on the line. However, by the method of least squares, the line that is found is that which minimizes the squared differences between the predicted Y' values and the actual Y values. It is therefore called a *goodness-of-fit measure*. Because we minimize the sum of the squared distances, the method is called least squares. Under the assumptions to be spelled out below, the Gauss-Markow theorem tells us that the line fitted by the method of least squares is the best linear, unbiased estimate of the population parameters α and β. And if the Y's are normally distributed, it also provides a maximum-likelihood estimation.

If there is no relationship between X and Y, the points on a scatter diagram will be completely dispersed, as in Figure 9-3, and the line that depicts this

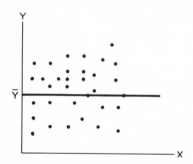

Figure 9-3.

will be parallel to the X axis and perpendicular to the Y axis at the mean of Y. Deviations from the line, then, will actually be deviations from $\bar{Y}$, and because there is no linear relationship between the two, the best estimate of Y, no matter what the value of X, is $\bar{Y}$. This, of course, depends on how the error is measured and also on the distribution of the population. Recall that in Chapter 2 it was noted that $\sum (Y_i - \bar{Y})^2 = \text{minimum}$. Hence, if the error is measured by squared deviations, the best estimate of each Y will be $\bar{Y}$. But if the error is measured as the sum of the absolute deviations, the best estimate is the median of Y because $\sum |Y_i - Md| = \text{minumum}$.

Note also that, if the frequency distribution is skewed, such as in income distribution, estimating the median will yield the smallest absolute error.

Now consider the following example. Suppose that we have a population consisting of 1,000 students who have taken the introductory political science course. Suppose that they are graded on a numerical scale, that the distribution of grades is normal, and that the mean is $\bar{Y} = 82$. Now suppose that we enter into a wager concerning the grade of each student who is selected at random from this population and the wager is that we lose an amount of money equal to the absolute amount of error we make. What numerical score should we guess for each student selected? The mean $\bar{Y}$, because our losses will be a minimum in the long run for such guesses (remember that for normal distributions, $\bar{Y} = Md$).

Now suppose that we know that the grades of students in introductory political science are related to the college entrance scores of each student. Suppose also that we are told the college entrance score of each student before we are asked to guess his grade. Now what grade should we guess? We should guess the mean introductory political science course grade for all the students who got the particular college entrance score of the student selected. For example, all students who have a college entrance score of 80 do not have exactly the same grade in the introductory political science course. The grades of all students who have a college entrance score of 80 form a frequency distribution with a mean $\bar{Y}_g$. We should guess the mean $\bar{Y}_g$ for each student selected. To the extent to which the two variables — introductory political science grade and college entrance score — are related, our predictions of Y will be improved over the former case, where we simply guessed the mean of all Y each time.

This is really another way of saying that, as two variables increase in strength of relationship, the line that best fits the coordinate points tilts on its axis, and deviations from the new line are less than deviations from the line defining the mean $\bar{Y}$; that is, $\sum (Y - Y')^2 < \sum (Y - \bar{Y})^2$.

We can find the line at which $\sum (Y - Y')^2$ is a minimum by drawing a series of lines visually and computing $\sum (Y - Y')^2$ each time, until we come to the one that is the best fit by least squares. This is an accepted method, and it is described in the book by Spurr and Bonini cited at the end of this chapter. But it also can be found mathematically by finding the point where the line passes through the Y axis and then its slope, or rate of rise relative to the X axis.

The point where the line passes through the Y axis is called the Y *intercept*, and it is given by the constant a in the regression equation. The slope of the line is given by the b coefficient in the regression equation. Thus, if we can solve for a and b, we can find the least-squares line.

Our objective is to find a and b such that the sum of the squared deviations of the predicted Y' values from the actual Y values is a minimum. This can be done with the differential calculus. We do not go into this in detail here, because we do not assume a knowledge of calculus in this book. However, a verbal description of what is done may be helpful.

Consider first the sum we want to minimize:

$$\sum (Y - Y')^2 \tag{9-1}$$

Recall now that

$$Y' = a + b(X) \tag{9-2}$$

We can substitute Equation (9-2) into (9-1), getting

$$\sum [Y - (a + bX)]^2 \tag{9-3}$$

as the sum to be minimized. When we set the partial derivative of (9-3) with respect to a and b equal to zero, we get the two normal equations in two unknowns a and b:

$$\sum Y_i = a \cdot n + b \cdot \sum X_i \tag{9-4}$$
$$\sum X_i Y_i = a \cdot \sum X_i + b \cdot \sum X_i^2$$

We can then solve these equations directly for a and b. To illustrate this, we substitute the values of the data in Table 9-1 into these normal equations, getting

$$348 = 7(a) + 47(b)$$
$$2,946 = 47(a) + 375(b)$$

We can now find a and b by elimination. To do so, we multiply one of the two equations by a number that enables us to eliminate one of the unknowns. In this case, if we multiply the first equation by 7.9787 (approximately) and then subtract the second from the first, we get[4]

$$2,777 = 56(a) + 375(b)$$
$$\underline{2,946 = 47(a) + 375(b)}$$
$$-169 = 9a$$

Therefore

$$a = -18.7$$

We can now solve for b by substituting -18.7 for a in the second equation:

$$2,946 = 47(-18.7) + 375(b)$$
$$2,946 = -878.9 + 375(b)$$
$$b = 10.19$$

Therefore, the values for a and b that ensure that $\sum (Y - Y')^2 = $ minimum are -18.7 and 10.19. The equation of the least-squares line for these data is therefore

$$Y' = -18.7 + 10.19(X)$$

[4] We have rounded in this example for ease of illustration.

Before interpreting this, consider another method of computing a and b. As with other statistics, there is a short method for computing a and b that can be used instead. Short-cut equations can be derived from the normal equations. When the normal equations are solved by substitution the result is

$$b = \frac{n \sum XY - (\sum X)(\sum Y)}{n \sum X^2 - (\sum X)^2} \tag{9-5}$$

$$a = \bar{Y} - b\bar{X} \tag{9-6}$$

Using the data in Table 9-1 as an illustration, we have

$$b = \frac{7(2,946) - 47(348)}{7(375) - (47)^2}$$

$$= 10.25$$

$$a = 49.7 - 10.25(6.7) = -18.96$$

The equation is thus $Y' = -18.96 + 10.25(X)$. (The slight difference in answers using the normal equations is due to rounding error.) This says, in effect, that the line that fits the data by least squares better than any other line has $a \doteq -18.96$ and $b \doteq 10.25$.

The equation of the regression line gives the predicted mean value of Y for any specified X. The constant a is the point at which the line crosses the Y axis (see Figure 9-4). The constant b is the slope, or the amount by which the estimated value of Y changes for unit changes in X.

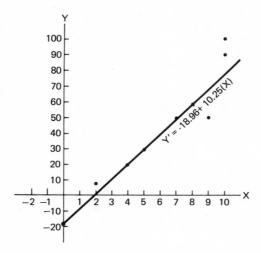

Figure 9-4. Regression line for data in Table 9-1.

Now consider what this means in terms of predictions. Imagine taking an infinite number of samples of congressmen and constituents, and for each sample computing the constituents' attitudes and the voting behavior of the congressmen; the latter variable will have a distribution for each value of constituents' attitudes and a mean for this distribution. The line of association for the mean voting behavior for each value of X, $Y' = a + b(X)$, is really a line of predicted means, or $(Y'|X) = a + b(X)$.

This is shown in Figure 9-5. For each value of X in the table, a distri-

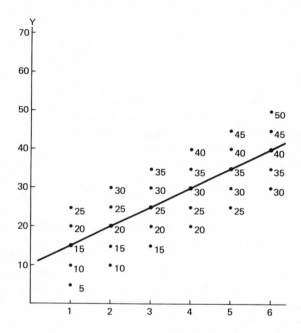

Figure 9-5. Populations of Y for given X.

bution of Y values is given. The Y' expected for each X is the mean of each distribution. Hence, the line can be defined as $E(Y'|X) = \bar{Y}_g$. The actual Y value will be different unless a straight line fits the data exactly.

The regression equation just computed is the sample regression equation. It can be used to estimate the population regression equation $Y = \alpha + \beta(X)$ if certain assumptions are met. These are (*a*) that the population relationship is linear, (*b*) that the distribution of each Y for given X values is normal, (*c*) that the variances are equal, and (*d*) that the errors are independent. If these assumptions are not met, the errors we observe, $\sum (Y - Y')^2$, may be due to more than sampling error.

Let us consider these assumptions briefly. If the assumption about linearity is not met, part of the difference between the predicted and actual

Y values may be due to the fact that a curve rather than a straight line describes the data better. We consider methods of finding curvilinear relationships later in the book. Note that, if the data do not approximate a straight line, we can conclude only that a linear model does not describe them. We cannot conclude that there is no relationship between the variables. The second assumption is based on the fact that the regression line can be defined as $E(Y'|X) = \bar{Y}_g$. Because we are predicting mean *Y* values for *X* values, if the distributions of various Y_i or X_i are not normal, some of the error will be the result of the skewness of the data. We saw in Chapter 2 that the mean for skewed data is pulled in the direction of the skewness. The assumption of equal variances means that the scatter of points about the regression line is uniform. If this assumption is not met, a falsely high or low correlation can result. This is considered later in the book. The assumption of independence means that the deviations of one point are not related to those of any other point. If this assumption is not met, the data are dependent, and different methods must be used (see Chapter 12). If all these assumptions are met, the sample *a* and *b* may be used to estimate the population parameters α and β. Otherwise, the regression equation may be used only as a description of the data and not to infer population relationships.

The ability to estimate the true mean depends, as in all estimation, on the variance within each subpopulation. If it is very large, there is a greater chance of error. But in any event, there will be some sampling fluctuation. For example, suppose that the constants found above apply to actual data. Suppose that we selected another congressional district at random and computed the attitude of voters concerning civil rights and found it to have a score of 9. What should we predict the congressman's score who represents this district to be? The answer is $Y' = -18.96 + 10.25(9) = 73.3$. But, of course, the actual *Y* is likely to be somewhat different.

Note that in the sample data in Table 9-1 the constituent with a score of 9 has a congressman with a score of 50. Because the relationship in this case is not a direct linear relationship, each predicted Y' value will be somewhat off. In some districts where the attitude of the voters is a 9, the actual voting score of the congressman might be 50, in others 80, and so on. The regression equation tells us to expect the score to be 73.3. How far off this prediction will be is a function of how badly a straight line fits the data as well as a function of sampling error.

The size of the *b* coefficient tells us, in terms of the units we are using to measure the variables, how much change in *Y* will occur for changes in *X*. Assume the following two regression equations, each with the same intercept:

$$Y' = -18 + 2X \tag{1}$$

$$Y' = -18 + 20X \tag{2}$$

Assume a unit increase in X for each equation:

$$Y' = -18 + 2(10) = 2 \tag{1}$$
$$Y' = -18 + 2(11) = 4 \tag{2}$$
$$Y' = -18 + 20(10) = 182 \tag{2}$$
$$Y' = -18 + 20(11) = 202 \tag{2}$$

In the first case, a unit increase in X produced an increase in Y' of 2, and in the second case, it produced an increase of 20. Thus, given the same intercept and the same unit of measurement, the b coefficient tells us how large a change in Y is produced by a unit change in X.

Since the size of b depends on the size of the unit being used to measure the data, it is not a good measure of strength of relationship. A better measure of strength is one that has the same range no matter what the size of the units being measured. This is called a *correlation coefficient*, and we may define it here as

$$r = \sqrt{b_{XY} b_{YX}} \tag{9-7}$$

Equation (9-7) tells us that a constant measure of the strength of the relationship between two variables can be obtained by taking the square root of the product of their b coefficients. The result is called the correlation of X and Y and is symbolized as r.

We shall come back to this question in a moment. First, let us continue with the question of inference concerning a and b. The sample a and b may be considered as estimates of the population coefficients α and β.[5] We can therefore construct confidence intervals around a and b. As with other estimates, if a large number of samples from a given population is drawn and the b for each of them computed, the sampling distributions of b can be approximated with a normal curve with a mean equal to β and a standard deviation given by

$$\sigma_b = \frac{\sigma_{Y'|X}}{\sqrt{\sum (X_i - \bar{X})^2}} = \frac{\sqrt{\sum (Y_i - \bar{Y})^2 - b \sum (X_i - \bar{X})(Y_i - \bar{Y})}}{\sqrt{\sum X_i^2 - (\sum X_i)^2/(n-2)}} \tag{9-8}$$

Hence, a 95 percent confidence interval for b is given by

$$\hat{\beta} = b \pm 1.96 \frac{\sigma_{Y'|X}}{\sqrt{\sum X_i^2 - (\sum X_i)^2/(n-2)}} \tag{9-9}$$

The term in the numerator of Equation (9-8) is called the *standard error of the estimate Y'* for X. It measures the scatter of the actual observations about the regression line and can be defined as

$$\sigma_{Y'|X} = \sqrt{\frac{\sum (Y - Y')^2}{N}} = \sigma_Y \sqrt{1 - \rho^2}$$

---

[5] Of course, they can only be used to estimate α and β if random sampling has been used and the assumptions mentioned above hold.

For samples, divide by the appropriate degrees of freedom, which, as will be explained below, are $n - 2$ in this case, getting

$$s_{Y'|X} = \sqrt{\frac{\sum (Y - Y')^2}{n - 2}}$$

$$= \sqrt{\frac{\sum (Y_i - \bar{Y})^2 - b \sum (X_i - \bar{X})(Y_i - \bar{Y})}{n - 2}}$$

For the data in Table 9-1, we have

$$s_{Y'|X} = \sqrt{\frac{7,163.43 - 10.25(609.43)}{7 - 2}} = 13.54$$

Hence, we may predict that a congressman with a constituency having an attitude score on civil rights of 9 should have a voting record of 73.3 ± 13.54, and we have two chances out of three of being correct.

The variance of the sampling distribution of a is

$$\sigma_a{}^2 = \frac{\sigma_{Y'|X^2} \left(\sum X^2 \right)}{n \sum (X - \bar{X})^2}$$

and the standard error of a is the square root of this.

SIMPLE LINEAR CORRELATION

Although the regression equation is used for prediction, correlation gives the strength of relationship between variables. A correlation coefficient is actually a measure of how good a straight line fits the points in a scatter diagram. It is thus called a *goodness-of-fit measure*. It varies between zero (no correlation, complete scatter) to ± 1 (perfect correlation, all coordinates falling on the line).

There are various ways we can measure how well the straight line fits the data, one of which is to take the sum of the squared deviations of the predicted Y' values from the actual Y values: $\sum (Y_i - Y_i')^2$.

When the coordinate points are closely bunched around the line, the fit is fairly good and the sum of the squared differences will be small. When the points are completely scattered, the fit is bad and the sum of squared differences will be large. These situations are depicted in Figure 9-6. Of course, simply taking the sum of the squared deviations of the predicted Y' values from the actual Y values would not be a good measure of how well the data fit a straight line if we wanted to compare different kinds of data, because this sum will be influenced by the size of the sample and by the size of units the data are being measured by.

| A. no relationship | B. strong positive relationship | C. strong negative relationship |

Figure 9-6. Hypothetical relationships.

One possible way to make the measure comparable would be to take the sum of the squared deviations of the predicted from the actual Y values as a ratio to some fixed point. We can do this by comparing $\sum (Y_i - Y_i')^2$ with $\sum (Y_i - \bar{Y})^2$. The latter is the sum of the squared deviations of the actual Y's from the mean value of Y. If the fit of the points around a straight line is good, $\sum (Y_i - Y_i')^2$ will be smaller than $\sum (Y_i - \bar{Y})^2$. If the fit is poor, $\sum (Y_i - Y_i')^2$ will be the same as $\sum (Y_i - \bar{Y})^2$. This can be illustrated by examining Figure 9-6*a* and *b*. In Figure 9-6*a*, the sum of the squared deviations of the actual points from the best-fitting straight line is exactly the same as the sum of the squared deviations from the mean of Y. In this case, the best estimate of Y is the mean of Y no matter what X is. In Figure 9-6*b*, if we drew a line at the mean of Y and took the sum of the squared deviations from this, the resulting sum would be larger than the sum of the squared deviations taken from the least-squares line in Figure 9-6*b*. To illustrate this, in Figure 9-7 the same coordinates are used as in Figure 9-6*b*, but the distances are drawn to the mean of $\bar{Y}$.

Given these facts, it is possible to measure goodness of fit by the statistic

$$r = \pm \sqrt{1 - \frac{\sum (Y_i - Y_i')^2}{\sum (Y_i - \bar{Y})^2}} \qquad (9\text{-}10)$$

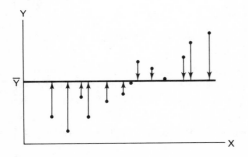

Figure 9-7. Identical plots given in Figure 9-6*b* represented as deviations from the mean of Y.

As we shall see in a moment, this gives us what is called the *Pearson product-moment correlation coefficient.* If the fit of the regression line is poor, the ratio of the two sums will be close to one, and r will be close to zero. If the fit is good, the ratio of the two sums will be close to zero, and r will be close to one.

To find r by means of this equation, we have to compute an estimated Y' for each value of X, subtract the estimated Y' from the actual Y, square, sum, and do the same for the squared deviations of the Y values from the mean of Y. Although we do not normally compute r in this manner, it is instructive to do so for the data in Table 9-1:

$$r = \sqrt{1 - \frac{914.00}{7,163.43}} = .93$$

As can be seen, the squared deviations from the estimated Y' are much smaller than the sum of the squared deviations from the mean $\bar{Y}$. Looked at from another point of view, we may say that the sum of squares from the least-squares line is about 12.7 percent of the sum of squares from the line drawn at the mean of Y. This is subtracted from one so as to have an index that is larger when the relationship is stronger, and its square root extracted to obtain an index that increases arithmetically as the relationship gets stronger.

Before going to shorter computing methods, it will be helpful to consider some other ways by which the correlation measure may be described. First, consider the difference in standard deviations of the estimated Y' and the actual Y values. If the estimated Y' values are all equal to the actual Y values, it is obvious that $\sigma_{Y'} = \sigma_Y$. But if the estimated Y' values are not equal to the predicted Y values, $\sigma_{Y'} < \sigma_Y$. Exactly how much smaller the standard deviation of the predicted Y' values will be depends on how good a straight line fits the data. If the fit is poor, $\sigma_{Y'}$ will be zero. And, as the fit improves, $\sigma_{Y'}$ will approach σ_Y. Thus, we may measure goodness of fit by taking the ratio

$$\frac{\sigma_{Y'}}{\sigma_Y} \tag{9-11}$$

Again, as an illustration, for the data above we have $29.9/32.0 = .93$, or the same result as before.

In variance terms, the variance of the Y values is equal to the variance of the predicted Y' values plus the error e variance, or $\sigma_Y^2 = \sigma_{Y'}^2 + \sigma_e^2$. We may say, therefore, that total variance is equal to the sum of explained and unexplained variance. The greater the $\sigma_{Y'}^2$ relative to σ_e^2, the better the prediction that can be made. Thus, we may define correlation as the square root of

$$\rho_{XY}^2 = \frac{\sigma_{Y'}^2}{\sigma_Y^2} = \frac{\text{explained variance}}{\text{total variance}}$$

This equation is really the *coefficient of determination*, because it tells us what proportion of variance in Y' is explained by X. We return to this point below.

A third way to view correlation is in terms of covariation; that is, in terms of deviations of Y_i and X_i from their respective means.[6] Let us adopt shorthand symbolism to explain this. We designate $\sum (Y_i - \bar{Y})^2$ by $\sum y^2$, and $\sum (X_i - \bar{X})^2$ by $\sum x^2$. If each deviation of Y from its mean is multiplied by the paired deviation of X from its mean, the result may be called *covariation* and symbolized as $\sum xy$. If the relationship between two variables is positive, high values of Y with respect to the mean of Y will be found with high values of X with respect to the mean of X, and low values of Y will be found with low values of X. The resulting $\sum xy$ will be positive. The stronger the relationship, the larger this amount will be. If the relationship

Table 9-2 Hypothetical Covariations

Strong positive			Weak positive		
$Y - \bar{Y}$	$X - \bar{X}$	xy	$Y - \bar{Y}$	$X - \bar{X}$	xy
100 − 50 = 50	10 − 5 = 5	250	100 − 50 = 50	8 − 5 = 3	150
90 − 50 = 40	9 − 5 = 4	160	60 − 50 = 10	8 − 5 = 3	30
80 − 50 = 30	8 − 5 = 3	90	60 − 50 = 10	7 − 5 = 2	20
70 − 50 = 20	6 − 5 = 1	20	40 − 50 = −10	6 − 5 = 1	−10
40 − 50 = −10	3 − 5 = −2	20	20 − 50 = −30	4 − 5 = −1	30
		$\Sigma xy = 540$			$\Sigma xy = 220$

No relationship		
$Y - \bar{Y}$	$X - \bar{X}$	xy
100 − 50 = 50	1 − 5 = −4	−200
80 − 50 = 30	4 − 5 = −1	−30
80 − 50 = 30	9 − 5 = 4	120
60 − 50 = 10	10 − 5 = 5	50
60 − 50 = 10	8 − 5 = 3	30
		$\Sigma xy = -30$

Strong negative			Weak negative		
$Y - \bar{Y}$	$X - \bar{X}$	xy	$Y - \bar{Y}$	$X - \bar{X}$	xy
100 − 50 = 50	1 − 5 = −4	−200	100 − 50 = 50	4 − 5 = −1	−50
90 − 50 = 40	2 − 5 = −3	−120	90 − 50 = 40	3 − 5 = −2	−80
70 − 50 = 20	3 − 5 = −2	−40	60 − 50 = 10	2 − 5 = −3	−30
40 − 50 = −10	7 − 5 = 2	−20	20 − 50 = −30	6 − 5 = 1	−30
20 − 50 = −30	9 − 5 = 4	−120	10 − 50 = −40	6 − 5 = 1	−40
		$\Sigma xy = -500$			$\Sigma xy = -230$

[6] Yet another way to define correlation will be considered when we discuss multiple correlation below.

between the variables is inverse, high positive values of Y will be associated with high negative values of X, and vice versa. The resulting $\sum xy$ will be negative. The stronger the relationship between the two variables, the larger this amount will be. When there is no relationship between the variables, the positive and negative values will tend to cancel out, and $\sum xy$ will be zero. All this is shown in Table 9-2, which shows that if, (a) $\sum xy$ is a large positive number, the relationship is positive and strong; if (b) $\sum xy$ is a large negative number, the relationship is negative and strong; and if (c) $\sum xy$ is near zero, there is no relationship.

But because the size of $\sum xy$ depends on the size of the units by which we are measuring X_i and Y_i and by sample size, it is not a good measure of *relative* strength. To overcome this difficulty, we can divide $\sum xy$ by a number that is influenced by the unit used in exactly the same way as covariation.[7] This was done by Karl Pearson when he defined the coefficient of correlation as

$$r_{xy} = \frac{\sum xy}{\sqrt{(\sum x^2)(\sum y^2)}} \tag{9-12}$$

This is a *pure number*, a number that does not depend on the units of measurement and can therefore be used to compare correlations of data no matter what units are used to measure the variable.

A little simple algebra can take us from Equation (9-12) to a computational equation. Let us take the numerator first, $\sum xy$. By definition, $\sum xy = \sum [(X - \bar{X})(Y - \bar{Y})]$, and carrying out the indicated multiplication in the parentheses,

$$\sum xy = \sum [XY - \bar{X}Y - X\bar{Y} + \bar{X}\bar{Y}]$$

$$= \sum \left[XY - \frac{(\sum X)}{n} Y - X \frac{(\sum Y)}{n} + \left(\frac{\sum X}{n}\right)\left(\frac{\sum Y}{n}\right) \right]$$

because $\bar{X} = \dfrac{\sum X}{n}$ and $\bar{Y} = \dfrac{\sum Y}{n}$

$$= \sum XY - \frac{(\sum X)(\sum Y)}{n} - \frac{(\sum X)(\sum Y)}{n} + n\left(\frac{\sum X}{n}\right)\left(\frac{\sum Y}{n}\right)$$

summing across terms

$$= \sum XY - 2\frac{(\sum X)(\sum Y)}{n} + \frac{(\sum X)(\sum Y)}{n}$$

$$= \sum XY - \frac{(\sum X)(\sum Y)}{n} \tag{9-13}$$

[7] The population correlation coefficient can be defined as $\Sigma Z_Y Z_Y / n$, where Z_X and Z_Y are the standard-score equivalents of the original X and Y. This measure is a pure number, and it can be seen that it is the mean of the covariation of standard scores.

Next, consider the denominator.

$$\sqrt{(\sum x^2)(\sum y^2)}$$

$$= \sqrt{\sum (X_i - \bar{X})^2} \sqrt{\sum (Y_i - \bar{Y})^2}$$

$$= \sqrt{\sum (X_i^2 - 2\bar{X}X_i + \bar{X}^2)} \sqrt{\sum (Y_i^2 - 2\bar{Y}Y_i + \bar{Y}^2)}$$

$$= \sqrt{\sum X_i^2 - 2\frac{(\sum X_i^2)}{n} + \frac{(\sum X_i^2)}{n}} \sqrt{\sum Y_i^2 - 2\frac{(\sum Y_i^2)}{n} + \frac{(\sum Y_i^2)}{n}}$$

$$= \sqrt{\sum X_i^2 - \frac{(\sum X_i^2)}{n}} \sqrt{\sum Y_i^2 - \frac{(\sum Y_i^2)}{n}} \qquad (9\text{-}14)$$

Putting the two parts together again, we have

$$r_{xy} = \frac{\sum XY - \frac{(\sum X)(\sum Y)}{n}}{\sqrt{\sum X^2 - \frac{(\sum X)^2}{n}} \sqrt{\sum Y^2 - \frac{(\sum Y)^2}{n}}} \qquad (9\text{-}15)$$

which is only one of the computing equations that can be used. It is quite simple to derive the following equivalent equation.

$$r_{xy} = \frac{n(\sum XY) - (\sum X)(\sum Y)}{\sqrt{n \sum X^2 - (\sum X)^2} \sqrt{n \sum Y^2 - (\sum Y)^2}} \qquad (9\text{-}16)$$

Equation (9-16) is the easiest to use in computing a correlation.

Because r is a pure number, to simplify computation, we can add, subtract, multiply, and divide X_i and Y_i by any constant without affecting r. This is also a good way to check for error, which is usually a good thing to do if the calculations are being done by hand. It is easy to make computational errors with correlation and therefore wise to check results before proclaiming them. For example, suppose that you are working with values that have several negative numbers and the largest negative number is -350. If you simply add 350 to all observations, you will eliminate the negative numbers but not affect the correlation coefficient at all. (The same thing applies to all variables.) Similarly, if you have Y or X observations that are decimals, such as 0.02, 0.05, etc., to get rid of the decimals, you can simply multiply each observation by 100. Again, the correlation coefficient will not be affected by such a linear transformation.

We computed r for the data in Table 9-1 by the two definitional equations (9-10) and (9-11). In order to show the comparability of the computing equation for arriving at the same result, we now compute r by the computing equation (9-16). To do so, we need only the quantities $n = 7$, $\sum X = 47$,

$\sum Y = 348$, $\sum XY = 2946$, $\sum X^2 = 375$, and $\sum Y^2 = 24{,}464$. Therefore

$$r = 7(2{,}946) - (47)(348)/\sqrt{[7(375) - (47)^2][7(24{,}464) - (348)^2]}$$

$$= 4{,}266/4{,}547$$

$$= .94$$

or the same result as before except for rounding error.

The coefficient r can be viewed as a measure of our ability to predict Y with knowledge of X on the basis of the least-squares regression line. If r is exactly ± 1, knowledge about X enables us to specify Y exactly. If r is somewhere between zero and ± 1, our prediction will not be perfect. If it is zero, our best guess of Y is $\bar{Y}$, because we shall come closer a larger number of times to the actual Y in the long run. When this is so, the amount of variation around the least-squares line is completely unexplained. If there is a linear relationship between the two variables, some of the total variation is explained, and the variation around the least-squares line is less than the variation around $\bar{Y}$. The amount of improvement in prediction is thus called the amount of explained variation in Y accounted for by X. r^2 is a measure of this explained variation. It is often called the *coefficient of determination*. For example, if $r = .6$, then 36 percent of the total variation is explained by the least-squares regression and $1 - .36$, or 64 percent is unexplained.

It is difficult to interpret r by itself. A correlation of .6 is not twice as strong as a correlation of .3. For, when $r = .6$, 36 percent of the variation in the dependent variable is explained by the independent variable. But when $r = .3$, only 9 percent of the variation in the dependent variable is accounted for by the independent variable. It is apparent, therefore, that an r of .8 is needed before we can say that 64 percent of the variance in the dependent variable is accounted for by the independent variable, and even in this case there is still 36 percent left unaccounted for by the independent variable.

Furthermore, as we shall see below, any simple correlation by itself is not too meaningful. Even with an r of .99, where it might now be said that 98 percent of the variation in the dependent variable is accounted for by the independent variable and only 2 percent is unaccounted for, we cannot infer that the two variables are causally related. It may be that a third, unidentified variable is related to both the independent and dependent variables and produces changes in both of them. For example, suppose that the correlation between constituency attitudes and representatives' attitudes is .99. We cannot conclude that the constituency causes representatives' attitudes or that the reverse is true. Suppose that we identify a third variable, such as socioeconomic level of district involved, and find that it is highly correlated with both constituency attitude and representatives' attitude. It is possible that this variable causes the correlation in attitudes of both constituents and representatives. What we may say is that both the representatives and his constituents agree, because they come from identical socioeconomic environments in which their respective attitudes were obtained. To begin to tackle the problem of sorting out cause, we must move on to partial and multiple correlation and regression analysis. We consider this in Chapter 10.

Effect on Correlation of Different Populations

Suppose that we are interested in whether people generally vote in accordance with their self-interest.[8] We want to know if people will vote for bond issues only if they think they will benefit from them. Specifically, suppose that we test whether or not there is a tendency for people with higher income to vote against a proposition to increase the county hospital facilities, which will not benefit them, because they do not use them. Suppose that the sample is 25 suburban cities and the data are those in Table 9-3 and Figure 9-8.

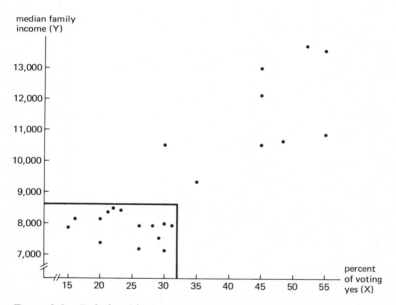

Figure 9-8. Relationship between votes on the proposition to increase county hospital facilities and median family income. Adapted from James Q. Wilson and Edward C. Banfield, "Voting Behavior on Municipal Public Expenditures: A Study in Rationality and Self-interest, in Julius Margolis (ed.), *The Public Economy of Urban Communities*, The Johns Hopkins University Press, Baltimore, Resources for the Future, Inc., 1965, pp. 74–92.

[8] This example is based on the work of James Q. Wilson and Edward C. Banfield, "Voting Behavior on Municipal Public Expenditures: A Study in Rationality and Self-Interest" in Julius Margolis (ed.), *The Public Economy of Urban Communities*, The Johns Hopkins Press, Baltimore, 1965, pp. 74–92. The conclusions reached here are not to be considered identical to those of the authors. The example is used illustratively. In general, the authors of the study did conclude that, under certain conditions, people do not vote for their own self-interest narrowly conceived.

Table 9-3   Percentage Voting Yes on County Hospital Proposal and Median
            Family Income of 25 Suburban Cities*

City	Median family income	Percentage voting yes
A	$ 7,900	15
B	7,400	20
C	8,200	16
D	8,100	20
E	8,500	22
F	8,400	23
G	8,300	21
H	7,900	26
I	7,200	26
J	7,900	28
K	7,500	29
L	7,100	30
M	8,000	30
N	7,900	31
O	10,500	30
P	9,300	35
Q	9,600	45
R	11,000	36
S	10,500	45
T	10,600	48
U	11,000	55
V	12,100	45
W	13,000	45
X	13,700	52
Y	13,600	55

*Adapted from: James Q. Wilson and Edward C. Banfield, "Voting Behavior on Municipal Public Expenditures: A Study in Rationality and Self-interest," in Julius Margolis (ed.), The Public Economy of Urban Communities, The Johns Hopkins Press, Baltimore, Resources for the Future, Inc., 1965, pp. 74–92.

The correlation for these data is .83. We may conclude that there is a strong positive relationship and that it appears that people do not vote in accordance with their narrow self-interest in the case of county hospital facilities.

Now suppose that instead of the sample of 25 suburban cities, the sample is only cities A to N (or n = 14). The correlation for this sample is −.3932. Now we conclude that people tend to vote in accordance with their narrow self-interest but that the relationship is not very strong. This is shown clearly if we focus only on the pairs in the lower left-hand corner of the scattergram in Figure 9-8 (the coordinates enclosed in the box).

The lesson to be learned is that the range of scores in data can greatly influence the correlation coefficient. If the sample includes extreme values, it may greatly increase or decrease the value of r. It is thus very important to be perfectly clear about what the population is that is being considering. Very misleading results can be reached if we do not consider the entire range of the variables.

But, at the same time, a few extreme values in the data can distort the correlation, and in some cases it might be desirable to leave them out. The decision as to what the population is that is being considered, and whether or not to exclude extreme values, must be based in the theory of the analysis. Statistics has nothing to say about this.

Significance Tests for r

The Pearson product-moment correlation coefficient can be used as a descriptive statistic or as an estimate of the correlation in a wider population we know nothing about. When it is used as a descriptive statistic, we cannot say that the relationship found to hold in a particular group of data is in any way true of a broader population. Because the objective of science is to generalize, the theoretically more interesting use of correlation is in terms of inferences.

To make inferences about a wider population on the basis of the sample r, we use a sampling distribution of r, as in other estimating problems. If we repeated the process of drawing random samples from the same population, each time computing the correlation between X and Y in the sample, we should find that the value of r for each sample would be slightly different. In other words, it would form a sampling distribution, with a mean and a standard deviation.

The sampling distribution of r has a form that varies with the value of the parameter ρ, similar to the situation of the binomial distribution. For example, when $\rho = .9$, the distribution will be negatively skewed, but when $\rho = -.8$, it will be positively skewed. There are two distributions we can use that take this into account: (a) Fisher's Z, which is not the same as the Z score in the normal distribution, and (b) the F distribution. The following assumptions must be met:

1. The relationship in the population is linear.
2. The two population distributions are normal (the bivariate normal assumption).
3. The distribution of Y values for a given X is normal, and the distribution of X values for a given Y is normal, that is, equal variance or homoscedasticity.
4. Random sampling has been used.
5. The errors (residual variance) are independent.

To use Fisher's Z, we first convert r to Z, draw an interval around Z, and then change Z back to r. We can convert r into Z by use of logarithms with the equation

$$Z = 1.151 \log \frac{1 + r}{1 - r}$$

But because tables are available for this, we can go directly to the table of Z values itself. For example, suppose that we have a correlation of .91 for $n = 7$. We want to draw a 95 percent confidence interval around $r = .91$. Table 5 in the Appendix shows that an r of .91 has a Z value of 1.5275. The standard error of Z is $1/\sqrt{n-3}$, which in this case is .50. A 95 percent confidence interval for Z therefore is

$$Z \pm 1.96\,(.50) = 1.5275 \pm 1.96\,(.50)$$

$$= .5475 \text{ to } 2.5075$$

We then reenter the table and convert these Z values back to r, obtaining a 95 percent confidence interval for r of .499 to .987. The interpretation is that we are 95 percent confident that the population correlation is within this interval.

The F distribution can be used to test the significance of a particular r. This distribution is described in Chapter 11. Here we simply note that the F distribution is a ratio of two variances:

$$\frac{\text{Explained variance}}{\text{Unexplained variance}}$$

Recall that r^2 is the proportion of total variation in Y explained by X, and $1 - r^2$ is the proportion left unexplained. Therefore, the explained sum of squares is $r^2 \sum y^2$, and the unexplained sum of squares must therefore be $1 - r^2 \sum y^2$. In order to use the F ratio, we must make two estimates of variance. As explained in Chapter 11, the sum of squares divided by the respective degrees of freedom gives the estimate of variance. There are $n - 2$ degrees of freedom associated with the unexplained sum of squares, because we had to estimate the two coefficients a and b in computing the least-squares line; and one degree of freedom associated with the explained, because once we have computed one of the sum of squares, the other is given. Therefore, the F ratio for the two estimates of variance is:

$$
\begin{aligned}
F_{1,n-2} &= \frac{(r^2 \sum y^2)/1}{[(1 - r^2) \sum y^2]/(n - 2)} \\
&= \frac{r^2 \sum y^2}{1} \cdot \frac{n - 2}{(1 - r^2) \sum y^2} \\
&= \frac{r^2(n - 2)}{(1 - r^2)}
\end{aligned}
\tag{9-17}
$$

For the data in the example above we have

$$F_{1,5} = \frac{(.94)^2(5)}{(1 - .94^2)}$$

$$= 36.66$$

This is significant at the .01 level. We can therefore reject the null hypothesis that $\rho = 0$. The subscripts under the F are the degrees of freedom for the explained and unexplained estimates of variance. The F distribution and how to read the table in the Appendix is explained in Chapter 11.

Critical Value of r

It often is convenient to compute the critical value of r for sample sizes rather than testing the significance of any particular value of r. This can be done by using the t distribution values for the levels of significance and the various sample sizes we are interested in. The equation is

$$r^2 = \frac{t^2}{t^2 + n - 2} \tag{9-18}$$

Suppose that we want to know what level of r is needed to be significant at the .05 level for a sample size of 12. First, look in the table of the t distribution for the t value at the .05 level of significance for a two-tailed test, and $n - 2$, or 10, degrees of freedom. This value is 2.228. We therefore have

$$r^2 = \frac{(2.228)^2}{(2.228)^2 + 10} = .329$$

and $$r = .5735$$

In other words, an r of about .57 or above is required for significance at the .05 level when $n = 12$.

Regressed Standard Scores and Correlation

It is instructive to show how correlation coefficients and regression coefficients compare. The equations we have been discussing all pertain to the data as measured. However, we may also compute correlation coefficients for standardized scores rather than for the raw scores themselves.

We said earlier that each individual score might be normalized or standardized by

$$Z = \frac{Y_i - \bar{Y}}{\sigma_Y} \tag{9-19}$$

Let us symbolize the resulting standardized scores as Z_Y. When the observations of both variables are expressed in the form of standard scores, the line fitted to the data will always pass through the origin of the axis system; (that is, the Y intercept coefficient a will be zero). If we first converted all sample observations to standardized scores before computing the correlation coefficient, we should find that any particular observation on the Y variable could be converted into an observation on the X variable by

$$Z_Y = r(Z_X) \tag{9-20}$$

Of course, the Y observation found in this manner is the predicted Y' and not the actual Y. Thus, for an r of .50, an observation that is one standard deviation above its mean has as its most probable associate on the other dimension an observation that is .50 above the mean.

For example, consider the data used above after substituting the appropriate definitions in Equation (9-20) and simplifying a little by algebra:

$$Z_Y = r(Z_X)$$

$$\frac{(Y' - \bar{Y})}{\sigma_Y} = \frac{r\,(X_i - \bar{X})}{\sigma_X}$$

$$Y' - \bar{Y} = r\frac{\sigma_Y}{\sigma_X}\,(X_i - \bar{X})$$

$$Y' = \bar{Y} + r\frac{\sigma_Y}{\sigma_X}\,(X_i - \bar{X}) \tag{9-21}$$

Using the data in Table 9-1 again, for the observations (9,50), we have

$$Y' = 49.7 + (.94)\frac{32}{2.91}\,(9 - 6.7)$$

$$= 73.40$$

or about the same answer we arrived at using the regression equation. From this it can be seen that $r(\sigma_Y/\sigma_X)$ is the regression coefficient b. The numerical value of $r(\sigma_Y/\sigma_X)$ for the data in Table 9-1 is 10.99, or almost the same thing as when the normal equations were used (the difference may be attributed to rounding error). We make use of this line or reasoning in Chapter 10.

READINGS AND SELECTED REFERENCES

Blalock, Hubert, Jr. *Social Statistics*, McGraw-Hill Book Company, New York, 1960, Chaps. 17, 18.

Diamond, Solomon. *Information and Error*, Basic Books, Inc., Publishers, New York, 1959, Chaps. 10, 11.

Freund, John. *Modern Elementary Statistics*, Prentice-Hall, Inc., Englewood Cliffs, N. J., 1960, Chaps. 14, 15.

Games, Paul, and George Klare. *Elementary Statistics*, McGraw-Hill Book Company, New York, 1967, Chaps. 13, 14.

Graybill, Franklin A. *An Introduction to Linear Statistical Models*, McGraw-Hill Book Company, New York, 1961, vol. 1.

Hammond, Kenneth, and James Householder. *Introduction to the Statistical Method*, Alfred A. Knopf, Inc., New York, 1963, Chap. 6.

Janda, Kenneth. *Data Processing: Applications to Political Research*, Northwestern University Press, Evanston, Ill., 1965.

Parl, Boris. *Basic Statistics*, Doubleday & Company, Inc., Garden City, N. Y., 1967, Chaps. 24, 25.

Spurr, William A., and Charles P. Bonini. *Statistical Analysis for Business Decisions*, Richard D. Irwin, Inc., Homewood, Ill., 1967, Chap. 22.

Yamane, Taro. *Statistics: An Introductory Analysis*, Harper & Row, Publishers, New York, 1967, Chaps. 14, 15.

10
Partial and Multiple Correlation and Regression

PARTIAL CORRELATION

The development of methods of analysis usually runs hand in glove with the development of theory. The simple deterministic theories of the relatively recent past have slowly been giving way to more complex systems theories, including the use of probability models.[1] We shall have more to say about probability models later in this book. In the present section we consider the methods of analysis that assume a deterministic theoretical framework but attempt to include many variables.

In general, if we find a relationship between two variables, it is not likely to be very interesting from a theoretical perspective. This is because of the ever-present possibility that a particular relationship may be "spurious." We define this term more rigorously below. At this point it suffices to give an example. Suppose that we find a strong correlation between constituency attitudes and the voting behavior of congressmen. We cannot infer that a causal relationship exists, because a third variable we have not discovered may account for the variance in both of these variables. It may be that the socioeconomic backgrounds of both constituents and representatives are similar, and this accounts for the tendency for the attitudes of each to vary in the same manner.[2]

In order to learn whether this is so, we may resort to multiple and partial correlation and regression analysis. The example from Chapter 9 concerning constituency attitude and voting behavior is retained, and a third variable — representative's attitude — introduced. The dependent variable is still the voting behavior of congressmen in the area of civil rights. We are interested

---

[1] See Karl Deutsch's excellent summary of this development, *The Nerves of Government*, The Free Press of Glencoe, New York, 1963, especially Chaps. 1–4.

[2] There may be a tendency for upper-income people to elect representatives from the same socioeconomic background, and so on. Of course, to the extent that congressional districts include people of mixed socioeconomic background, this is not so. The possibility offered is only illustrative; however, the author does not know of any studies that try to relate the two factors.

in the relationship between this dependent variable and each of the independent variables, the combined effect of the two independent variables as well as the independent effect of each, and the relationship among all three. The simple correlations for this problem are shown in the Figure 10-1.

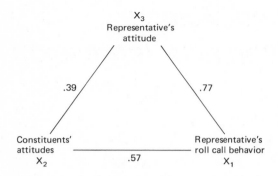

Figure 10-1. Intercorrelations for civil rights. Adapted from Warren E. Miller and Donald F. Stokes, "Constituency Influence in Congress," *American Political Science Review*, vol. 57, pp. 45–56, March, 1963.

Our first questions concern the relationships between X_1 and X_2 controlling for X_3 and between X_1 and X_3 controlling for X_2; that is, we want to determine whether or not the correlations we have found remain the same when we control for a third variable. To put it in slightly different terms, we want to know whether or not the third variable is responsible for some of the correlation we have found in the first two.

One way we may control is to take all representatives with constituents whose attitudes toward civil rights are exactly the same, and see whether or not there is a relationship between representatives' attitudes and voting behavior within these groups. Analysis of variance is the method to use. For the relationship controlling for representatives' attitudes, we may take all representatives who have the same attitudes toward civil rights, and see whether or not there is a relationship between constituency attitudes and voting behavior within these groups. But, of course, we are not likely to find many cases in each of these categories, and it may not be possible, therefore, to do this. We can resort to mathematical manipulation of the variables instead as a means of controlling.

First let us clarify notation. The standard notation for partial correlation is to put the variables we are trying to relate on the left side of a dot in the subscript and the one we are trying to control on the right side of the dot. Thus, $r_{12.3}$ means the correlation between variables 1 and 2 controlling for 3; and $r_{13.2}$ means the correlation between variables 1 and 3 controlling for 2; and $r_{13.2456}$ means the relationship between variables 1 and 3 controlling for variables 2, 4, 5, and 6; and so on. In the example we are using here, the numbers 1, 2, and 3 stand for representative's voting behavior on civil rights, constituency attitude, and representative's own attitude toward civil rights, respectively.

We said that we are going to control by mathematical manipulation. Let us first consider $r_{13.2}$, or the correlation between the representative's attitude and his voting behavior, controlling for constituency attitude. The total correlation between the representative's attitude and his voting behavior is .77. This says that representatives tend to vote on civil-rights legislation in accordance with their beliefs. However, we want to know if some of this positive correlation is "spurious." That is, when we hold constituency attitude constant, is there still some covariation between the representative's attitude and his voting behavior? Another way to put this question is whether all the covariation found between the representative's attitude and his voting behavior has been produced by constituency attitude.

We shall explain briefly what we do when we hold a third variable constant and give a more precise explanation below when multiple regression analysis is considered. In the present example we want to find the partial correlation between X_1 (representative's voting behavior on civil rights) and X_3 (representative's attitude toward civil rights). To do so, we have to remove the effects of X_2 (constituency attitude toward civil rights) on each of these two by deducting it from each. Of course, we do not subtract each X_2 from each X_1 and X_3 directly; we subtract the improvement brought about by the regression line of X_2 on X_1 and X_3. The remaining sum of squared deviations around the regression line is the amount of variation in X_1 and X_3 left unexplained by X_2. We then see if this residual sum of squares is correlated.

Recall that in Equation (9-21) we defined the b coefficient as $r(\sigma_Y/\sigma_X)$. This is the amount of improvement in Y for values of X. To bring the symbolic notation into line with partial-correlation notation, let $Y = X_1$ and $X = X_2$. Hence, $b_{X_1X_2} = r(\sigma_{X_1}/\sigma_{X_2})$.

Now consider each individual score. When we remove from x_1 (where $x_1 = X_1 - \bar{X}$) the improvement brought about by the regression of x_2 (where $x_2 = X_2 - \bar{X}$), we have

$$x_1 - b_{12}x_2 = x_1 - r_{12}\frac{\sigma_1}{\sigma_2}x_2$$

and when we remove from x_3, the improvement brought about by the regression of x_2, we have

$$x_3 - b_{32}x_2 = x_3 - r_{32}\frac{\sigma_3}{\sigma_2}x_2$$

In each case, the quantity designated on the right-hand side of the equation is the sum of the squares around the regression line left unexplained by X_2. To get the partial correlation between X_1 and X_3 controlling for X_2, we correlate these residuals in somewhat the same way we obtain a total correlation; but in this case the covariation is that of the residuals, and we divide by a slightly different correction factor:

$$r_{13.2} = \frac{\sum [x_1 - r_{12}(\sigma_1/\sigma_2)x_2][x_3 - r_{32}(\sigma_3/\sigma_2)x_2]}{\{\sum [x_1 - r_{12}(\sigma_1/\sigma_2)x_2]^2 \sum [x_3 - r_{32}(\sigma_3/\sigma_2)x_2]^2\}^{1/2}} \qquad (10\text{-}1)$$

$$= \frac{r_{13} - r_{12}r_{32}}{\sqrt{1 - r_{12}^2} \sqrt{1 - r_{32}^2}} \qquad (10\text{-}2)$$

In simpler terms, what we have done in Equation (10-2) is to subtract from the total correlation between 1 and 3 the product of the correlation of the controlled variable (2) with each, and divide by a factor that takes account of how much the controlled variable leaves unexplained in each of the correlated variables. Using the data from Figure 10-1 we have

$$r_{13.2} = \frac{.77 - (.57)(.39)}{\sqrt{1 - .57^2} \sqrt{1 - .39^2}} = .72$$

This says that the partial correlation between the representative's attitude and his voting behavior, controlling for constituents' attitudes, is .72. Note that it has been reduced from .77 to .72. The square of the partial correlation is a coefficient of determination. Therefore, $r_{13.2}^2 = .518$, or about 52 percent of the representative's voting behavior is explained by his own attitude when we allow for the influence of constituents' attitudes. This is in contrast to the $r_{13}^2 = .593$ for the total correlation.

At this point, we may say that only a small part of the .77 correlation between the representative's attitude and his voting behavior contains the effect of constituents' attitudes on his attitude. Now consider the partial correlations between constituents' attitudes and the representative's voting behavior and between constituents' attitudes and the representative's attitude. We have for the former

$$r_{12.3} = \frac{.57 - (.77)(.39)}{\sqrt{1 - .77^2} \sqrt{1 - .39^2}} = .46$$

and for the latter

$$r_{23.1} = \frac{.39 - (.57)(.77)}{\sqrt{1 - .57^2} \sqrt{1 - .77^2}} = -.09$$

Figure 10-2 summarizes these partial correlations.

We now come to a rather complex question concerning the interpretation of correlation data. Let us first consider the problem from the point of view of causal terminology. Our objective is to say something about whether or not constituents' attitudes cause the representative's attitude, which then

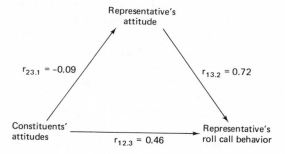

Figure 10-2. Partial correlations for civil rights.

causes his vote, and so on.[3] We want to discover something about the "forcings" among these variables in order to discover which is the antecedent factor and if there is independence between the variables of constituents' attitudes and the representative's attitude.

In order to do so, we must make two major assumptions. First, we must assume that only certain relationships among the three variables are plausible, so that we may restrict the number of possible models we must investigate. For example, we might assume that constituents' attitudes influence the representative's attitude but that the reverse causal sequence does not occur. We might also assume that constituents' attitudes influence the representative's voting behavior but the reverse does not occur and that the representative's attitude can influence his own voting behavior but the reverse does not occur. In other words, we first assume that there is only one-way causation involved.

It is obvious that of the three assumptions only the last seems theoretically plausible. It is possible that the representative's attitude may influence those of his constituents, assuming that the representative can convert some of his voters, and it is also possible that his voting behavior on civil rights may influence his constituents' attitudes, although this seems less plausible. It does not seem theoretically possible for the representative's voting behavior to influence his own attitude. For the moment, however, we make the simplifying assumption of one-way causation, and thus the arrows can run only in the direction shown in Figure 10-2. We must make this simplifying assumption, because if we allow the possibility of two-way causation, we cannot separate the independent effect of each variable, and partial-correlation analysis cannot be used.

The second major assumption we must make is that any other variables that may influence the dependent variable are not correlated with the two

[3] See Hubert Blalock, *Causal Inferences in Non-experimental Research*, The University of North Carolina Press, Chapel Hill, N. C., 1964; Herbert Simon, *Models of Man*, John Wiley & Sons, Inc., New York, 1957, Chaps. 1–3; Hayward Alker, *Mathematics and Politics*, The Macmillan Company, New York, 1965.

independent variables we are investigating. We admit, therefore, that there are uncontrolled factors which may also explain the representative's voting behavior and which may be related to the two independent variables. For example, the sociocultural background of the individuals involved might influence all three variables. We must assume that this variable or any other has a random effect on all three variables. If we cannot make this assumption, the other variables that might be related must be brought into the model and controlled. It becomes immediately apparent that there is no end to the process, for there may be many variables that are related. At some point, we must consider the variables we have identified as part of a closed system. For the present example we say that the three variables do form a closed system.

Having made these two broad assumptions, we can proceed to eliminate some of the possible relationships in the three-variable model through partial-correlation analysis. There are three different models that may hold when we consider that there is a relationship among the three variables and that the causal arrows run in only one direction (see Figure 10-3).[4]

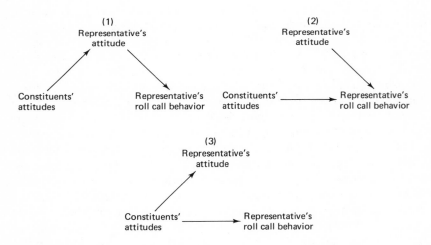

Figure 10-3.

The first model suggests that constituents' attitudes influence the representative's attitude, which influences his voting behavior. We consider constituents' attitudes the independent variable in Figure 10-3a and the representative's attitude the intervening variable. By intervening variable we mean that the influence of constituents' attitudes on the representative's voting behavior is carried through, and modified by, the representative's attitude

---

[4] There are actually many more than 3 models than can be applied in this case. We consider the rest below.

and that constituents' attitudes are only indirectly related to the representative's voting behavior. What effect they have is carried through the representative's attitude.

Figure 10-3*b* suggests that constituents' and the representative's attitudes each have an independent effect on voting behavior and that the simple correlation between constituents' attitudes and the representative's attitude is spurious. By spurious we mean that the simple *r* is produced by the fact that both independent variables are related to the same dependent variable and act in the same direction.

Figure 10-3*c* suggests that the correlation between the representative's attitude and his voting behavior is spurious, being produced by the action of constituents' attitudes on both these variables. In Figure 10-3*a* we have one independent variable, one intervening variable, and one dependent variable. In Figure 10-3*b* we have two independent variables, each with an effect on the dependent variable. And in Figure 10-3*c* we have one independent variable and two dependent variables.

We may use partial-correlation (and regression) analysis to learn which of the three models is most likely correct. When the relationship between two variables is spurious, the partial correlation between them, when a control for the third variable is introduced, will be reduced to zero (or close to zero, allowing for sampling error). The hypothetical model in Figure 10-4 illus-

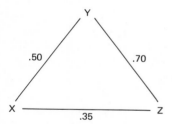

Figure 10-4. Simple correlations for three-variable relationship.

trates this situation. The partial correlation between X and Z controlling for Y is zero, because we subtract from the total correlation between X and Z the product of the correlations between X and Y times Y and Z. Therefore, when a correlation is spurious, $r_{XZ} = r_{XY}r_{YZ}$, because

$$r_{XZ \cdot Y} = \frac{.35 - (.7)(.5)}{\sqrt{1 - .5^2} \sqrt{1 - .7^2}} = 0$$

There are two possible interpretations of this situation. The correlation .35 between X and Z is produced by the action of Y on both X and Z, as in Figure 10-5*a*, or it is the result of the effect of X on Y and Y on Z, as in

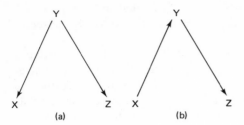

Figure 10-5.

Figure 10-5*b*. Which of the two models is correct can be determined only on the basis of an investigation of the data and the compatibility of the interpretation with theory. Let us return to the data in the previous example to illustrate this.

In the model in Figure 10-2 it appears that the correlation between constituents' attitudes and the representative's attitude is spurious because the partial correlation between these variables, controlling for the representative's voting behavior, is reduced almost to zero. Also, it does not seem plausible to argue that the representative's voting behavior should be treated as an intervening variable so that the causal arrows run from constituents' attitudes to the representative's voting behavior and thence to the representative's attitude. Hence, it seems more likely that we can treat each variable, constituents' attitudes and the representative's attitude, as independent variables and conclude that each variable explains a part of the representative's voting behavior. We may therefore conclude that constituents' attitudes explain 21 percent of the variance in voting behavior and the representative's attitude explains a separate 52 percent. In a moment we shall consider how to determine the combined effect of the two independent variables on the representative's voting behavior. It is apparent that of the two independent variables, the representative's attitude contributes the most and that constituents' attitudes and the representative's attitude are not related by a linear model. We return to this illustration in a moment.

In addition to the models just considered, there are others that might apply in this situation. In general, for a three variable case, if we assume one-way causation and allow each of the variables to be the dependent variable, there will be six simple causal chains ($_3P_3 = 3!$) as follows:

$$a.\ X \rightarrow Y \rightarrow Z$$
$$b.\ X \rightarrow Z \rightarrow Y$$
$$c.\ Y \rightarrow X \rightarrow Z$$
$$d.\ Y \rightarrow Z \rightarrow X$$
$$e.\ Z \rightarrow X \rightarrow Y$$
$$f.\ Z \rightarrow Y \rightarrow X$$

In the simple causal chain the middle variable is the intervening variable and the relationship between the two end variables is indirect (not spurious). Notice that models (*a*) and (*f*) both give the same predictions: $r_{xz.y} = 0$, and $r_{zz.y} = 0$. In this case we can only distinguish which is correct by using the partial regression coefficients since $b_{xz} \neq b_{zx}$ even though $r_{xz} = r_{zx}$. We shall consider partial regression coefficients later in this chapter. In addition to these six models, for each there are two possible interpretations:

In the first there is one independent and two dependent variables. If $r_{yz.x} = 0$, then we say that r_{yz} is spurious. Similarly for (B); if $r_{xy.z} = 0$, then r_{xy} is said to be spurious. Notice, however, that it is impossible to distinguish statistically between models (A) and (*c*); they both predict $r_{yz.x} = 0$, $b_{yz.x} = 0$. Hence, we cannot use the regression equation in this case. This is the problem called "identification" in econometrics, and is too complex to consider here.[5] Which of these various models is correct is as much a matter for theory as for statistics. It is not sufficient to simply run every possibility to see which partials reduce to zero. In some cases sampling error will not enable us to determine which model is correct. But, more importantly, it may be better to apply an indeterminate model (such as those to be considered in Chapter 14) rather than the determinate models of correlation and regression analysis.

Partial correlation can be extended to more than two independent variables. The formula for the second-order partial is

$$r_{12.34} = \frac{r_{12.3} - (r_{14.3})(r_{24.3})}{\sqrt{1 - r_{24.3}^2}\ \sqrt{1 - r_{14.3}^2}}$$

and for the third-order partial it is

$$r_{12.345} = \frac{r_{12.34} - (r_{15.34})(r_{25.34})}{\sqrt{1 - r_{25.34}^2}\ \sqrt{1 - r_{15.34}^2}}$$

Thus, to compute the third-order partial correlation it is necessary to compute the second-order first, and to compute the second-order partial, it is necessary to compute the first-order first.

The interpretation of causal models for many variables becomes very complex. The number of alternative models that must be evaluated increases rapidly, especially if we cannot assume that two-way causation does not exist. Of course, we may also take the standard position here. Correlation does not mean causation, and causation cannot be inferred from correlation. We may

[5] See Dennis J. Palumbo, "Causal Inference and Indeterminacy in Political Behavior," in Mike Haas (ed.), *Statistics and Quantitative Methods*, forthcoming.

reach causal statements with experimental design, because the independent variable can be physically manipulated and its effect on the dependent variable observed directly. In this case the experimenter acts like a puppeteer. But in nonexperimental design, the investigator is an observer. He can simply observe the play, where nature is the puppeteer and is pulling a thousand strings.[6] In addition, it should be noted that if r or n is small, sampling fluctuations can be quite great. For example, a sample of $n = 10$ would yield an r of $+0.60$ or higher 5 percent of the time even if the population correlation is zero. Therefore, it would be dangerous to try to make causal inferences with correlations that are small (i.e., ≤ 0.50). Of course, if r is large, then there is less danger of getting wide fluctuations in samples. But if all the variables are highly intercorrelated, then a control for any one of the variables will always reduce the correlation between any two variables to zero. For example, if $r_{12} = 0.80$, $r_{13} = 0.90$, and $r_{23} = 0.90$, then

$$r_{12.3} = \frac{0.80 - (0.9)(0.9)}{\sqrt{1 - 0.9^2}\,\sqrt{1 - 0.9^2}} \doteq 0$$

The problems involved in this question are too complex to consider in detail here. The interested reader should consult the references cited.

MULTIPLE REGRESSION

Multiple linear regression is essentially an extension of simple linear regression. It enables us to measure the combined effect of any number of independent variables on a dependent variable. The objective is the same in fitting the regression line. The coefficients for each independent variable are found so as to minimize $\sum (Y - Y')^2$, or the sum of the squared deviations of the predicted from the actual Y values. The multiple-regression equation may be written

$$Y' = a + b_1(X_1) + b_2(X_2) + \cdots + b_n(X_n)$$

where X_1 represents the first independent variable, X_2 the second, and the b's the respective regression constants (the subscripts have been omitted for the moment). The equation is linear because it assumes a constant rate of change for the relationships between each of the independent variables and the dependent variables.

The multiple-regression equation cannot be represented in geometrical form, because the coordinate points would require as many dimensions as there are independent variables. For two independent variables and one dependent variable we can get an intuitive geometrical sense with a two-

---

[6] See Lee J. Cronbach, "The Two Disciplines of Scientific Psychology," *American Psychologist*, vol. 12, pp. 671–684.

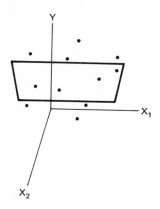

Figure 10-6. Geometrical representation of regression plane for two independent variables.

dimensional figure like that in Figure 10-6. The coordinates of X_1 and X_2 against Y are fitted by a plane rather than a line. The method of least squares can be used to fit the plane to the coordinate points. Each b coefficient found by this method represents the slope of the regression line for its respective variable, holding the effect of all the others constant. The a coefficient is simply the value of Y' when all the X's are zero.

Let us first consider the three-variable case. The example is the same as before — the relationship among constituents' attitudes, the representative's attitude, and his voting behavior. We change the notation in this case, however, designating the dependent variable by X_1 (representative's voting behavior) and the independent variables by X_2 (constituents' attitudes) and X_3 (representative's attitude) respectively. We want to find the constants in the equation $X_1 = a_{1.23} + b_{12.3}(X_2) + b_{13.2}(X_3)$.

Before proceeding, a word on the notation is necessary. The notation $a_{1.23}$ designates the intercept point on the Y axis for both X_2 and X_3. It may be read, as we shall see in regard to multiple correlation, as the combined effect of variables 2 and 3 on 1. The notation $b_{12.3}$ designates the partial regression coefficient of variables 1 and 2, holding variable 3 constant. Thus, for the general case, the equation can be written

$$X_i = a_{i.jk} + b_{ij.k}(X_j) + b_{ik.j}(X_k)$$
$$X_i = a_{i.jkl} + b_{ij.kl}(X_j) + b_{ik.jl}(X_k) + b_{il.jk}(X_l)$$

Because our objective is to fit the coefficients so as to minimize $\sum (Y - Y')^2$, we can use the normal equations to find the least-squares values. For example, in the case of three variables, we want to estimate the coefficients

a, b_1, and b_2 so that $\sum [Y - (a + b_1X_1 + b_2X_2)^2] = $ minimum. Recall that the normal equations for the two-variable case are

$$\sum Y = na + b \sum X$$
$$\sum XY = a \sum X + b \sum X^2$$

The normal equations for the three-variable case are a straightforward extension of this.

$$\sum Y = na + b_1 \sum X_1 + b_2 \sum X_2$$
$$\sum X_1Y = a \sum X_1 + b_1 \sum X_1^2 + b_2 \sum X_1X_2$$
$$\sum X_2Y = a \sum X_2 + b_1 \sum X_1X_2 + b_2 \sum X_2^2$$

These equations can be solved directly, but it is simpler to use the following computing equations:

$$b_{12.3} = \frac{b_{12} - (b_{13})(b_{32})}{1 - (b_{23})(b_{32})} \tag{10-3}$$

$$b_{13.2} = \frac{b_{13} - (b_{12})(b_{23})}{1 - (b_{32})(b_{23})} \tag{10-4}$$

Note that b_{32} is not the same as b_{23}. The b coefficient is not symmetrical, as is the correlation coefficient. The coefficient of the Y intercept is also straight-forward:

$$a_{1.23} = \bar{X}_1 - b_{12.3}(\bar{X}_2) - b_{13.2}(\bar{X}_3) \tag{10-5}$$

The equations for three independent variables are

$$b_{12.34} = \frac{b_{12.3} - (b_{14.3})(b_{42.3})}{1 - (b_{24.3})(b_{42.3})} \tag{10-6}$$

$$a_{1.234} = \bar{X}_1 - b_{12.34}\bar{X}_2 - b_{13.24}\bar{X}_3 - b_{14.23}\bar{X}_4 \tag{10-7}$$

To find the regression constants by Equations (10-3) to (10-7), we first find the simple regression coefficients. Thus, we first find b_{12} by Equation (9-5) and then substitute it into Equations (10-3) and (10-4). These in turn are substituted into Equation (10-6) if three independent variables are involved. If more variables are involved, it is better to use short computing methods that use matrix algebra. These are described in Yamane.

We now illustrate the computation of the coefficients for the multiple-regression equation, using Equations (10-3) to (10-5). Recall from Chapter 9 that $b_{12} = 10.25$ (the b coefficient for the representative's voting behavior in civil rights and constituent's attitudes — see page 184). Suppose that b_{13} (representative's voting behavior and his own attitude) is 12, b_{23} (constituents' attitudes and the representative's attitude) is 1.5, and b_{32} is 4 (these are all computed the same way as b_{12}).

Then

$$b_{12.3} = \frac{10.25 - (12)(4)}{1 - (1.5)(4)} = 7.55$$

$$b_{13.2} = \frac{12 - (10.25)(1.5)}{1 - (4)(1.5)} = .68$$

To illustrate the computation of a, recall that the mean vote on civil rights (for the hypothetical data in Table 9-1) was 49.7, the mean score of constituent's attitudes was 6.7, and suppose that the mean representative's attitude (on a scale of 0 to 10, with 10 most favorable) is 4.5; then

$$a_{1.23} = 49.7 - 7.55(6.7) - .68(4.5)$$
$$= -3.95$$

The complete multiple-regression equation would therefore be:

$$X_1 = -3.95 + 7.55(X_2) + .68(X_3)$$

What is the predicted voting behavior of a congressman who comes from a constituency with an attitude score of 2 (highly unfavorable) and an attitude himself of 8 (highly favorable)? It would be

$$X_1 = -3.95 + 7.55(2) + .68(8)$$
$$= 16.59$$

or not very favorable. In this hypothetical example the constituents' attitudes were set up to have a stronger influence than the representative's own attitude, contrary to the actual data discussed above.

The partial b coefficient gives the slope of the regression line for each independent variable, controlling for all others. The multiple-regression equation would be used for predicting the dependent variable from the combined independent variables. However, it is sometimes useful to get an idea of the relative importance of each of the independent variables in predicting the dependent variable. The b coefficient cannot give us this information, because, as we noted before, the size of the b coefficient is affected by the size of the unit used to measure the variables. We can standardize the b coefficient so that it will tell us the relative importance of each independent variable on a standard scale by multiplying it by the ratio of the standard deviation of the variable not controlled to the standard deviation of the dependent variable. For the three-variable case we should have

$$X_1 = a_{1.23} + \left(b_{12.3}\frac{s_2}{s_1}\right)(X_2) + \left(b_{13.2}\frac{s_3}{s_1}\right)(X_3)$$

The resulting coefficients are called *beta weights*, symbolized as β (beta) coefficients.[7] The beta weights can be used in various ways. We may take the

---

[7] Note that these β coefficients or beta weights are not the same as the population β in the multiple regression equation.

square of the beta weight as the relative percentage of variance accounted for by each of the independent variables. For example, suppose that we have the following respective beta weights for the independent variables, constituents' attitudes, representative's attitude, and socioeconomic status, with the dependent variable X_1, representative's voting behavior: $X_1 = a + (.60)(X_2) + (.26)(X_3) + (.7)(X_4)$. We may say that constituents' attitudes account for 36 percent, representative's attitude another 4 percent, and socioeconomic status another 49 percent, for a total of 89 percent of the variance in the dependent variable accounted for by the three independent variables. We shall see in a moment how this is related to the multiple-correlation coefficient.

Another way to interpret the beta weights is as follows. Using the same hypothetical figures given above, we say that for every increase of one standard deviation in constituents' attitudes, the representative's voting behavior increases by .60 standard deviations, and for one standard deviation increase in the representative's attitude, his voting behavior increases by .26 standard deviations. Because the betas are pure numbers, they can be compared. Of the three independent variables in the hypothetical example above, we conclude that socioeconomic status is most important, being slightly more important than constituency attitude and almost three times as important as the representative's own attitude. Note, however, the artificiality of this conclusion, because it assumes that it is possible to separate the "independent" contribution of each of these variables. This assumption must have theoretical plausibility to be meaningful. In addition, it is necessary to consider the problem of multicollinearity if all of the "independent" variables are related. This is considered in the next section.

MULTIPLE CORRELATION

The multiple-correlation coefficient R is a measure of the goodness of fit of the multiple regression between the dependent variable and the various independent variables. The square of the multiple-correlation coefficient R^2 is called the *coefficient of multiple determination*. It tells us the percentage of variance explained by the combined effect of all the independent variables. The reasoning behind the derivation of the multiple R^2 is similar to that for the simple r. We can measure the sum of squared deviations of the predicted Y' values from the actual Y values $\sum (Y - Y')^2$. This is the deviation that is left unexplained after fitting the regression plane. The sum of the squared deviations of the individual Y values from the mean of Y, $\sum (Y - \bar{Y})^2$, is the total deviation. Therefore, the deviations of the predicted Y' values from the mean of $\bar{Y}$, $\sum (Y' - \bar{Y})^2$, is the explained deviation. Thus we have

$$\sum (Y - \bar{Y})^2 = \sum (Y - Y')^2 + \sum (Y' - \bar{Y})^2$$

Total SS   Unexplained SS   Explained SS

and the ratio of the explained to the total is the coefficient of multiple determination:

$$R^2 = \frac{\sum (Y' - \bar{Y})^2}{\sum (Y - \bar{Y})^2}$$

It would be tedious to compute R^2 by this equation. Fortunately, there is another way to define the multiple R^2 that simplifies computations greatly:

$$R^2_{1.23} = r^2_{12} + r^2_{13.2}(1 - r^2_{12})$$

As an example we use the data in Figure 10-1:

$$R^2_{1.23} = (.57)^2 + (.72)^2(1 - .57^2) = .68$$

This says that 68 percent of the variation in the representative's voting behavior in civil rights is accounted for by the combined effect of constituents' attitudes and the representative's own attitude. Note that, when we did partial-correlation coefficients for each and squared them, the added effect was 73 percent. This is almost the same. Part of the difference is due to rounding error and part to the possibility that the effect of each variable is not additive. Of course, if the correlation between the two independent variables were zero, $R^2_{1.23} = r^2_{12} + r^2_{13}$.

When the independent variables are highly correlated with each other, the regression coefficients are unreliable, and it is meaningless to consider the "independent" contribution of each variable.[8] This situation is referred to as multicollinearity. The regression coefficients are unreliable because, as the correlation for the independent variables approaches one, the standard error of the regression coefficient gets very large. This can be seen easily from the equation for the standard error of the regression coefficient for two independent variables:

$$s_{b_1} = \frac{s_{Y|X_1X_2}}{\sqrt{\sum (X_1 - \bar{X}_1)^2(1 - r_{12}^2)}}$$

The numerator is the standard error of the estimate of Y based on X_1 and X_2.[9] It can be seen that the standard error of b_1 is smallest when the coefficient of determination r^2_{12} in the denominator is smallest and that it becomes very large when r^2_{12} approaches one. Thus if X_1 and X_2 move together, it is difficult

[8] As an example, see Dennis J. Palumbo and Oliver P. Williams, "Predictors of Public Policy: The Case of Local Public Health," *Urban Affairs Quarterly*, vol. II, pp. 75–92, June, 1967.

[9] It is defined as

$$s_{Y|X_1X_2} = \sqrt{\frac{\Sigma(Y - Y')^2}{n - k}}$$

$$= \sqrt{\frac{\Sigma(Y - \bar{Y})^2 - b_1\Sigma(X_1 - \bar{X}_1)(Y - \bar{Y}) - b_2\Sigma(X_2 - \bar{X}_2)(Y - \bar{Y})}{n - k}}$$

to distinguish their separate effects on Y. It is also not very meaningful to add the independent contributions of each variable. For example, suppose that the correlation between X_1 and X_2 is .90 and that the correlation of each of these with the dependent variable Y is .9 and .9 respectively. Then it is possible to say that variable X_1 explains 81 percent of the variance in Y. But it is not very meaningful to say that variable X_2 explains an additional 81 percent. In fact, the partial correlation between any two of these is almost zero, indicating something false in the relationships.

One solution is to drop the X's that are not important and retain only those which are. Another solution is to approach the question through canonical correlation or factor analysis. The later is considered in Chapter 12.

Multiple correlation can be extended to more than two independent variables. For three independent variables we have

$$R_{1.234}^2 = r_{12}^2 + r_{13.2}^2(1 - r_{12}^2) + r_{14.23}^2[(1 - r_{12}^2 - r_{13.2}^2)(1 - r_{12}^2)]$$

$$= R_{1.23}^2 + r_{14.23}^2(1 - R_{1.23}^2)$$

It is clear that using this method of computing the multiple R^2 for many independent variables can be laborious. There are short methods. One computing equation for the three-variable case that uses only simple correlations is

$$R_{1.23}^2 = \frac{r_{12}^2 + r_{13}^2 - 2r_{12}r_{13}r_{23}}{1 - r_{23}^2} \qquad (10\text{-}8)$$

However, computing multiple-regression coefficients and the multiple-correlation coefficient by hand often involves more labor than the result warrants. Fortunately, computer programs for multiple regression and correlation (and many other statistics) are usually readily available in "canned" form. Many of these canned programs are adaptations of the Biomedical Computer Programs.[10] For example, the BMDO2R performs a stepwise regression program in which the variables are taken into the regression equation in the order of their importance in explaining variations in the dependent variable. The program first takes the independent variable with the highest correlation with the dependent variable and computes the simple regression for these two, giving the a and b coefficients, along with the multiple R, standard error, and the partial-correlation coefficients of each variable not yet included in the equation. It then selects, as the next variable, the one that makes the greatest additional contribution to explained variance in the dependent variable, and so on. Whatever the canned program used, however, the interpretations given here apply. See Janda for a further description of the use of computers in political research (cited at the end of the chapter).

[10] See BMD *Biomedical Computer Programs*, Health Services Computing Facility, School of Medicine, University of California, Los Angeles, Jan. 1, 1964. Also William W. Cooley and Paul R. Lohnes, *Multivariate Procedures for the Behavioral Sciences*, John Wiley & Sons, Inc., New York, 1962; Kenneth Janda, *Data Processing: Applications to Political Research*, Northwestern University Press, Evanston, Ill., 1965.

SIGNIFICANCE TESTS FOR THE MULTIPLE AND PARTIAL CORRELATIONS

To test significance of the multiple R and partial r, we test the null hypothesis of no correlation, H_0: $\rho = 0$. The assumptions are the same as in simple regression, namely: (a) the relationship between variables in the population is linear, (b) the residual errors are independent of each other, (c) the variances are equal, and (d) the values of the dependent variable are normally distributed (this can be ignored for large samples). We can use the F distribution, as follows:

For Multiple R:
$$F_{k,n-k-1} = \frac{R^2}{1 - R^2}$$

For Partial, $r_{12.3}$:
$$F_{1,n-3} = \frac{r_{12.3}^2}{1 - r_{12.3}^2} (n - 3)$$

CURVILINEAR RELATIONS

So far we have been discussing how to approximate relationships assumed to be linear. Most variables in political science research studied to date seem to fit the linear model very well. Of course, if they do not, we cannot conclude that there is no relationship between the variables but only that a linear model does not seem to fit the data. There is always the possibility that better predictions can be afforded by a curved line, in which case we say that the relationship is curvilinear.

There have not been a great number of cases in social science research where curved lines afford better predictions. In many cases, a curvilinear relationship can be more accurately described by a straight line after the variables have been logarithmically transformed (this is considered in Chapter 13). However, it is still useful to understand some of the simpler curved lines. We consider two such equations here and additional ones in Chapter 13.

A function whose values are given by a quadratic polynomial, $aX^2 + bX + c$, where $a \neq 0$, is called a polynomial function of degree two, or a quadratic function. To illustrate, let us first consider the simplest of quadratic functions, those specified by equations of the form $Y = aX^2$. If we set a at 1, the graph of this function is that shown in Figure 10-7.

The graph of the quadratic function $Y = aX^2$ is said to be symmetric with respect to the Y axis. If we folded the graph at the Y axis, the part of the graph in the first quadrant would coincide with that in the second quadrant.

The second-degree polynomial, or quadratic function, specified by $Y = a(X - h)^2 + k$, would take a similar form when graphed, but because

Y = X²		
X	X²	Y
−3	(−3)²	9
−2	(−2)²	4
−1	(−1)²	1
0	0²	0
1	1²	1
2	2²	4
3	3²	9

Figure 10-7. Graph of $Y = aX^2$.

we are subtracting the constant h from X, its graph is symmetric with respect to the line whose equation is $X = h$, and its vertex is at the point (h,k). If $a > 0$, the vertex is a minimum point of the graph (see Figure 10-8); and if $a < 0$, the vertex is a maximum point.

The Parabola

We now consider the curve of the form $Y' = a + bX + cX^2$, which is the parabola. a gives the Y intercept, b the slope, and c the degree of curvature. To fit such a curve, we solve the normal equations:

$$\sum xy = b \sum x^2 + c \sum x^3$$
$$\sum x^2y = b \sum x^3 + c \sum x^4$$

where x and y are deviations from their respective means.

To illustrate this, suppose that a campaign organizer is interested in determining the effects of increasing the number of volunteer workers on getting people to register to vote. He selects 20 election districts. In 5 of these he uses only 5 workers; 5 districts are allocated 20 workers each, 5 are allocated 40 workers each, and 5 are allocated 80 workers each. The results of this are shown in Table 10-1.

These data are plotted in Figure 10-9, and a parabola is fitted by free-hand drawing. It appears that, as the number of voluntary workers increases the voter turnout increases; but the increase is successively smaller, which seems logical.

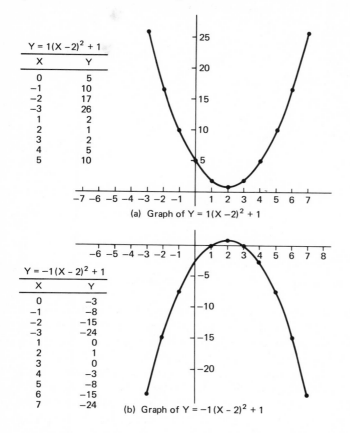

Y = 1(X −2)² + 1

X	Y
0	5
−1	10
−2	17
−3	26
1	2
2	1
3	2
4	5
5	10

(a) Graph of Y = 1(X −2)² + 1

Y = −1(X − 2)² + 1

X	Y
0	−3
−1	−8
−2	−15
−3	−24
1	0
2	1
3	0
4	−3
5	−8
6	−15
7	−24

(b) Graph of Y = −1(X − 2)² + 1

Figure 10-8. (A) Graph of $Y = 1(X - 2)^2 + 1$. (B) Graph of $Y = -(X - 2)^2 + 1$.

Table 10-1 Voter Turnout and Voluntary Workers

Number of workers	Voter turnout (thousands) in district number					Total
	1	2	3	4	5	
5	4	10	12	6	8	40
20	40	50	40	60	55	245
40	70	90	100	110	95	465
60	100	110	115	120	105	550

This can also be seen if we compare the mean turnout for each of the different categories of workers. These are 8, 49, 93, and 110 respectively. The increases are 41, 44, and 17 respectively.

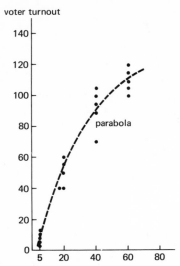

Figure 10-9. Voter turnout (thousands) and voluntary workers.

It is possible to test whether or not a curvilinear relationship, such as a parabola, can better explain the data than a linear relationship. First, let us note that the reader will not be able to follow this discussion without having read Chapter 11. We consider the matter here, because it is part of regression analysis, but analysis of variance techniques and notation, which are covered in Chapter 11, are used to test for curvilinear relationships. We simply describe here how it is done and do not explain the various terms. They are all explained in Chapter 11. The reader should return to this section after reading Chapter 11.

Consider again the equation $Y = a + bX + cX^2$. Suppose that we try to fit a linear equation to these data, $Y' = a + bX_1 + cX_2$. There will be an amount of deviation from this equation, and this deviation will contain not only sampling error but error depending on the squared value of X. The basic model for the sum of squares for curvilinear regression can therefore be represented as

Total SS = linear regression SS + deviations from linear regression SS

$$+ \text{error } SS$$

To test whether or not some curvilinear regression exists, we can use the F distribution, which is described in Chapter 11.

$$F_{j-2, n-j} = \frac{MS \text{ deviations from linear regression}}{MS \text{ error}}$$

The null hypothesis is that there is no curvilinear regression. We illustrate how to compute this F ratio, using the data in Table 10-1. We need the following sums of squares:

$$SS \text{ total} = (4)^2 + (10)^2 + \cdots + (105)^2 - \frac{(1,300)^2}{20} = 33,060$$

$$SS \text{ between groups} = \frac{(40)^2 + \cdots + (550)^2}{5} - \frac{(1,300)^2}{20} = 31,570$$

$$SS \text{ error} = SS \text{ total} - SS \text{ between groups} = 1,490$$

Finally, we need the SS linear regression, which can be obtained by

$$SS \text{ lin. reg.} = \frac{n\left[\sum_j \sum_i X_j Y_{ij} - \left(\sum_j n_j X_j\right)\left(\sum_j \sum_i Y_{ij}/\right)n\right]^2}{n\left(\sum_j n_j X_j^2\right) - \left(\sum_j n_j X_j\right)^2}$$

Because there are 5 in each group ($n_j = 5$), and the total number of districts is $n = 20$, we have

$$\sum_j \sum_i X_j Y_{ij} = \sum_j X_j \left(\sum_i Y_{ij}\right) = 5(40) + 20(245) + 40(465)$$
$$+ 60(550) = 56,700$$

$$\sum_j 5X_j = 5(5) + 5(20) + 5(40) + 5(60) = 625$$

$$\sum_j 5X_j^2 = 5(25 + 400 + 1,600 + 3,600) = 28,125$$

and $\quad SS \text{ lin. reg.} = \dfrac{20 \; 56,700 - 625(1,300/20)^2}{20(28,125) - (625)^2} = 30,069.01$

The deviations from linear-regression sum of squares are

$$\text{Dev. from lin. reg. } SS = SS \text{ between groups} - SS \text{ lin. reg.}$$
$$= 31,570 - 30,069$$
$$= 1,501$$

To obtain the respective mean squares (or estimates of variance, as explained in Chapter 11) for the F ratio, we divide the respective sum of squares by the degrees of freedom, which are $j - 2 = 2$ and $n - j = 16$. The F ratio is:

$$F_{2,16} = \frac{1,501/2}{1,490/16} = 8.05$$

The table of the F distribution in the Appendix tells us that for 2 and 16 degrees of freedom we need an F of 6.23 or better for significance at the .01 level. Because the F here is larger, we can reject the null hypothesis that there is no curvilinear correlation, and we conclude that a curve can afford some description of these data. Note that the respective sum of squares are

$$\text{Total } SS = \text{lin. reg. } SS + \text{dev. from lin. reg. } SS + \text{error } SS$$
$$33,060 = 30,069 + 1,501 + 1,490$$

It appears that most of the variance can be explained by linear regression and only a relatively small part by curvilinear regression. To get an estimate of how much of the total variance is accounted for by curvilinear regression, we use the following:

$$\text{Est. } \rho_c{}^2 = \frac{\text{dev. from lin. reg. } SS - (j-2) \, MS \text{ error}}{SS \text{ total} + MS \text{ error}}$$

which for the above data is

$$\text{Est. } \rho_c{}^2 = \frac{1,501 - (2)1,490/16}{33,060 + 1,490/16} = .03$$

and the estimate of curvilinear correlation is the square root of this, .17. We can also estimate the linear correlation by

$$\text{Est. } \rho^2 = \frac{SS \text{ lin. reg.} - MS \text{ error}}{SS \text{ total} + MS \text{ error}}$$

$$= \frac{30,069 - 93.1}{33,060 - 93.1}$$

$$= .90$$

and the estimate of the correlation in the population is .95. We conclude that most of the variance can be accounted for by a straight line and that a curve adds only a little additional. In general, unless the phenomenon we are studying is related in a complex manner, best described by a curve with a high degree of curvature, a straight line usually suffices.

READINGS AND SELECTED REFERENCES

Blalock, Hubert, Jr. *Social Statistics*, McGraw-Hill Book Company, New York, 1960, Chap. 19.

Freund, John. *Modern Elementary Statistics*, Prentice-Hall, Inc., Englewood Cliffs, N. J., 1960, Chap. 16.

Graybill, Franklin A. *An Introduction to Linear Statistical Models*, McGraw-Hill Book Company, New York, 1961, Vol. 1.

Janda, Kenneth. *Data Processing: Applications to Political Research*, Northwestern University Press, Evanston, Ill., 1965.

Parl, Boris. *Basic Statistics*, Doubleday & Company, Inc., Garden City, N.Y., 1967, Chaps. 26–28.

Spurr, William A., and Charles P. Bonini. *Statistical Analysis for Business Decisions*, Richard D. Irwin, Inc., Homewood, Ill., 1967, Chap. 23.

Yamane, Taro. *Statistics: An Introductory Analysis*, Harper & Row, Publishers, New York, 1967, Chap. 22.

11
Analysis of Variance

The analysis of variance is a direct extension of the difference of means. In the difference of means test, the simplest kind of experiment is involved. A single independent variable is used and one experimental and control group. Most experiments are much more complex than this, involving a number of groups and different treatments. But no matter how complex the design, the objective is still the same: the means of the different groups are compared simultaneously.

When more than two means are to be compared, the analysis of variance is the appropriate method. The null hypothesis is H_0: $\mu_1 = \mu_2 = \cdots = \mu_k$. The principal objective is to determine whether or not there is a significant difference among the several different means. By "significant difference" we mean whether or not there is a difference among the means of the population. A statistically significant difference is not necessarily a strong relationship. This latter question is considered later in this chapter.

The analysis of variance test is used in research designs similar to those described in Chapter 7. It is used frequently in agricultural experiments. A typical agricultural problem analyzed by the analysis of variance is the differences in yield from plots of land to which fertilizer is added. The fertilizer is the *treatment* and the differences in yield are the *effects* of this treatment. We use the term "treatment" synonomously with the term "independent variable" and the term "effects" synonomously with the differences in means of the dependent variable.

In recent years, the analysis of variance has become a very broad and technical subject. It is used extensively in disciplines in which experimental design is most appropriate, such as experimental psychology. In political science, of course, researchers rely more heavily on nonexperimental research design, and the analysis of variance, therefore, is not so appropriate a statistical method. However, the underlying principles are central to some of the logic of methods like correlation analysis, and the essential aspects of the analysis of variance are therefore considered in this chapter.

An analysis of variance test would be appropriate for a study like the Guetzkow-Simon study described briefly in Chapter 7. Recall that three groups were involved, and that they might be viewed roughly[1] as falling on a single dimension, running from the most open communications structure to the most restricted, in the following order: (a) all channel, (b) wheel, and (c) circle. Recall also that the mean time it took each group to perform certain tasks was (a) 24.38, (b) 19.12, and (c) 29.45 minutes, respectively. We might infer from this that the intermediate communications structure is best. By the analysis of variance we test whether there is a significant difference in the population means.

It would be possible to answer this question by using difference of means tests on each group separately, but suppose that we found a significant difference between groups (b) and (c) and not between groups (a) and (c). Should we then conclude that there is a significant difference among all three groups in the population or only between groups (b) and (c)? This question cannot be answered by a difference of means test. The *t* test of a difference between means, and the analysis of variance test, produce exactly the same results when there are only two groups. If there are more than two groups, or treatments, the analysis of variance test should be used rather than the *t* test.

The assumptions made in the analysis of variance test are (a) the populations from which the samples are drawn are normally distributed (or at least similar in distribution), (b) the samples are randomly drawn, (c) the observations in each sample are independent (hence the test is not appropriate for paired observations), and (d) a common variance exists among the populations from which the samples are drawn (this is also called *homoscedasticity*). The last assumption says, in effect, that, although there may be a difference among means in the population produced by the independent variable, the dispersion of scores has not been affected.

For some models of analysis of variance (discussed below) the first assumption may be violated if the samples are large, and so may the fourth assumption if the number of cases in each sample is the same. But, in general, if the assumptions are clearly not being met, the analysis of variance test will yield false results, and we shall be misled in making inferences.

THE GENERAL MODEL

We said that the objective of the test is to determine if $\mu_1 = \mu_2 = \cdots = \mu_k$. If the means are equal, we may treat them as one large population having

---

[1] The word "roughly" is used here because the conditions of the communications nets are more complex than implied by the notion of "openness" of communications. For the full description of the conditions imposed on each group the reader should consult Harold Guetzkow and Herbert Simon, "The Impact of Certain Communications Nets upon Organization and Performance in Task-oriented Groups," in Albert H. Rubenstein and Chadwick Haberstroh, *Some Theories of Organization*, The Dorsey Press, Homewood, Ill., 1966, pp. 425–443.

a single mean and variance. The samples drawn from this population may then be conceived of as having been drawn from the same population. Therefore, estimates of the population variance based on each of the samples should be the same except for chance variation due to sampling.

In analysis of variance we make separate estimates of σ^2 based on the samples. The separate estimates of σ^2 are not made from each of the samples separately. One estimate is based on the variance within each of the categories of the independent variable and is called the within-column estimate; the other is based on the variance between the means of each of the categories of the independent variables and is called the between-columns estimate.

For example, consider the data in Table 11-1. Suppose that this represents an experimental situation in which we want to determine if three dif-

Table 11-1 Scores on Political Values Test for Three Different Teaching Methods

Student	Method 1	Method 2	Method 3
1	3	4	6
2	6	7	7
3	5	7	8
4	4	4	7
5	7	8	7
Total	25	30	35
Group mean $\bar{X}$ =	5	6	7

Grand mean $\bar{\bar{X}}$ = 6

ferent methods of teaching the introductory political science course to undergraduate students has an influence on their political values. A random sample of 15 students has been selected from the total enrollment (normally a much larger sample would be used). These 15 students are then assigned randomly to each of 3 different courses. At the end of the term a test of some political value, say, confidence in the rule of law, is administered to each group of 5. Suppose that a higher score means a greater degree of this value.

Table 11-1 seems to indicate that those exposed to method 3 have the greatest degree of this value at the end of the course. We want to know if the differences are real, that is, also exist in the population, or if they can be attributed to sampling error.

The null hypothesis is $\mu_1 = \mu_2 = \mu_3$. If this hypothesis is true, $\mu_1 = \mu_2 = \mu_3 = \mu$, and the three populations can be considered as one large population having a variance σ^2. We make two separate estimates of σ^2 based on the three samples. The first is a pooled estimate of the variances within each column.[2] The variance within each column is

$$s_1{}^2 = \frac{\sum\limits_{i=1}^{n_1} (X_{i1} - \bar{X}_{.1})^2}{n_1}$$

$$s_2{}^2 = \frac{\sum\limits_{i=1}^{n_2} (X_{i2} - X_{.2})^2}{n_2}$$

$$s_3{}^2 = \frac{\sum\limits_{i=1}^{n_3} (X_{i3} - \bar{X}_{.3})^2}{n_3}$$

A pooled estimate of σ^2 based on these three within-column variances is

$$\sigma^2 = \frac{n_1 s_1{}^2 + n_2 s_2{}^2 + n_3 s_3{}^2}{n_1 + n_2 + n_3}$$

$$= \frac{\sum\limits_{i=1}^{n_1} (X_{i1} - \bar{X}_{.1})^2 + \sum\limits_{i=1}^{n_2} (X_{i2} - \bar{X}_{.2})^2 + \sum\limits_{i=1}^{n_3} (X_{i3} - \bar{X}_{.3})^2}{n - 3}$$

$$= \frac{\sum\limits_{i=1}^{n_i} \sum\limits_{j=1}^{3} (X_{ij} - \bar{X}_{.j})^2}{n - 3} \tag{11-1}$$

[2] The concept of degrees of freedom has been used earlier in the book, and in this chapter we make estimates of the population variance by dividing the sum of the squared deviations by the degrees of freedom rather than by n. Therefore, a short explanation of this is necessary at this point.

In general the degrees of freedom are $n - 1$. If we deal with population data, we do not have to make any assumptions about the various quantities we are trying to compute, and we are not concerned with degrees of freedom. But when we estimate an unknown, like σ^2, with a known quantity like s^2, we lose one degree of freedom, because when we accept s^2 as our estimate, any other computations we make are determined by this quantity. When we are dealing with an analysis of variance table like Table 11-1, we no longer make an estimate of σ^2 based on individual observations but on the variance within each of the columns and between each of the means. In the case of the within-column estimate, we lose as many degrees of freedom as there are columns. Hence, the within-column-estimate degrees of freedom are $n - k$, where k is the number of columns. In the case of the between-columns estimate, the general rule applies except that the individual column means are treated as individual observations, and the degrees of freedom in this case are therefore $k - 1$. The total degrees of freedom are $n - 1$. Given these definitions, we can see that the following equality holds: $n - 1 = (k - 1) + (n - k)$, or total degrees of freedom equals the between- plus the within-column degrees of freedom. For example, in Table 11-1 we have $15 - 1 = (3 - 1) + (15 - 3)$. The significance of this additive relationship is considered later in this chapter. For the moment it is important to note that degrees of freedom are used to compute the estimates of variance rather than n, because estimates computed in this manner are unbiased.

The notation being used in this chapter is standard. A dot in place of a subscript is used to distinguish row and column means. Thus, $\bar{X}_{.j}$ indicates the mean of the jth column, $\bar{X}_{.1}$ that of the first column; $\bar{X}_{i.}$ indicates the mean of the ith row, $\bar{X}_{1.}$ that of the first row; and so on. $\bar{\bar{X}}$ indicates the grand mean, or the mean of all the observations in the table regardless of row or column.

Because we are dividing by the degrees of freedom, this within-column estimate is an unbiased estimate of σ^2.

A second way of estimating the population variance σ^2 is by using the column means. Recall that the standard error of the mean is

$$\sigma_{\bar{X}} = \frac{\sigma}{\sqrt{n}}$$

then

$$\sigma^2 = n\sigma^2_{\bar{X}}$$

Therefore, we can estimate σ^2 by estimating $\sigma^2_{\bar{X}}$. For each of the samples we have

$$\hat{\sigma}^2 = n_1\sigma_{\bar{X}_1}{}^2 = n_1(\bar{X}_{.1} - \bar{\bar{X}})^2$$

$$\hat{\sigma}^2 = n_2\sigma_{\bar{X}_2}{}^2 = n_2(\bar{X}_{.2} - \bar{\bar{X}})^2$$

$$\hat{\sigma}^2 = n_3\sigma_{\bar{X}_3}{}^2 = n_3(\bar{X}_{.3} - \bar{\bar{X}})^2$$

$$3\hat{\sigma}^2 = \sum_{j=1}^{3} n_i(\bar{X}_{.j} - \bar{\bar{X}})^2$$

$$\hat{\sigma}^2 = \frac{\sum_{j=1}^{3} n_i(\bar{X}_{.j} - \bar{\bar{X}})^2}{k - 1} \tag{11-2}$$

In less mathematical terms, we make two separate estimates of σ^2. One is a pooled estimate based on the variance within each of the three columns. The other is a weighted estimate based on the variance among the means of the three columns. The first is an unbiased estimate of σ^2. The second will also be an unbiased estimate of σ^2 provided that the null hypothesis is true. If the null hypothesis is not true, the second will tend to be larger than the first.

If the null hypothesis is true, we should expect that $\bar{X}_{.1} = \bar{X}_{.2} = \bar{X}_{.3}$. But, of course, there will be some variation in sample means due to sampling error. Hence, the between-columns estimate becomes larger by chance, and if the amount of variation among the three means is too large to attribute to sampling fluctuation, we reject the null hypothesis. The within-column estimate is not affected by the sampling fluctuation of $\bar{X}_{ij}$. As the between-columns estimate becomes larger than can be attributed to sampling error, so does the ratio V_b/V_w, where V_b is between-columns variance and V_w within-column variance.

How large must this ratio become for us to reject the null hypothesis? In order to answer this question, it is necessary to set up a sampling distribution of all the possible differences in estimates we may get if we draw samples of size n_1, n_2, and n_3 respectively from the same populations an infinite number of times. The ratio of these differences forms a known distribution, called the F distribution. We can use this distribution as an approximation of the sampling distribution we should get if we continually sampled from the same population and each time computed the two estimates of variance.

We shall consider this distribution in a moment. First, let us consider the ratio V_b/V_w for the data in Table 11-1 so that the reader does not get lost in the large dose of symbols characteristic of the analysis of variance.

Recall that V_b/V_w is

$$F = \frac{\left[\sum_{i=1}^{3} n_i(\bar{X}_{.j} - \bar{\bar{X}})^2\right]/(3 - 1)}{\left[\sum_{i=1}^{n_i} \sum_{j=1}^{3} (X_{ij} - \bar{X}_{.j})^2\right]/(n - 3)} \qquad (11\text{-}3)$$

The numerator of Equation (11-3) is the sum of the weighted, squared differences between the column means and grand mean, or

$$\frac{5(5 - 6)^2 + 5(6 - 6)^2 + 5(7 - 6)^2}{3 - 1} = 5$$

The denominator is simply the sum of the squared deviations of the individual observations from the mean of their column, or

$$(3 - 5)^2 + (6 - 5)^2 + (5 - 5)^2 + (4 - 5)^2 + (7 - 5)^2$$
$$+ (4 - 6)^2 + (7 - 6)^2 + (7 - 6)^2 + (4 - 6)^2 + (8 - 6)^2$$
$$+ (6 - 7)^2 + (7 - 7)^2 + (8 - 7)^2 + (7 - 7)^2 + (7 - 7)^2$$
$$= 26/16 = 2.17$$

and the F ratio[3] in this case is $F = 5/2.17 = 2.3$. How large must this ratio be for us to reject H_0. When the two estimates σ^2 are the same, the ratio is 1. As the differences in means increase, the between-column estimate increases, and F becomes larger than 1. Let us now consider this distribution in general.

THE F RATIO

It was noted in Chapter 3 that the standard deviation of a sample is not an unbiased estimate of the population parameter σ. The standard deviation of the sample consistently underestimates σ. However, the central limit theorem states that, if random samples ($n > 30$) are taken from a normally distributed universe, the standard deviation of the samples will be distributed normally, and the standard deviation of the sampling distribution will be

$$\sigma_s = \frac{\sigma}{\sqrt{2n}} \qquad (11\text{-}4)$$

[3] We do not ordinarily compute the F ratio in this manner. We shall consider in a moment a more detailed example and a short method for doing so.

For large samples drawn from a normal universe, we can estimate σ with s, getting

$$\hat{\sigma}_s = \frac{s}{\sqrt{2(n-1)}} \tag{11-5}$$

Therefore, a 95 percent confidence interval for σ can be obtained by

$$\sigma = s \pm 1.96 \frac{s}{\sqrt{2(n-1)}}$$

For example, suppose that a sample of 400 has been drawn from the universe of all voters in the United States and the problem is to estimate the standard deviation of education for this universe. Suppose also that the sample standard deviation is 4. The 95 percent confidence interval for σ is

$$\sigma = 4 \pm 1.96 \frac{4}{\sqrt{2(399)}}$$

$$= 3.72 - 4.28$$

If the universe we are sampling from is not normal or if the sample is small, the sampling distribution of standard deviations will not be normally distributed. In this case, the sample variance, multiplied by the number of degrees of freedom, will be distributed as a χ^2 (read "chi square") variable when expressed in the form of a ratio to the variance:

$$\chi^2 = \frac{(n-1)s^2}{2}$$

It is therefore possible to draw a confidence interval around s and s^2 by converting to χ^2 values. For example, assume a sample size of 25 and a sample standard deviation of 10. To draw a confidence interval around s using the χ^2 distribution, we find the χ^2 values corresponding to the width of the interval. For a 90 percent confidence interval we find the χ^2 values for $p = 0.05$ and $p = 0.95$ for 24 degrees of freedom. These are 13.85 and 36.42 respectively. Thus, the 90 percent confidence interval in this case is

$$36.42 = \frac{(25-1)10^2}{\sigma^2}$$

$$\sigma^2 = \frac{2,400}{36.42} = 65.9 \qquad \sigma = 8.118$$

$$13.85 = \frac{(25-1)10^2}{2}$$

$$\sigma^2 = \frac{2,400}{13.85} = 173.3 \qquad \sigma \neq 13.164$$

The respective 90 percent confidence intervals run

$$\sigma^2 : 65.9 \text{ to } 173.3$$

$$\sigma : 8.118 \text{ to } 13.164$$

We can now recall that the F ratio is the ratio of two estimates of variance: V_b/V_w. Because the sample variance can be converted to a χ^2 value by

$$\chi^2 = \frac{(n-1)s^2}{2}$$

then

$$s^2 = \frac{\chi^2 \sigma^2}{n-1}$$

and

$$F = \frac{s_1^2}{s_2^2} = \frac{\chi_1^2 \sigma_1^2 / (n_1 - 1)}{\chi_2^2 \sigma_2^2 / (n^2 - 1)}$$

If the null hypothesis is true, then $s_1^2 = s_2^2 = \sigma_1^2 = \sigma_2^2$, and the F ratio becomes

$$F = \frac{\chi_1^2 / (n_1 - 1)}{\chi_2^2 / (n_2 - 1)}$$

This equation says that the F distribution is a ratio of two χ^2 variables. An examination of the equation shows that the shape of this distribution depends on the degrees of freedom and thus upon sample size. In general, it is skewed positively, but it approaches the normal distribution for very large values of n (see Figure 11-1).

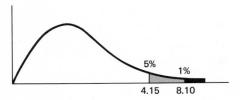

Figure 11-1. The F distribution, where L and e are constants, and $V =$ degrees of freedom.

There is no need to consider the mathematical equation that defines the F distribution. The values of the F distribution have been tabled, and we can simply use the existing tables. Some of these are included in Table 4 in the Appendix. Let us consider how to use this table.

Suppose that $F = (u/\phi_1)/(v/\phi_2)$, where $u = \sum_{j=1}^{n_i} n_i(\bar{X}_{.j} = \bar{\bar{X}})^2$, $v = \sum_{i=1}^{3} \sum_{j=1}^{n_i} (X_{ij} - \bar{X}_{.j})^2$, and ϕ_i are the respective degrees of freedom. Suppose that the degrees of freedom in this case are 8 and 6 respectively. The distribution of F for these values is shown in Figure 11-1. The F table shows that for 8 and 6 degrees of freedom the area of the curve beyond $F = 4.15$ is 5 percent of the total area and that the area beyond $F = 8.10$ is 1 percent of the total area. In probability terms, this says that if the F ratio is larger than 4.15 for 8 and 6 degrees of freedom, we may safely reject the null hypothesis that $\mu_1 = \mu_2 = \mu_3$ with less than 5 percent chance of making error type 1.

Note that the smallest value of F in the table is 1. The F ratio is equal to 1 when the between and within degrees of freedom approach ∞. When the F ratio is calculated, the greater variance is put in the numerator, which is the between-columns variance in the analysis of variance test. It is possible for the F value to be less than 1 and even 0. But this should not occur and may signal something amiss in the design of the experiment or in the computations.

TYPES OF ANALYSIS OF VARIANCE

There are many potential applications for the analysis of variance. Here we consider only two simple situations. The first is where there are one nominal-scale variable and one quantitative dependent variable. This is called *one-way analysis of variance*. The second is where there are two (or more) independent variables and a single dependent variable. This is called *two-way analysis of variance*.

For each of these experimental situations, two different sampling situations are possible. The first is usually called the *fixed-effects model* (or model 1), and the second the *random-effects model* (or model 2). Let us use the Guetzkow-Simon study in order to clarify the difference between these two.

Recall that in the experiment only two variables were involved: communications structure (the independent variable) and performance (the dependent variable). Hence, one-way analysis of variance is appropriate. The independent variable has three different stages. A random sample of individuals was selected, and the assignment to each of the groups was also random. The three groups were compared on the dependent variable, which is the mean time it took each group to complete given tasks.

An experiment performed in this manner is called a fixed-effects model. Inferences made in this model can be made only about the particular treatments (in this case, communications nets) administered, and not about any other kinds of treatments that might have been administered. If a significant difference among the three means does exist in this case, we can infer that it also exists in the population for these three communications nets; we cannot infer that the differences apply to other nets. The population is all individuals who might be given similar tasks under the same three communications structures: all channel, wheel, and circle.

Now think of these three communications structures as three points along a scale of communications structures, ranging from a completely open structure to a completely closed one. Suppose that we label each point along the continuum and that we draw a random sample of three communications nets from this large group of all possible nets. In this case, we are able to make inferences not only about the population for the three that have been drawn but also about all possible communications nets. This is called the random-effects model.

The random-effects model applies where an inference is desired to a large set of possible treatments. The treatments applied in an experiment are a random sample taken from all possible treatments. In this case, two populations are involved: the population of all possible communications nets and the population(s) from which each of the particular groups being tested has been drawn. Note that both would be analyzed by one-way analysis of variance. There are differences in the *F* ratios for the fixed-effects and random-effects models, and these are noted below. At this point we want to emphasize that the notion of one-way and two-way analysis of variance is a more general classification and that we might have a fixed-effects or random-effects model for each.

The fixed-effects model is the one used most extensively in research today. However, computationally, both the fixed-effects and the random-effects models are identical in most instances. In the discussion that follows, we describe the fixed-effects model. Where the random-effects model requires a different treatment, or different assumptions, these are noted. A general discussion of some of the differences is presented toward the end of the chapter.

Before proceeding to a discussion of one-way analysis of variance, let us note one more important classification in research design: the distinction between a completely randomized design and a randomized block design. If a study has one or more treatments and if every other factor influencing the dependent variable is randomized, the design is called a *fully randomized design*. In this design, a random sample of subjects is drawn, and they are allocated among treatments in a completely random manner. If some other factor is being controlled in addition to the treatment, such as age or sex, this will also be completely randomized.

However, if we want to introduce controls for the factors that may influence the dependent variable other than the treatments, individuals paired for each level of the factor to be controlled are selected within each of the control groups for each treatment to be tested. This is called a *randomized block design*.

For example, consider the following situation. Suppose that we want to determine if the role perceptions of city councilmen are related to their power aspirations, but we know that both age and occupation of councilmen are strongly related to the dependent variable, power aspirations. We want to control for these factors, but we cannot consider all occupation and age-group combinations. Hence, we set up an exhaustive classification of occupation and age combinations. Two subjects that are alike in each age-occupation group are selected, and they are assigned at random to the treatment, that is, differences in role perceptions. Each treatment gets one subject. Suppose that there are only two role perceptions involved: delegate and trustee. Then, within each age-occupation group, there would be two councilmen, one delegate and one trustee. The difference in means of the various groups would then be analyzed by two-way analysis of variance. This is a randomized block design.

The problem of research design is too complicated to discuss in detail here. The reader should consult the references given at the end of this chapter for a more extended discussion of this question if he intends to use more complicated analysis of variance techniques. In the sections that follow we first discuss one-way analysis of variance for the fixed-effects made. The computations for the random-effects model are the same; the inferences are different. Following this, two-way analysis of variance is considered. The random-effects computations are exactly the same, but the F ratios are different; these differences are noted.

ONE-WAY ANALYSIS OF VARIANCE (FIXED-EFFECTS MODEL)

The example used to describe the procedure for one-way analysis of variance is the Guetzkow-Simon experiment discussed before. We use hypothetical data (see Table 11-2) to make the computations easier to follow.

Table 11-2 Task Trial Times for Different Communications Nets (Minutes)*

All channel	*Wheel*	*Circle*
20	10	45
30	15	25
35	30	35
20	25	15
20	20	30
125	100	150
$\bar{X}_{ij} =$ 25	20	30

$$\bar{\bar{X}} = 25$$

*Note that, although the same size of sample is in each group, this does not have to be so for one-way analysis.

We suppose that the 15 members are a random sample and that they have been allocated randomly among the three groups. We are testing the null hypothesis that $\mu_1 = \mu_2 = \mu_3$. To do so, we make two estimates of σ^2. The first is the within-column estimate, which we can obtain by means of Equation (11-1):

$$\sigma^2 = \frac{1}{n-k} \sum_{i=1}^{n_i} \sum_{j=1}^{3} (X_{ij} - \bar{X}_{.j})^2.$$ (11-1)

$= 1/12\,[(20-25)^2 + (30-25)^2 + (35-25)^2 + (20-25)^2 + 20-25)^2]$

$+ [(10-20)^2 + (15-20)^2 + (30-20)^2 + (25-20)^2 + (20-20)^2]$

$+ [(45-30)^2 + (25-30)^2 + (35-30)^2 + (15-30)^2 + (30-30)^2]$

$= 950/12 = 79.17$

The second estimate is the between-columns estimate, and it can be obtained by Equation (11-2):

$$\hat{\sigma}^2 = \frac{1}{k-1} \sum_{j=1}^{3} n_j(\bar{X}_{.j} - \bar{\bar{X}})^2$$ (11-2)

$= 1/2[5(25-25)^2 + 5(20-25)^2 + 5(30-25)^2]$

$= 250/2 = 125$

Therefore, the F ratio for these data is

$$F_{2,12} = 125/79.17 = 1.58$$

The degrees of freedom have been entered as a subscript of F here. The degrees of freedom for the F ratio are $(k-1)$ for the between-columns estimate and $(n-k)$ for the within-column estimate. The table of F values shows that an F of 3.88 or larger is needed for significance at the $\alpha = 0.05$ level, and we are therefore not able to reject the null hypothesis in this case.[4] We conclude that there does not seem to be any relationship between communications structure and performance if these were actual data.

Let us consider the mathematical model of analysis of variance before considering a short method of computing the various quantities needed for the F ratio. If the null hypothesis is true, the overall population mean is

$$\mu = \frac{1}{k} \sum_{i}^{k} \mu_i = 1/3 \sum_{i}^{3} \mu_i$$

Deviations from this overall mean may be defined as

$$\alpha_i = \mu_i - \mu \qquad i = 1, 2, 3$$ (11-6)

---

[4] Because the data in Table 11-2 were invented for this example, this is not the same conclusion as reached by Guetzkow and Simon.

The deviations α_i, if any exist, are the effects due to the treatments. But if $\mu_1 = \mu_2 = \mu_3$, then

$$\sum_i^k \alpha_i = \sum_i^k (\mu_i - \mu) = 0$$

Consider now the population of individuals that might be exposed to the all-channel communications net. The scores of these individuals are assumed to be normally distributed with a mean μ_i and a variance σ_i^2. The individual scores X_{ij} will be randomly distributed around this mean. We can define these deviations as

$$\in_{ij} = (X_{ij} - \mu_i) \qquad (11\text{-}7)$$

And we may consider them to be random errors.

Given these assumptions, the following is true:

$$V(\in_{ij}) = V(\bar{X}_{ij} - \mu_i)$$
$$= V(X_{ij})$$
$$= \sigma^2,$$

where V = variance.

We may also expect any value X_{ij} selected at random from this population to equal the mean μ of the population it is drawn from. (This is the maximum-likelihood expectation.) The sum of the deviations of the random error $\in_{ij}$ is expected to equal zero. Given the fact that we are sampling at random from this population, we therefore expect the *j*th score of the *i*th person to be

$$X_{ij} = \mu + \in_{ij}$$

And if there is any treatment effect, we expect this score to be

$$X_{ij} = \mu + \alpha_{ij} + \in_{ij} \qquad (11\text{-}8)$$

The latter is the basic linear model of the analysis of variance. In terms of the samples, the model says that we may expect each observation X_{ij} drawn into the sample to deviate from the grand mean by a given amount $(X_{ij} - \bar{\bar{X}})$. This deviation is composed of two parts: the amount of deviation of the individual score from the mean of its column $(X_{ij} - \bar{X}_{.j})$ plus the deviation of the column mean from the grand mean $(\bar{X}_{.j} - \bar{\bar{X}})$. The amount of deviation of the score from the mean of its column was defined as random error in Equation (11-7), and for samples, it may be written e_{ij}. Note that this is the within-column variation. The amount of deviation of the column mean from the grand mean was defined as treatment effect in Equation (11-6), and for samples it may be written a_{ij}. This is sometimes called the *explained deviation*. Hence, the model for the samples may be written

$$X_{ij} = \bar{\bar{X}} + a_{ij} + e_{ij}$$

In terms of deviations, the total amount of deviation equals the deviation within each group plus the deviation between the groups:

$$(X_{ij} - \bar{\bar{X}}) = (X_{ij} - \bar{X}_{.j}) + (X_{.j} - \bar{\bar{X}})$$

If we square this and sum for all deviations across all individuals i, we can derive the following sum of squares:

$$\sum_i \sum_j (X_{ij} - \bar{\bar{X}})^2 = \sum_i \sum_j (X_{ij} - \bar{X}_{.j})^2 + \sum_i \sum_j (\bar{X}_{.j} - \bar{\bar{X}})^2 \quad (11\text{-}9)$$

Total sum of	Within-column sum	Between-column sum
squares	of squares	of squares
	(error)	(explained)

This equality is usually called the *partition of the sum of squares*.

For the random-effects model the basic mathematical model is the same:

$$X_{ij} = \mu + \alpha_{ij} + \in_{ij}$$

except that α_{ij} is a random variable rather than a fixed variable. In the random-effects model we thus deal with two populations: the distribution of treatments from which X_{ij} was selected and the distribution of observations within each group. The partition of the sum of squares is carried out in the same way as in the fixed-effects model, and the estimates of the between and within variances are identical.

Let us now consider a short method of computing the variance components in Equation (11-9). Recall that variance, by definition, is the mean squared deviation of a series of scores from their mean. Therefore, if we divide each of the sums of squared deviations in Equation (11-9) by their respective degrees of freedom, we can obtain the respective estimates of variance (sometimes called mean squares and designated *MS*). However, we can easily derive a short way of computing the sums, just as for computing the variance of individual scores. To make it easier to follow, we drop the double summation notation and subscripts:

$$\sum (X - \bar{X})^2 = \sum (X^2 - 2\bar{X}X + \bar{X}^2)$$

$$= \sum X^2 - \sum 2\bar{X}X + n\bar{X}^2$$

$$= \sum X^2 - 2\bar{X} \sum X + n\left(\frac{\sum X}{n}\right)^2$$

$$= \sum X^2 - 2\frac{(\sum X)^2}{n} + \frac{(\sum X)^2}{n}$$

$$= \sum X^2 - \frac{(\sum X)^2}{n}$$

When we reintroduce double summation notation, we have

$$\sum_i \sum_j X_{ij}^2 = \frac{\left(\sum_i \sum_j X_{ij}\right)^2}{n} \quad (11\text{-}10)$$

This is the equation for obtaining the total sum of squares, which is computed by taking the sum of the squares of each score and subtracting from this the sum of the individual scores squared and divided by n.

For example, consider Table 11-2 again.

$$\sum_i \sum_j X_{ij}{}^2 = 10{,}575 \qquad \text{the sum of the square of each score}$$

$$\left(\sum_i \sum_j X_{ij}\right)^2 = 140{,}625 \qquad \text{the sum of the individual scores squared}$$

$$\frac{\left(\sum_i \sum_j X_{ij}\right)^2}{n} = 9{,}375 \qquad \begin{array}{l}\text{the sum of the individual scores squared} \\ \text{and divided by } n\end{array}$$

The total sum of squares is therefore $10{,}575 - 9{,}375 = 1{,}200$. Because the total sum of squares is equal to the within-column sum of squares plus the between-columns sum of squares, we need to compute only one of the latter two in order to obtain all three quantities. Consider the between-columns sum of squares, which is $\sum_i \sum_j (\bar{X}_{.j} - \bar{\bar{X}})^2$. The following short computing formula can be derived:

$$\frac{\left(\sum_j X_{ij}\right)^2}{n_j} - \frac{\left(\sum_i \sum_j X_{ij}\right)^2}{n} \qquad (11\text{-}11)$$

Note that the right term in (11-11) is identical to the right term in Equation (11-10). We therefore need only one additional quantity, the left term in (11-11), which is the sum of the column totals squared and divided by the respective column n_j. For the data in Table 11-2, we have

$$\frac{(125)^2}{5} + \frac{(100)^2}{5} + \frac{(150)^2}{5} = 9{,}625$$

And to obtain the between-columns sum of squares by (11-11), we have

$$\text{Between } SS = 9{,}625 - 9{,}375 = 250$$

The within-column sum of squares can be obtained by simple subtraction:

$$\text{Total } SS - \text{between } SS = \text{within } SS$$
$$1{,}200 \quad - \quad\;\; 250 \quad\;\; = \quad\;\; 950$$

As we said before, variance is defined as the mean squared deviation of a series of scores from their mean. We have just found the respective squared deviations. To obtain the estimates of variance, we divide these by their respective degrees of freedom.

MS between: $\qquad V_b = \dfrac{\text{between } SS}{k-1} = \dfrac{250}{2} = 125$

MS within: $\qquad V_w = \dfrac{\text{within } SS}{n-k} = \dfrac{950}{12} = 79.17$

Finally the F ratio, using the short method is

$$F_{2,12} = \frac{125}{79.17} = 1.58$$

which is what we obtained with the longer method.

Now it should be apparent that, the larger the variance within each category as compared with that among the categories, the smaller the F ratio. If we think of the variance within each group as unexplained or error variance and that among groups as explained variance, then as the ratio of explained variance to error variance increases, the less likely is it that we have chance difference between means. Figure 11-2 shows the two situations where the ratio of the two estimates of variance are reversed.

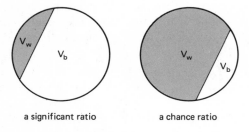

a significant ratio   a chance ratio

Figure 11-2.   Two hypothetical F ratios.

Consider for a moment why the within-column variance is called error variance. If the variance within a category is large, it is much more likely that samples drawn from such a population will produce large differences of means. For example, suppose that we had two different populations as follows:

$$P_1 = (10,4,100,40,500,10,1,40,60,50,40,30,90,20,120,2)$$
$$P_2 = (40,60,50,45,60,50,70,40,30,65,76,60,50,55,40,65)$$

P_1 has a much larger variance than P_2. Hence, two successive samples of size 5 drawn at random from the first population (each time with replacement) will yield means that will be more different than in the second population. As an experiment, two random samples of size $n = 5$ drawn from the first population produced the following two means: $\bar{X}_{P_1} = 64$; $\bar{X}_{P_1} = 32.8$. Two random samples of size $n = 5$ drawn from the second population produced the following two means: $\bar{X}_{P_2} = 58$; $\bar{X}_{P_2} = 58$. The means in the two samples drawn from the second population were the same, but the means of the two samples drawn from the first population were quite different. Note that, although first two samples were drawn from a population with a mean

$\mu = 69.81$, each sample mean was quite different. As a matter of fact, a difference of means test for the two sample means is

$$t = \frac{64 - 32.8}{\sqrt{[5(2064) + 5(367.32)]/(5 + 5 - 2)} \sqrt{1/5 + 1/5}} = 1.27$$

which is not significant at the 0.05 level. The lesson is clear; what appears to be a large difference of means is not, because of the large variation in the scores of this population. As a final note on this hypothetical example, we observe that the variance of the first sample drawn from P_1 was 2,064 and for the second sample drawn from P_1 was 367.32. The true population variance for the first population is 13,558.77. Thus, each random sample greatly underestimated the variance of the population.

Strength of Relationship

A statistically significant relationship is not necessarily strong. In fact, a very weak relationship may be statistically significant. It is worth emphasizing that statistical significance is simply another way of saying that what was found in a set of samples is also likely to be true of the population.

The strength of a relationship between variables can be measured in one-way analysis of variance by means of an index called the *coefficient of determination*. It may be symbolized by the Greek omega ω^2 in this case. We want to measure the variance in the dependent variable Y_i accounted for by the independent variable X_i. In the Guetzkow-Simon example, this would mean the variation in performance accounted for by communications nets.

Let $\sigma_Y{}^2$ represent the variance in Y_i and $\sigma^2{}_{Y|X}$ the conditional distribution of Y_i given a fixed X_i category. If the variance in Y_i is reduced by knowing X_i, we are better able to predict Y_i values knowing X_i values than if we do not know the associated values of X_i. Hence, we speak of X_i explaining or accounting for variance in Y_i.

The absolute amount of reduction in Y_i, knowing X_i, is $\sigma_Y{}^2 - \sigma^2{}_{Y|X}$. And the relative amount of reduction is

$$\hat{\omega}^2 = \frac{\sigma_Y{}^2 - \sigma^2{}_{Y|X}}{\sigma_Y{}^2} \qquad (11\text{-}12)$$

This is the population index. However, we can estimate ω^2 from sample data by

$$\omega^2 = \frac{\text{between } SS - (k - 1)V_w}{\text{total } SS + V_\omega} \qquad (11\text{-}13)$$

If the resulting quotient is negative, it is set at zero. The upper limit is one. Hence, decimals in between can be interpreted as increasing strength. For the hypothetical data in Table 11-2 we have

$$\omega^2 = \frac{250 - (3 - 1)79.16}{1,200 - 79.16} = .08$$

If these were actual data, we should be able to conclude that the independent variable, communications structure, explains about 8 percent of the variation in the dependent variable (performance), leaving 92 percent to be explained by other factors. Hence, we conclude that the relationship is not very strong.

We shall have more to say about measures of strength of relationship after considering two-way analysis of variance.

TWO-WAY ANALYSIS OF VARIANCE (FIXED-EFFECTS MODEL)

In most research we are interested in the relationship among more than two variables. Two-way analysis of variance enables us to measure the effect of more than one independent variable on the dependent variable and also whether there is an interaction effect.

Two-way analysis of variance can become very complex. We consider here only the orthogonal design; that is, the observations in each combination are sampled at random and independently from a normal population, and the number of observations in each combination (cell) is the same. If unequal numbers are assigned to the cells, the computational procedure for two-way analysis of variance changes.

The mathematical model for two-way analysis states that any particular observation X_{ijk} deviates from the grand mean, and this deviation is composed of a column variance, row variance, interaction variance, and error variance; that is,

Total SS = column SS + row SS + interaction SS + error SS

To illustrate, we use data concerning a relationship between party competition, type of city, and welfare effort.[5] Party competition is assumed to be measured on a scale that classifies communities as Democrat if the percentage of registered voters who are Democrats is greater than 60 percent, Republican if the percentage of registered voters who are Democrats is less than 40 percent, and competitive if the percentage of registered voters who are Democrats is between 40 and 60 percent. Type of city is classified as manufacturing and nonmanufacturing, where a city is classified as manufacturing if the per-

---

[5] These data are hypothetical, although the idea in the example is patterned after a study by John Fenton, *People and Parties in Politics*, Scott, Foresman and Company, Glenview, Ill., 1966, especially pp. 34–36.

centage of its total labor force in manufacturing occupations is more than one standard deviation higher than the percentage of the national labor force engaged in manufacturing occupations. Welfare effort is assumed to be measured by total expenditures per capita for social-welfare payments. This is shown in Table 11-3.

Table 11-3   Analysis of Variance for Three-Variable Relationship

Type of city	Political Composition			
	Republican	*Democrat*	*Competitive*	*Total*
Nonmanufacturing	2, 4, 3 $\Sigma X = 9$ $\bar{X} = 3$	10, 8, 15 $\Sigma X = 33$ $\bar{X} = 11$	20, 40, 30 $\Sigma X = 90$ $\bar{X} = 30$	$\Sigma\Sigma X_{1j} = 132$ $\bar{X}_{1j} = 14.6$
Manufacturing	4, 8, 6 $\Sigma X = 18$ $\bar{X} = 6$	20, 15, 10 $\Sigma X = 45$ $\bar{X} = 15$	40, 50, 30 $\Sigma X = 120$ $\bar{X} = 40$	$\Sigma\Sigma X_{2j} = 183$ $\bar{X}_{2j} = 20.3$
Total	$\Sigma\Sigma X_{i1} = 27$ $\bar{X}_{i1} = 4.5$	$\Sigma\Sigma X_{i2} = 78$ $\bar{X}_{i2} = 13$	$\Sigma\Sigma X_{i3} = 210$ $\bar{X}_{i3} = 35$	$\Sigma\Sigma\Sigma X_{ijk} = 315$ $\bar{\bar{X}} = 17.5$

The numbers in the cells represent hypothetical per capital expenditures for three cities in each cell, the sum of these, and the mean for each cell. (The trends indicated in these data are not to be considered empirical trends. The data were set up to produce an increase in welfare expenditures from Republican to competitive, and a larger expenditure in manufacturing cities.) There are three different null hypotheses to be tested in this case. We want to know if there is a relationship between the party composition of a city and its welfare effort, and the null hypothesis in this case is that there is no significant difference among the means of the columns, symbolized as H_0: $\mu_{C_1} = \mu_{C_2} = \mu_{C_3}$. We also want to know if there is a relationship between city type and welfare effort. The null hypothesis in this case is that there is no significant difference between the rows, symbolized as H_0: $\mu_{r_1} = \mu_{r_2}$. And, finally, we want to know if the combined effect of party composition and type of city produces a multiplicative rather than an additive effect, that is, whether there is interaction.

These three null hypotheses are tested by taking the ratio of the estimates of variance for each one to the error variance. The error variance is the within-subclass variance. The within-subclass estimate is an unbiased estimate of the population variance. The other estimates are unbiased if the null hypotheses are true. Thus, the respective tests are:

1. For the relationship between party composition and welfare effort:

$F = V_b/V_e$, where V_b = the between-columns estimate and V_e = the within-subclass estimate, or error variance.

2. For the relationship between city type and welfare effort: $F = V_r/V_e$, where V_r is the between-rows estimate of variance and V_e is as before.

3. For whether there is an interaction effect: $F = V_i/V_e$, where V_i is the interaction estimate and V_e is as before.

Each of the estimates may be computed by first obtaining the sum of squares and then dividing by the respective degrees of freedom, as in one-way analysis. To do so, we compute the total SS, the between-columns SS, the between-rows SS, and the between-subclass SS. The other two quantities can be derived from these.

The computing equations are essentially the same as one-way analysis of variance except that, because we now have rows and columns we need triple summation notation. This can be illustrated best by way of the example using the data in Table 11-3:

$$\sum_j \sum_k \sum_i X_{ijk}^2 = 9,159.0 \qquad \text{the sum of the square of each observation in the cells}$$

$$\left(\sum_j \sum_k \sum_i X_{ijk}\right)^2 = 99,225 \qquad \text{the sum of all observations squared}$$

$$\frac{\left(\sum_j \sum_k \sum_i X_{ijk}\right)^2}{n} = 5,512.5$$

The total sum of squares, therefore, is $9,159.0 - 5,512.5 = 3,646.5$.

The between-columns sum of squares is computed in the same manner as for one-way analysis of variance:

$$\frac{27^2}{6} + \frac{78^2}{6} + \frac{210^2}{6} - 5,512.5 = 2,973$$

The between-rows sum of squares is

$$\frac{132^2}{9} + \frac{183^2}{9} - 5,512.5 = 144.5$$

The between-subclass sum of squares can be obtained in the following manner:

$$\frac{9^2}{3} + \frac{33^2}{3} + \frac{90^2}{3} + \frac{18^2}{3} + \frac{45^2}{3} + \frac{120^2}{3} - 5,512.5 = 3,160.5$$

The error sum of squares is the difference between the total sum of squares and the between-subclass sum of squares: $3,646.5 - 3,160.5 = 486$. The interaction sum of squares is the between-subclass sum of squares less the sum of the between-columns and between-rows sums of squares: $3,160.5 - (2,973 + 144.5) = 43$. This is all summarized in Table 11-4.

Table 11-4 Two-Way Analysis of Variance with Interaction Test

	Sum of squares	Degrees of freedom	Estimate of variance (mean square)	F
Total	3,646.5	$N - 1 = 17$		
Between subclass	3,160.5	$kr - 1 = 5$		
Between column	2,973	$k - 1 = 2$	1,486.5	
Between row	144.5	$r - 1 = 1$	144.5	
Interaction	43	$(k-1)(r-1) = 2$	21.5	.53
Error	486	$N - kr = 12$	40.5	

The test for interaction is always made first. It is $F_{2,12} = V_i/V_e = 21.5/40.5 = .53$, which is not significant at the 0.05 level, and therefore we cannot reject the null hypothesis that there is no interaction. Having decided that there is no interaction between political composition and type of city, we put the interaction back into the error term, getting the new Table 11-5.

Table 11-5 Two-Way Analysis of Variance

	Sum of squares	Degrees of freedom	Estimate of variance (mean square)	F	p
Total	3,646.5	17	214.5		
Between column	2,973	2	1,486.5	39.3	<.001
Between row	144.5	1	144.5	3.8	>.05
Error	529	14	37.8		

The table shows that the relationship between political composition and welfare effort is significant beyond the 0.001 level but that the relationship between type of city and welfare effort is not significant at the 0.05 level. We therefore reject the null hypothesis that the means of the columns are the same but accept the null hypothesis that the means of the rows are the same. In general, if these data were real, we should conclude that political composition is related to welfare effort but that type of city is not.

A Note on the Random-effects Model for Two-way Analysis of Variance

For both the one-way analysis and the two-way analysis of variance, the actual arithmetic computations for the sum of squares and estimates of variance are the same for the random-effects models as for the fixed-effects models. The inferences are different. For the fixed-effects model, we are inferring something about the differences in population means based on the samples. For the random-effects model, we are inferring something about the variances of the effects actually sampled by the experimenter. Thus, in the random-effects model, we make two inferences: one about the population of observations and another about the population of treatments.

As we said above in the discussion of one-way analysis of variance, there is no difference in the computation of the sums of squares nor in the F ratio used to make the test in the two models. However, for two-way analysis of variance, the F tests are different (the computation of the components of variance are the same). In the random-effects model, the sum of squares for rows is assumed to have both row and interaction effect, and the sum of squares for columns is assumed to have both column and interaction effect. Therefore, the F ratios to test the null hypotheses change. The hypothesis of no column effect is tested by

$$F = \frac{MS \text{ columns}}{MS \text{ interaction}}$$

where MS represents mean squares or the estimate of variance, with $c - 1$ and $(r - 1)(c - 1)$ degrees of freedom. The no rows effect is tested by

$$F = \frac{MS \text{ between rows}}{MS \text{ interaction}}$$

with $r - 1$ and $(r - 1)(c - 1)$ degrees of freedom. The no interaction effect is tested by

$$F = \frac{MS \text{ interaction}}{MS \text{ error}}$$

with $(r - 1)(c - 1)$ and $rc(n - 1)$ degrees of freedom.

It may also be noted that it is possible to have a design that combines both the fixed-effects and the random-effects assumptions, which is sometimes called a *mixed model*. In this case some of the treatments are assumed to be fixed and some sampled randomly from a universe of possible treatments. The two-way analysis of variance computations is used, and it is exactly as described above. However, the F tests are slightly different again. The no column effects are tested by the same F ratio as for the random-effects model, and the no rows and no interaction tests are the same as for the fixed-effects model. The reasons for these differences, and a further description of their application, can be found in the references cited at the end of this chapter.

A Test of Homoscedasticity

In Chapter 7 we noted that the *t* test to be used in a particular case depends on whether or not $\sigma_1^2 = \sigma_2^2$. We are now in a position to test whether or not this is likely to be so. An *F* test of whether or not $\sigma_1^2 = \sigma_2^2$ should always be made before proceeding with a *t* test.

In using the *F* ratio to test the homoscedasticity assumption, the larger estimate of variance is always put in the numerator. Consider again the example in Chapter 7. In that example, $s_1 = 5.08$ and $s_2 = 3.09$, with $n_1 = 21$ and $n_2 = 15$. We test whether or not $\sigma_1^2 = \sigma_2^2$ by means of the *F* ratio as follows:

$$F_{20,14} = 25.806/9.548 = 2.7$$

The table of the *F* distribution shows that this is significant at the .05 level, and hence we reject the hypothesis that $\sigma_1^2 = \sigma_2^2$. The *t* test that is appropriate in this case is the one where separate estimates of the population variances are made.

Measures of Strength of Relationship for Two-way Analysis of Variance

A statistically significant relationship is not necessarily strong. Hence, sometimes the more important question in research is whether the relationship between the variables is strong and not whether the relationship found in the sample also holds in the population. Which of the questions is more important depends, naturally, on what the research question is. For certain questions, statistical significance, or confidence intervals, may be more important. For other questions, they may be only secondary to the question of how strong the relationship is.

There are several ways we may measure the strength of relationships of variables cast in an analysis of variance table. One logical way to do so is to use the ratio of explained variance to total variance. When the explained variance is a large part of the total variance, we may reasonably conclude that the relationship is fairly strong. Also, it is useful if the measure of the strength we are using has the same upper and lower limits under all conditions, so that we may compare the relative strength of different kinds of data and different variables. The two standard measures used for this purpose are the correlation ratio and the interclass correlation coefficient.

The correlation ratio has two forms. If the sample is large, we may measure it directly by comparing the explained and total sum of squares:

$$E^2 = \frac{\text{explained } SS}{\text{total } SS}$$

It is obvious that, if there is no within-column variance in the data, the total sum of squares will be the same as the between (or explained) sum of squares, and $E^2 = 1$. Conversely, when all the variance is within columns, and there is no variance between columns (the column means are all the same), $E^2 = 0$, and we may say that there is no relationship. Thus, ratios between zero and one might be interpreted as increasing in strength as the ratio approaches one. For the data used in the two-way analysis of variance concerning the relationship between political composition and welfare above, we have

$$E^2 = \frac{2{,}973}{3{,}646.5} = .81 \qquad E = .9$$

We may say that the relationship between the two variables is very strong. 81 percent of the variance in the dependent variable is explained by the independent variable; only 19 percent is unaccounted for by the independent variable, because there is 100 percent to be accounted for in total.

There are two drawbacks to this measure. In the first place, because the between-columns variance must always be smaller than the total variance, this ratio cannot take on negative values, and we have therefore to consider the data in the table to tell whether the relationship is positive or inverse. More important, however, is the fact that the sum of squares consistently underestimates the population sum of squares. It is therefore somewhat better to work directly with an estimate of the population variance by taking the comparable estimates of variance rather than the sum of squares. This measure is called the unbiased correlation ratio, given by

$$CR^2 = 1 - \frac{V_w}{V_t}$$

which is the same as V_b/V_t in the case of simple analysis of variance. Using the same example, we see that in this case the result is similar to the previous measure:

$$CR^2 = 1 - \frac{37.8}{214.5} = .82 \qquad CR = .91$$

However, because the lower limit of this is not zero and it can take on various negative values depending on n and k, it is difficult to interpret and not a very useful measure.

The second measure is the interclass correlation coefficient. It takes advantage of the difference between the explained and unexplained variances:

$$r_i = \frac{V_b - V_w}{V_b + (\bar{n} - 1)V_w}$$

where $\bar{n}$ represents the average number of cases in a class. Using the same example, we have

$$r_i = \frac{1,486.5 - 37.8}{1,486.5 + (2)37.8} = .9$$

The interclass correlation coefficient is equal to one when there is no within-column variance, and the relationship can be said to be perfect in this case. When the variance between columns is equal to the variance within columns, $r_i = 0$. However, this does not designate a case of no relationship. Where all the variance is within columns and there is no between-columns variance, the interclass correlation coefficient takes on negative values, but the lower limit is not stable. When there is no between-columns variance, the interclass correlation coefficient becomes

$$r_i = \frac{-V_w}{(\bar{n} - 1)V_w} = \frac{-1}{(\bar{n} - 1)}$$

and it is obvious that this is -1 when there is an average of two cases in each class. But when there is an average of three cases in each subclass, the lower limit is -0.50; and so on. Thus, it is difficult to interpret the interclass correlation coefficient because it is not independent of the number of cross classifications being used.

READINGS AND SELECTED REFERENCES

Blalock, Hubert, Jr. *Social Statistics*, McGraw-Hill Book Company, New York, 1960, Chap. 16.

Campbell, Donald T., and Julian Stanley. "Experimental and Quasi-Experimental Designs for Research on Teaching," in N. L. Gage (ed), *Handbook of Research on Teaching*, Rand McNally & Co., 1963, Chicago.

Cochran, W. G., and G. M. Cox. *Experimental Designs*, 2nd ed., John Wiley & Sons, Inc., New York, 1957.

Edwards, Allen L. *Expected Values of Discrete Random Variables and Elementary Statistics*, John Wiley & Sons, Inc., New York, 1964.

Federer, W. T., *Experimental Design Theory and Application*, The Macmillan Company, New York, 1955.

Games, Paul, and George Klare. *Elementary Statistics*, McGraw-Hill Book Company, New York, 1967, Chap. XV.

Hays, William L. *Statistics for Psychologists*, Holt, Rinehart and Winston, Inc., New York, 1963, Chaps. 12, 13.

Kerlinger, Fred N. *Foundations of Behavioral Research*, Holt, Rinehart and Winston, Inc., New York, 1965, Chaps. 11–13.

Parl, Boris. *Basic Statistics*, Doubleday & Company, Inc., Garden City, N. Y., 1967, Chap. 22.

12

Time Series Analysis

In Chapter 6 a distinction was made between experimental and nonexperimental research design. It was noted there that the principal difference between the two might be characterized in terms of whether the researcher sets up a controlled situation and physically manipulates the variables he is interested in (experimental design) or whether he observes things "in the world" as they happen and tries to determine the effect of variables by mathematically manipulating values of the variables (by partial correlation and regression, for example) he is interested in (nonexperimental design). It is useful to introduce a further distinction in research design at this point. In either experimental or nonexperimental research design the variable of time can be considered an important independent variable — something whose effects must be controlled or something that is a part of a dynamic as opposed to a static analysis of relationships among variables. If time is considered in the research, the design is called *longitudinal*. If it is not, the design is called *cross-sectional*. Cross-sectional research involves an attempt to understand relationships among variables by observing their values at a particular point in time. Longitudinal research involves an attempt to understand the nature of changes in these variables during the course of time. The statistics discussed so far in this book have focused for the most part on cross-sectional research, or statics. The concern of this chapter is statistics that enable us to say something about changes that occur during the course of time; that is, longitudinal research, or dynamics.

When we observe political behavior during the course of time, we observe a flow of activity. Through some method — content analysis, survey research, record inspection, and so on — we might collect data about this activity at different points of time, that is, hour, day, month. A theory that incorporates time as a variable is concerned with explaining the flow of activity during the course of time. If we are able to do so using quantitative measures, the result may be called a time series.

There are a number of different things we might try to describe about this flow of activity. We might be interested in the rate of increase or decrease of a particular kind of activity; for example, violence, revolution, voting participation, party preferences, and so on. We might be interested in determining if there is a regular fluctuation, or periodicity, in certain kinds of political behavior; for example, changes in party preferences, magnitude of majorities in elections, kinds of legislation considered by state legislatures, differences in themes of presidential acceptance speeches. Of course, whether or not we are interested in these aspects (and more) of time series depends on the theory we are using to study the data. The presence or absence of regular fluctuations in a time series is either (a) not detected in the absence of a theory predicting their existence or (b) obvious when charted on a graph but totally devoid of interest or meaning in the absence of a theory. This point cannot be stressed too much, because almost all analyses in time series have been done by economists, who have a special interest in this form of analysis. This special interest might be considered to stem from theoretical concerns that lead us to expect such things as long-term growth (or decline) and cyclical changes in business activity. As we consider various aspects of time series in this chapter, it will become apparent that some of them do not have much relevance to politics, at least in the present state of the development of political theory.

However, certain methods of fitting trend lines, different components of time series, and the different kinds of curves that may be described are of both immediate and potential relevance to political science. We therefore consider them in this chapter. We consider four well-known components of time series and some methods of fitting lines for them. Then we consider some of the different shapes that trend lines may take. The four components of time series discussed here are:

1. Secular trend
2. Cyclical variation
3. Seasonal variation
4. Irregular variation

Not all time series contain all four components. As we shall see shortly, for political data it is usually only the secular trend that has a theoretical rationale.

SECULAR TREND

The secular trend can be defined as the long-term movement in a series of numbers. Long-term changes in the size of the population — of the country, cities, a city, and so on — per capita income, votes for the Democratic or Republican party, the amount of money spent on campaigns for national, state, or local office — all are examples of data that may have a long-term secular trend.

Generally we cannot speak of a trend unless we have a long enough period of time (at least forty to fifty years). Some of these may be simple, linear increases during the course of time, with some kind of periodic fluctuation around this straight line. Others may be more complex movements. We consider the simple linear situation first.

We can describe a trend in terms of either an equation or a graph. As with any other data, we can plot a graph of trend data and then try to fit a line or curve to them in order to describe the long-run trend. As an example,

Table 12-1   Percentage of Turnout in Election for President, 1920–1960*

Year	Percent	Three-year moving total	Three-year moving average
1920	44.2		
1924	44.3	140.8	46.9
1928	52.3	149.5	49.8
1932	52.9	162.7	54.2
1936	57.5	170.1	56.7
1940	59.7	173.5	57.8
1944	56.3	167.5	55.8
1948	51.5	169.8	56.6
1952	62.0	173.6	57.9
1956	60.1	185.9	62.0
1960	63.8		

*Adapted from: U.S. Bureau of the Census, *Statistical Abstract of the United States: 1962,* 8th annual edition, 1967.

we shall use the percentage of turnout in presidential elections in the years 1920 to 1960 given in Table 12-1 and plotted in Figure 12-1 (ignore, for the moment, the last two columns in Table 12-1).

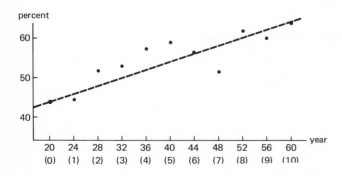

Figure 12-1.   Graph of voter turnout for presidential elections, 1920–1960.

An examination of the points in Figure 12-1 reveals a steady upward movement, although with a slight downswing in the years 1944 and 1948. We might try to fit the trend line to these points by a freehand drawing (dashed line — it is called freehand because it is nonmathematical). Of course, we have already studied (in Chapter 9) more precise ways of fitting a line to a set of coordinates than freehand drawing, that is, the method of least squares. We can fit a line to these data using the same method. However, it is useful to consider some other methods which are simpler and which also reveal the way the different methods influence projections.

We can begin by focusing on the coordinate points on the line drawn freehand for the first and last years in the series. The freehand line has been drawn through the coordinate points of these two years. First we change the scale of the X axis from years to integers, making the year 1920 equal to 0 and increasing in units of 1 for each Election (see the numbers in parentheses in Figure 12-1). The coordinates for the years 1920 and 1960 are (0,44.2) and (10,63.8) respectively.

Recall now that the general equation for a straight line is $Y = a + b(X)$. We consider the percentage of turnout on the Y axis the dependent variable. We then substitute into the general equation for a straight line the coordinate points for the first and last years in the series. For the first, when $X = 0$, $Y = 44.2$; and for the second, when $X = 10$, $Y = 63.8$:

$$44.2 = a + 0(b) \tag{12-1}$$

$$63.8 = a + 10(b) \tag{12-2}$$

As shown in Chapter 9, it is possible to solve for a and b in these equations by subtracting one from the other so as to eliminate one of the unknowns:

$$\begin{aligned} 63.8 &= a + 10(b) \\ -44.2 &= a + 0(b) \\ \hline 19.6 &= 10b \\ b &= 1.96 \end{aligned}$$

Having found b, we can substitute it in Equation (12-2) to find a:

$$63.8 = a + 10(1.96)$$

$$a = 44.2$$

Therefore, the equation for the straight line going through points (0,44.2) and (10,63.8) is

$$Y' = 44.2 + 1.96(X) \tag{12-3}$$

The origin is 1920, and the units are four years. Note that the Y' values computed by this equation are estimated and not actual. For example, when $X = 0$, that is, 1920, the estimated percentage of turnout for the presidential election is $Y' = 44.2 + 1.96(0) = 44.2$. But when $X = 4$, that is, 1936, $Y' = 44.2 + 1.96(4) = 52.04$. Table 12-1 shows that the turnout for the year 1936 is 57.5. The difference between the predicted and actual Y is attributable to the fact that the coordinates do not fit a straight line exactly.

Before turning to another method of fitting a trend line to these data, let us consider what happens if we use the trend equation found by the free-hand method to predict what the percentage of turnout for presidential elections will be in the year 1972.[1] Because we are using integers instead of years and incrementing in units, the X value for the year 1972 is 13. When we substitute this into Equation (12-3), we get $Y' = 44.2 + 1.96(13) = 69.68$; that is, the estimated turnout for the year 1972 is 69.68 percent. But this estimate may be quite wrong, for a number of reasons. First, it is apparent that there is some fluctuation around the trend line that should be taken into consideration. We discuss this later in this chapter. Second, it is possible that we do not have a long enough time period to identify the long-term trend. This is also considered later. Third, the freehand method of fitting the trend line is not necessarily the best. Let us now consider additional methods.

Another method of fixing the trend line is called the method of semi-averages. In this case, we divide the time series into two parts, find the average of each, and then fit a trend line through these averages. The average percentage of turnout for the first part, that is, the first five elections, is $(44.2 + 44.3 + 52.3 + 52.9 + 57.7)/5 = 50.3$. The average percentage of turnout for the second part is $(59.7 + 56.3 + 51.1 + 62.0 + 60.1 + 63.8)/5 = 58.9$. Next, we can plot the average percentage of turnout for the first five elections at their midpoint, which is 1928, and the average percentage of turnout for the last six elections at their midpoint, which, in this case, is 1950, because there is an even number of elections. This has been done in

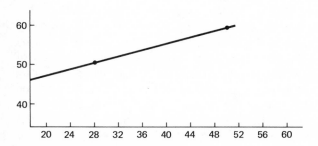

Figure 12-2. Semiaverages trend line.

[1] The question of using time series for forecasting is considered in more detail below. A relatively short period was selected to illustrate forecasting, because it is hazardous at best to predict beyond a few years.

Figure 12-2, and a straight line has been drawn connecting the two points. The coordinate points are (2,50.3) and (7.5,58.9). The equation of the straight line going through these two points can be computed in the same way as in the freehand method:

$$58.9 = a + b(7.5)$$
$$\underline{-50.3 = a + b(2)}$$
$$8.6 = 5.5b$$
$$b = 8.6/5.5$$

and
$$50.3 = a + 8.6/5.5 \ (z)$$
$$a = 47.2$$

The equation of the line fitted by the method of semiaverages is thus

$$Y' = 47.2 + 1.6(X) \tag{12-4}$$

As a means of comparing this method with the last, we compute the estimated turnout for the years 1920 and 1972 with Equation (12-4):

1920: $Y' = 47.2 + 1.6(0) = 47.2$

1972: $Y' = 47.2 + 1.6(13) = 68$

The estimated turnout for the year 1920 is higher than that given by the free-hand method (and also higher than the actual figure). The estimated turn-out for the year 1972 is a little lower than that of the freehand method. Hence, the method of semiaverages has, in this case, tended to tilt the line to the right, moderating the projected or estimated turnout for the year 1972. Of course, this is not due to the different methods of fitting the line, because a freehand line might have been drawn even more horizontal than that in Figure 12-2 and still have seemed to be a good fit to the actual coordinate points. The point that should be emphasized here is that neither method necessarily gives a better estimate for the year 1972. We can easily determine which method gives the better estimate for the actual years by simply computing the estimated turnout for each year by each method of fitting the line, subtracting the estimates from the actual figures, squaring and summing. Which will be more accurate for the actual data depends on how well one fits a freehand line. Which will be more accurate for future years depends on whether the other sources of fluctuation in the time series have been identified and taken into account. We return to these questions later in this chapter. For the moment let us consider additional methods of fitting the trend line.

Another method of fitting the trend line is called the method of moving averages. To compute the trend line by the method of moving averages, we must first obtain moving totals and then moving averages. A moving total is the sum of any number of successive years entered in sequence in the table, that is, two, three, four years, and so on. If we use a three-year[2] total, we sum the value for three successive election years, beginning with the first, and enter the result next to the middle year in the period. If we use a five-year total, we sum the values for five successive years, and so on. We use a three-year moving total here, because there are only eleven elections in all.[3]

The moving total for the first three elections — 1920, 1924, and 1928 — is 44.2 + 44.3 + 52.3 = 140.8. This figure is placed next to the middle year, 1924 (see Table 12-1). The next moving total is for the years 1924, 1928, and 1932, which is 44.3 + 52.3 + 52.9 = 149.5. The other totals have been computed in the same manner. To find the three-year moving averages, we divide each of the moving totals by 3 and enter the results next to the moving totals. In effect, column 4 in Table 12-1 gives us the average turnout for each successive three elections. We can plot the three-year moving averages as has been done in Figure 12-3.

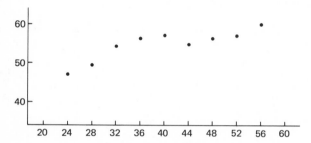

Figure 12-3. Moving averages trend line.

In contrast to the coordinate points in Figure 12-1, those in Figure 12-3 have less wide fluctuation. The long-run trend shown in Figure 12-3 is, graphically, a little more descriptive than the straight line in Figure 12-1. It shows not only the upward trend but also some of the up and down fluctuation. If there is a regular cycle, and we discuss below how it may be measured, it determines the length of the period to use for moving averages. If it appears that there is a two-year cycle, we use a two-year moving average to smooth out the line. If there is a four-year cycle, we use a four-year moving average, and so on. Hence, we first determine if a regular cycle exists before applying moving averages.

If the trend line is straight, the trend line obtained by the method of

---

[2] The term "year" here is to be interpreted as an election year rather than a calendar year.

[3] If we used an eleven-year moving total, we should have only one sum, and thus an eleven-year moving total is too much. Note that a three-year moving total does not refer to three calendar years. In the present case, because elections are held every four years, a three-year moving total actually refers to three elections and a time period of eight years.

moving averages will be straight. If the trend line is curved, the trend line will appear as a curve. If the trend line is curved, we want to fit it by a different equation than (12-3) or (12-4). We consider some equations for various curves later in this chapter.

Before turning to other components of time series, let us consider the method of least squares, described in Chapter 9, as a means of fitting the trend line. In Chapter 9, it was noted that we can find the equation of the line that minimizes the sum of the squared differences of the actual and estimated Y values by two methods. We can solve the normal equation:

$$\sum Y = na + b \sum X$$
$$\sum XY = a \sum X + b \sum X^2$$

or we can use Equations (9-5) and (9-6), which can be simplified for time series if we convert the years to intervals, with the origin as the midpoint if there is an odd number of years or not using an origin if there is an even number of years, as shown in Figure 12-4. The simplified equations for

										odd number of years
20	24	28	32	36	40	44	48	52	56	60
−5	−4	−3	−2	−1	0	1	2	3	4	5

									even number of years
20	24	28	32	36	40	44	48	52	56
−5	−4	−3	−2	−1	1	2	3	4	5

Figure 12-4.

finding the coefficients a and b are then

$$b = \frac{\sum X_i Y_i}{\sum X_i^2} \qquad (12\text{-}5)$$

$$a = \frac{\sum Y_i}{n} \qquad (12\text{-}6)$$

As an example, consider the data on percentage of turnout for presidential elections. The data needed to compute a and b are given in Table 12-2. Substituting the appropriate values in Equations (12-5) and (12-6), we get

$$b = \frac{186.3}{110} = 1.7$$

$$a = \frac{604.6}{11} = 55.0$$

and the equation for the trend line is

$$Y' = 55.0 + (1.7)X$$

Table 12-2   Percentage of Turnout in Election for President, 1920–1960

Year	X_i	Y_i	X_iY_i	X^2
1920	−5	44.2	−221.0	25
1924	−4	44.3	−177.2	16
1928	−3	52.3	−156.9	9
1932	−2	52.9	−105.8	4
1936	−1	57.5	−57.5	1
1940	0	59.7	0	0
1944	1	56.3	56.3	1
1948	2	51.5	103.0	4
1952	3	62.0	186.0	9
1956	4	60.1	240.4	16
1960	5	63.8	319.0	25
Total	. . .	604.6	186.3	110

For comparison with the other methods of computing the trend line, estimations of the percentage of turnout in presidential elections for the years 1920 and 1972, using the equation computed by least squares, are

1920: $Y' = 55 + (1.7)(0) = 55$

1972: $Y' = 55 + (1.7)(8) = 68.6$

(Note that 8 is used as the value for the year 1972 here, because the origin for the regression equation computed here is 1940.)

The estimates based on the equation computed by least squares are similar to those computed by the other methods. However, none of the estimates is necessarily correct, because there is some deviation from a straight-line trend. Figure 12-3 brings this out. It is possible that the line that best fits the data is curved, and this would give us a better estimate for future years. Or it is possible that the data can be described adequately in terms of a long-term straight line plus fluctuations around it.

It should be clear by now that fitting a curve or line to data is more than a mechanical operation. Some logic or theory is also required. In the present case, it might be argued that the increase in turnout for presidential elections will continue but at a diminishing rate. A theory concerning voter turnout would be necessary to support this. We might look at past figures to substantiate this, and with a theory we can make better projections of future developments. It is also important to choose a long enough period. If the period chosen is affected by an unusual series of events, trends projected forward will be quite wrong. The classic example of an error of this kind is that of the economist Alvin Hansen, who, in the late 1930's, predicted a gloomy economic future for the United States, because he thought that popu-

lation growth was slowing down. He was wrong because he based his predictions on the decline of births during the 1930's.

CYCLICAL AND SEASONAL VARIATION

Before considering cyclical variation, we must first consider seasonal variation. The reason will be apparent soon. The intuitive meaning of the concept
seasonal implies activity that fluctuates regularly as a result of changes in
climatic conditions during the course of a year. However, it can be used to
apply to any variations that occur during the course of a twelve-month period,
the underlying cause of which may be social custom and habit as well as
climate. Examples of seasonal fluctuations are sales in department stores
(which increase heavily each December), building construction, and various
farm crops. The important point is that "seasonal" refers to regular fluctuations in activity within a twelve-month period. In contrast, "cyclical"
refers to fluctuations that occur during a period of years and will vary in
the amount of upswing and downswing.

    A good illustration of how the concept of seasonal fluctuation might
be applied to political science comes from organization theory.[4] Suppose
that we are interested in the rate of innovation in a particular governmental
function, such as local public health. Suppose also that innovation can be
measured adequately as the percentage of total man-hours being devoted to
new as opposed to traditional activities. Without elaborating here as to why
this is so, we say that the measure focuses on the percentage of total man-
hours devoted to air pollution, chronic-disease control, narcotics, accident
prevention, and radiation control. Some of these have a seasonal component;
that is, if a health department is devoting any time at all to these things,
there will be fluctuations in the amount of time devoted that may be a function of changes in environmental conditions occurring in a twelve-month
period. This is true of air pollution, which tends to increase in the winter
months, and accidents also. Hence, if we are to say anything at all about
the time series, we must take account of this seasonal factor.

    As an example, hypothetical data have been set up in Table 12-3. Sup-
pose that the figures represent the average percentage of man-hours devoted
to air-pollution control.[5] In order to compute the seasonal variation, we
must eliminate the trend, cyclical, and irregular variations from the series.
We suppose that each of these components is additive, and thus we can say
that the total time series TS is composed of trend T plus cyclical C plus

---

[4] The principal reason for measuring seasonal variations in economic activity is to
eliminate them from the series so as to better study trend and cyclical behavior. In political
science the reason is somewhat the same, as the example brings out.

[5] The figures are not at all realistic, because the percentage of time actually being
devoted in the United States to air pollution is more in the order of a few percent.

Table 12-3   Hypothetical Data for Computing Seasonal Variation

Month	1 1965	2 1966	3 (1) + (2)	4 Average	5 T	6 (4) − (5)	7 S
January	68	46	114	57	. 0	57.000	.958
February	62	40	102	51	.375	50.625	.851
March	57	35	92	46	.750	45.250	.760
April	56	32	88	44	1.125	42.875	.720
May	53	33	86	43	1.500	41.500	.697
June	56	38	94	47	1.875	45.125	.758
July	63	49	112	56	2.250	53.750	.903
August	76	64	140	70	2.625	67.375	1.132
September	87	77	164	82	3.000	79.000	1.327
October	88	80	168	84	3.375	80.625	1.355
November	85	81	166	83	3.750	79.250	1.331
December	77	75	152	76	4.125	71.875	1.208
Total	. . .	. . .	. . .	. . .	. . .	714.250	

seasonal S plus irregular I variation, or $TS = T + C + S + I$. Therefore, $S = TS − (T + C + I)$. In other words, the seasonal fluctuation is equal to the total time series minus the sum of the trend, cyclical, and irregular fluctuation. There are several methods of arriving at S.

When using the method of simple averages, we can assume that both I and C disappear, because if there is a sufficiently long time, the averaging process cancels out both the irregular and cyclical variation.[6] Column 4 of Table 12-3 contains the average of columns 1 and 2, and it can be interpreted as containing only the trend and seasonal variation. If our assumption about the disappearance of C and I is correct, all we have to do to obtain S is to subtract T from column 4. Column 4 can thus be viewed as the time series without the cyclical and irregular fluctuations, which we assume have been eliminated by the averaging process. Although it is reasonable to assume that irregular fluctuations are canceled as a result of the averaging process, it is not reasonable to make this assumption about cyclical fluctuation unless the cycles are even in amplitude. Thus, the figures we derive through the method of simple averages are correct only if the assumptions hold. For this example, we assume that they do. Hence, all we must do is to remove T from column 4.

We know how to compute a trend line for yearly data. However, to do so for the hypothetical example we need a few more years. Hence, we set up a new set of hypothetical data in Table 12-4 that represent the average

[6] In this example we have used only two years, which is not sufficiently long for this assumption. Two years was selected for simplicity. The reader can readily visualize figures for an extended number of years. Usually at least seven to eight years is necessary to average out the pecularities in individual years. A longer period is needed for irregular data.

Table 12-4

Year	X	Y	XY	X^2
1962	−2	42	−84	4
1963	−1	48	−48	1
1964	0	44	0	0
1965	1	69	69	1
1966	2	54	108	4
Total	. . .	257	45	10

$$b = \frac{\Sigma XY}{\Sigma X^2} = \frac{45}{10} = 4.5$$

$$a = \frac{\Sigma Y}{n} = \frac{257}{5} = 51.4$$

$$Y' = 51.4 + 4.5X$$

percentage of total man-hours devoted to air pollution for the five-year period 1962 to 1966.[7]

The trend line has been fitted by the method of least squares. The figures in the Y column represent the average for the year, computed by taking the average of monthly figures. For example, for the year 1965 we sum the averages for the months January to December and divide by 12 to get a figure of 69 percent as the monthly average devoted during this twelve-month period to air pollution. The b coefficient in the linear equation tells us the amount of change taking place in these monthly averages each year. Hence, to obtain the amount of increase that occurs each month, we may divide the b by 12. In the example, we get $4.5/12 = .375$, or an increase of about .375 of a percent a month.

These trend values are entered in column 5 of Table 12-3. When we subtract column 5 from column 4, we get the seasonal index. In order to express it with a base of 100, we can compute the average of column 6 and divide this into each figure in column 6. The result, given in column 7, tells us the seasonal index in percentages.

Having found the seasonal fluctuation in the example above and recalling that $TS = T + C + S + I$, we may simply subtract the seasonal component from the time series. The activity that remains after S has been removed may be considered an average rate of growth. This is done, for example, for sales figures where S (the amount in column 6 in Table 12-3) is subtracted and the remainder is considered the average monthly sales. In the example above, if we were trying to compute a measure of innovation,

[7] Again we have simplified by using only five hypothetical years rather than the larger number of years that would be necessary with actual data.

we could first subtract seasonal fluctuations. We might take the remainder as the average amount of innovation for each department.

The *cyclical variation* in a time series may be measured as a residual after the trend, seasonal, and irregular fluctuations have been measured and extracted. However, it is not always necessary to subtract seasonal and irregular fluctuations. If we have yearly data, seasonal and irregular movements do not have to be removed, because they tend to cancel out. The notion of a cycle refers to up- and downswings in activity that vary with respect to length and amplitude. The concept has the most meaning in relationship to business activity. We are all familiar with the concepts of recession, depression, and inflation. They have occurred at irregular intervals and have varied in length and intensity, although, on the average, the expansion phases have been of longer duration than the contraction phases.

Because seasonal variations are not reflected in annual data, the cyclical component may be isolated simply by elimination of the trend from the data. Consider, for example, the data in Table 12-5 and Figure 12-5 concerning the Democratic percentage of the two-party vote for President in the years 1900 to 1964. Because these are annual data, there is no seasonal variation to consider. We also may assume that the irregular fluctuations have been averaged out. Thus, the cyclical variation is the difference between

Table 12-5   Democratic Percentage of Vote for President, 1900–1964

Year	Democratic Percentage	Trend	Cyclical Percentage
1900	45.5	40.0	113
1904	37.6	40.7	92
1908	43.1	41.7	103
1912	41.9	42.5	99
1916	49.3	43.3	114
1920	34.1	44.2	77
1924	28.8	45.0	64
1928	40.8	45.9	89
1932	57.4	46.7	123
1936	60.8	47.5	128
1940	54.7	48.4	113
1944	53.4	49.7	107
1948	49.5	50.1	99
1952	44.4	50.9	87
1956	42.0	51.7	81
1960	49.7	52.7	94
1964	61.1	53.4	114

$$Y' = 46.7 + \frac{345.9}{408} (X)$$

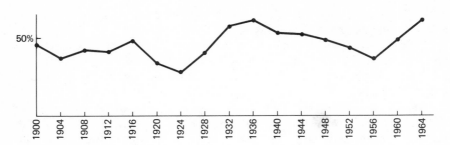

Figure 12-5. Democratic percentage of vote for President, 1900–1964.

the trend percentage and the actual Democratic percentage of the vote, expressed as a percent. In 1900, for example, the Democratic percentage was 5.5 percent higher than the trend. The cyclical percentage in 1900 was 45.5/40 = 113 percent. An examination of Table 12-5 shows that the years 1900, 1908, 1916, 1932 to 1944, and 1964 were above the normal Democratic percentage as expressed by the trend values and that the other years were below normal. Hence, it appears that there is some cyclical behavior in this series, at least in the sense that the upswings of Democratic percentage are irregular and of different duration. The long-term trend seems to be upward also.

A note of caution must be introduced here concerning cyclical fluctuations. Cyclical behavior in business activity cannot be studied by the use of statistics alone. Each business cycle has a different duration and is associated with different factors as opposed to the periodicity of a seasonal variation that has the same duration and is a function of the same factors.[8] Hence, an adequate study of business cycles requires, in addition to the use of statistics, the application of various theoretical models. Similarly, with political phenomena, whether or not cyclical behavior can be said to exist depends as much on the development of analytical models that tell us to look for this kind of behavior as on statistical techniques for measuring the cycles.

The use of the concepts of normal, expected, and predicted voter turnout points to some of the problems of using time-series data as a basis for making forecasts or for computing an average year. We might look at this as a sampling question in which the predicted Y' for a given year is one of the many possible predictions for that year. The annual series of data used to make the predictions might be considered a sample. For example, the percentage of turnout for presidential elections for the years 1920 to 1960 can be viewed as a sample of size 11 that is to be used as the basis for an inferential linear-regression problem, as described in Chapter 9. In this case we might test for the significance of b. We noted in Chapter 9 that the Gauss-Markoff

[8] See, for example, A. F. Burns and Wesley C. Mitchell, *Measuring Business Cycles*, National Bureau of Economic Research, Inc., New York, 1946.

theorem states that the method of least squares gives a linear unbiased estimate of Y. However, the assumptions we made for the regression problems in Chapter 9 may not hold for time-series data. The assumption of independence of observations may not be true, because what happens in one election may influence what happens in the next, particularly if there is an underlying factor, such as the Great Depression of the 1930's or World War II. Similarly, the assumption of normally distributed Y's may not be met, and hence the method of least squares can be used only in a questionable sense as a method for estimating parameters. These possibilities lead some writers to conclude that, when we use a trend equation to forecast an event, we are simply projecting an existing trend into the future. There is no statistical way of telling how much error is likely, because the standard error of the estimate cannot be validly computed. Moreover, if there is cyclical behavior in the time series as well, projections are likely to be off, because there may be a cyclical upswing or downswing in the projected year. Thus, it would be necessary to forecast the cycles as well as the trend to get a more accurate reading.

The question of forecasting cycles in economics is very difficult, involving complicated theoretical problems. One method is to correlate the phenomenon we want to predict with some basic measure, such as population growth, the projections for which may be more certain. The correlation can be a straight correlation of the actual two time series. We can then make predictions of the phenomenon we are interested in, using the projections of the basic measure. For example, suppose that we want to forecast the Democratic percentage of the presidential vote for the next few elections. If we know that there is a strong correlation of this variable with a more stable economic measure, such as GNP, we might use the regression of Democratic percentage of the presidential vote and GNP as a means of forecasting. Of course, if some unusual factors occur in the intervening period, the projections are likely to be very wrong.

There has been very little attempt to make forecasts of political behavior (outside of forecasting the possible election results just before the elections by means of polling). The problems involved are such that the question is best left unexplored in a book of this kind.

Non-linear Trend Curves

A possible reason that projections made on the basis of the equation $Y' = a + b(X)$ may be wrong is that the time series cannot adequately be described in terms of a straight line, and a curve of some kind must be fitted to the data. Let us consider two of the simpler nonlinear equations that could be used to fit a trend line (in fact there are an infinite number of such curves, but only some have possible practical significance at the moment.)

We considered a second-degree polynomial in Chapter 10. The equation for such a polynomial is $Y' = a + b(X) + c(X)^2$. A third-degree polynomial is defined as $Y' = a + b(X) + c(X)^2 + d(X)^3$. A fourth-degree polynomial is defined as $Y' = a + b(X) + c(X)^2 + d(X)^3 + e(X)^4$, and so on. The specific shape of any of these, when graphed, depends on the values of the coefficients $a, b, c, \ldots, n$. However, to get an idea of how these functions look when graphed, we assume the intercept value a to be at the origin and the coefficients $b, c, \ldots, n$ to have a constant value of one. Under these assumptions the second- and third-degree polynomials reduce to $Y' = X + X^2$ and $Y' = X + X^2 + X^3$. The graph of the second-degree polynomial for small integer values of X is shown in Figure 12-6. The second-degree polynomial

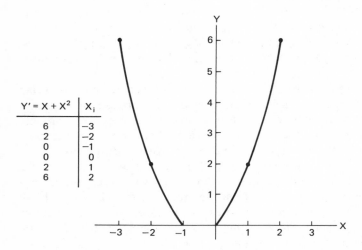

$Y' = X + X^2$	X_i
6	−3
2	−2
0	−1
0	0
2	1
6	2

Figure 12-6. Graph of $Y' = X + X^2$.

has the general shape of a parabola when graphed and can curve either up or down. The graph of the third-degree polynomial for small integer values of X is shown in Figure 12-7. Note that it takes an additional turn in direction as compared with the second-degree polynomial.

In general, the higher-degree curves have more changes in direction. Hence we can find a polynomial equation that describes a curve that fits empirical data accurately. To do so, we must solve for the coefficients a, $b, c, \ldots, n$ in the equation. This can be done for a second-degree (and higher-degree) polynomial by means of the normal equations:

$$\sum Y = na + b \sum X + c \sum X^2$$
$$\sum XY = a \sum X + b \sum X^2 + c \sum X^3$$
$$\sum X^2Y = a \sum X^2 + b \sum X^3 + c \sum X^4$$

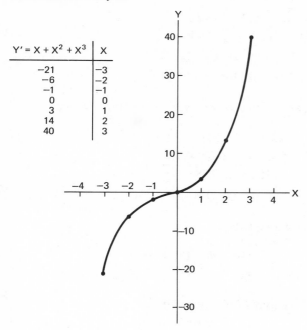

Y' = X + X² + X³	X
−21	−3
−6	−2
−1	−1
0	0
3	1
14	2
40	3

Figure 12-7. Graph of $Y' = X + X^2 + X^3$.

However, we can usually assume a to be at the origin, so that $\sum X = 0$, and the equations simplify to

$$\sum Y = na + c \sum X^2$$
$$\sum XY = b \sum X^2$$
$$\sum X^2Y = a \sum X^2 + c \sum X^4$$

To find the coefficients a, b, and c, we can start with b, because both $\sum XY$ and $\sum X^2$ are known. Then we can solve for c and a by the method described above for solving the normal equation for a straight line. As an example, suppose that we wanted to fit a parabola to the data in Table 12-2. First we find b by the middle term

$$186.3 = b(110)$$
$$\frac{186.3}{110} = b$$
$$b \doteq 1.7$$

as we found before. Then we find a and c by solving the other two equations simultaneously:

$$\sum Y = na + c \sum X^2$$
$$\sum X^2Y = a \sum X^2 + c \sum X^4$$
$$604.6 = 11(a) + c(110)$$
$$5{,}930.5 = 110a + 1{,}958c$$

Multiplying the first equation to equalize a and subtracting, we have

$$
\begin{aligned}
6{,}046 \ \ &= 110a + 1{,}100c \\
-5{,}930.5 &= 110a + 1{,}958c \\
\hline
115.5 &= -858c \\
c &= -.13
\end{aligned}
$$

Substituting this into the first equation,

$$604.6 = 11a + (-.13)(110)$$
$$a = 54$$

Hence the equation of the parabola for these data is

$$Y' = 54 + 1.7(X) - .13(X^2) \qquad \text{origin } 1940$$

Of course, the parabola is not a good fit for these data. The projected turn-out for the year 1972, using the parabola, is $Y' = 54 + 1.7(8) - .13(8^2) = 59.28$ percent. This is probably too low.

Just as we said in regard to the straight line, we cannot use a curve for estimating parameters, because the assumptions of independence of observations and normally distributed Y values may not be true. The method of least squares enables us to find a straight line or a curve that minimizes the differences between the observed Y values and the estimated Y' values. But sometimes visual examination and a good theory can afford a better basis for making forecasts than mechanical fitting of a line by mathematics.

We may describe a trend better in terms of a polynomial equation than in terms of a straight-line equation. But this does not mean that we are therefore in a better position to estimate parameter values. The methods of freehand drawing and semiaverages described above as a means of finding a straight-line equation can also be used to find the parabola trend line (second-degree polynomial). We do not consider these here, because these curves have limited practical research application.[9]

[9] For a description of mathematical and other methods of fitting such lines, see F. C. Croxton and D. J. Cowder, *Practical Business Statistics*, Prentice-Hall, Inc., Englewood Cliffs, N.J., 1960, Chap. 38; Jack Sherman and W. J. Morrison, "Simplified Procedures for Fitting a Gompertz Curve and a Modified Exponential Curve, *Journal of the American Statistical Association*, pp. 87–97, March, 1950.

The Exponential Trend Line

If the rate of growth in a time series follows a geometric progression, that is, grows at a constant rate, then the trend line may be represented by an exponential function $Y' = ab^X$. Certain phenomenon seem to be adequately described by this equation, such as population growth. As an example, a trend line of the form $Y' = ab^3$, for $a = 1$, $b = 2$, has a partial graph that is shown in Figure 12-8.

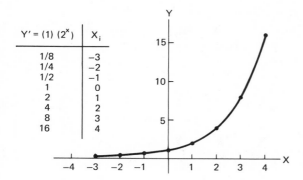

Figure 12-8. Graph of $Y' = ab^3$.

However, exponential data graph as a straight line if we either plot them on a logarithmic scale or work with the logarithms of the numbers. For example, the equation $Y' = ab^X$ becomes $\log Y' = \log a + \log b$ when we work with logarithms. The normal equations of this function[10] are

$$\sum \log Y = n \log a + \log b \sum X$$
$$\sum (X \log Y) = \log a \left(\sum X \right) + \log b \sum X^2$$

If the intercept is at the origin, $\sum X = 0$, and the equation becomes

$$\sum \log Y = n \log a$$
$$\sum (X \log Y) = \log b \sum X^2 \tag{12-8}$$

As an example, assume the following data:

Year	X	Y	$\log Y$	$X \log Y$	X^2
1965	−1	100	2	−2	1
1966	0	1,000	3	0	0
1967	1	10,000	4	4	1
Totals	...	11,100	9	2	2

[10] We can also find a and b by $a = (\Sigma \log Y)/n$ and $b = (\Sigma X \log Y)/\Sigma X^2$.

Substituting these into Equations (12-8), we have

$$(1) \qquad 9 = 3(\log a) \qquad (2) \qquad 2 = 2(\log b)$$
$$\log a = 9/3 = 3 \qquad \log b = 2/2 = 1$$

The logarithm trend equation is $\log Y' = 3 + 1$. This can be converted back into exponential form by looking up the antilogarithms for $\log a$ and $\log b$. In this example, we have $Y' = 1{,}000(10)^X$. Thus, if the origin is 1966 and $X = 1$, the estimate for the year 1967 is $Y' = 1{,}000(10)^1 = 10{,}000$.

There are many other equations that may describe trend lines better. For example, if the trend is convex rather than concave, this can be described by adding a constant to the equation as follows: $Y' = k + ab^X$. As an example, if $k = 16$, $a = -8$, and $b = .5$, the graph of the curve will be as shown in Figure 12-9. Note that the upper end of the curve approaches the

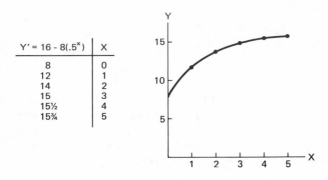

$Y' = 16 - 8(.5^X)$	X
8	0
12	1
14	2
15	3
15½	4
15¾	5

Figure 12-9.

value of k. In this case, it comes closer and closer to 16 but never exceeds it.

It is apparent that, by combining different values of a and b, we can make the curve slope convex upward or downward or concave upward or downward. A curve such as that shown in Figure 12-9 might be used to describe a flow of activity that grows rapidly in the early years and then reaches a maximum point at which there is no further growth or only a small amount. There are empirical examples that approximate this, such as the growth of industries like radio (which also have a decline after the maximum point has been reached) or television. This might also be true of voter turnout. We can use the method of freehand drawing or the method of semiaverages to fit the curve to actual data and find the values of the coefficients a and b. However, we do not go into these here (see Yamane, Chap. 23).

SERIAL CORRELATION

We said above that one of the assumptions that must be met before we may use a trend-line equation to compute parameter values is the assumption of independence of observations. We can determine if there is independence in a series by means of the serial correlation coefficient.[11] The serial correlation between successive terms in a series may be defined as

$$r_1 = \frac{\text{cov}(X_i)(X_{i+1})}{\sqrt{\text{var}(X_i)\text{var}(X_{i+1})}} \tag{12-9}$$

that is, the standardized covariation of the successive terms in the series. To compute the serial correlation, we can use

$$r_1 = \frac{\sum X_i X_{i+1} - (\sum X_i)^2/n}{\sum X_i^2 - (\sum X_i)^2/n} \tag{12-10}$$

Let us illustrate this with the data concerning the percentage of turnout in presidential elections in Table 12-1, but rounding off each percentage for easier computation. The necessary sums are given in Table 12-6.

The serial correlation is

$$r_1 = \frac{32,953 - 601^2/11}{33,267 - 601^2/11} = \frac{117}{431} = .27$$

Table 12-6 Serial Correlation for Percentage of Turnout in Presidential Elections, 1920–1964

X_i	X_i^2	$X_i X_{i+1}$
44	1,936	1,936
44	1,936	2,288
52	2,704	2,756
53	2,809	3,021
57	3,249	3,420
60	3,600	3,360
56	3,136	2,856
51	2,601	3,060
60	3,600	3,600
60	3,600	3,840
64	4,096	2,816
601	33,267	32,953

[11] Other methods are also available, such as Von Neuman ratio and the Durbin-Watson test. See J. Von Neuman, "Distribution of Ratio of the Mean Square Successive Difference to the Variance," *Annals of Mathematical Statistics*, vol. 12, pp. 367 ff., 1941; J. Durbin and G. S. Watson, "Testing for Serial Correlation in Least Squares Regression," *Biometrika*, vol. 37, pp. 409 ff., 1950.

Note that in this computing equation for the serial correlation coefficient, $X_{n+1} = X_i$. In this example, $X_{11+1} = X_1$. Thus, the product X_iX_{i+1} for
the last term is 64×44. Therefore, this definition is called the circular definition of the serial correlation coefficient. If there is a pronounced trend in the
data, the value of the first X_i will be much smaller than that of the nth, and
the correlation will be affected. The correction for this is called the noncircular definition:

$$r_0 = \frac{\sum X_iX_{i+1} - 1/(n-1)(\sum X_i)(\sum X_{i+1})}{[\sum X_i^2 - 1/(n-1)(\sum X_i)^2]^{1/2}[\sum X_{i+1} - 1/(n-1)(\sum X_{i+1})^2]^{1/2}}$$

(12-11)

Both the circular and noncircular definitions are regarded as sample correlation coefficients. We can test whether or not the coefficient is significant.
The distribution of the serial correlation coefficient has been developed by
R. L. Anderson.[12] The table is not reproduced in this book. The Anderson
table shows that for significance at the .05 level, for an n of 11, an r_1 of .353 or
better is needed. Because our correlation is .24, we may conclude that the
series of the percentage of turnout for presidential elections is random; that
is, the turnout for one year is not dependent on the preceding year.

Although we may find that a particular series is random, we should still
not use the trend equation to estimate parameter values unless we could
identify the cyclical and irregular components. We might try to do so by
"detrending" the series by subtracting the trend values. The next step would
be to test this detrended, or stationary, time series for randomness, using the
serial correlation coefficient. If it is not random, we might then look for the
cyclical component. But as was said several times above, this is as much a
theoretical as a statistical piece of work.

READINGS AND SELECTED REFERENCES

Freund, John. *Modern Elementary Statistics*, Prentice-Hall, Inc., Englewood Cliffs,
N. J., 1960, Chap. 17.
Key, V. O. *A Primer of Statistics for Political Scientists*, Thomas Y. Crowell Com-
pany, New York, 1954, Chaps. 2, 3.
Parl, Boris. *Basic Statistics*, Doubleday & Company, Inc., Garden City, N. Y.,
1967, Chaps. 29–32.
Spurr, William A., and Charles P. Bonini. *Statistical Analysis for Business Decisions*,
Richard D. Irwin, Inc., Homewood, Ill., 1967, Chaps. 19–21, 24.
Yamane, Taro. *Statistics: An Introductory Analysis*, Harper & Row, Publishers,
New York, 1967, Chaps. 12, 13, 23.

[12] See R. L. Anderson, "Distribution of the Serial Correlation Coefficient," *Annals of
Mathematical Statistics*, vol. 13, no. 1, pp. 1–13, 1942.

13
Factor Analysis

In Chapter 10 we considered the question of trying to analyze the relationship between a dependent variable and several independent variables. We saw there that it is difficult to sort out the contribution of each independent variable and that under conditions of multicollinearity it is not possible to do so. In Chapter 11 we considered a method appropriate to experimental design when there are two nominal-scale and one interval-scale variables. We considered a method for testing for interaction among the two independent variables to see if the combined effect of the two was more than the additive effect of each alone. In the present chapter we consider a completely different approach to this question. Instead of trying to sort out the contribution of each independent variable, factor analysis attempts to find out if a large number of independent variables really are measures of one or more underlying common variables.

Factor analysis originated with psychologists in an attempt to discover the factors underlying intellectual ability. Charles Spearman is credited with originating, in 1904, the theory that a general factor (inherited intellectual ability) and specific factors (learning and experience) might be found to underlie the multitude of tests used to measure different intellectual abilities. He originated the idea that this might be done by analyzing a matrix of correlations of the various tests mathematically. Factor analysis has gone through several changes since then, but its basic objective remains the same: to substitute a few concepts in place of many as explanations of behavior. This is another way of saying that its objective is to try to make complex phenomena simpler and, by doing so, increase our understanding of it.

Factor analysis has been used in many other fields than psychology. It has not grown so much as was envisioned by J. P. Guilford in 1936, who wrote then that it might be used some day to "...solve some of the remaining mysteries in the field of atomic physics." Perhaps this is partly a result of the fact that the use of factor analysis cannot proceed without the development of a theory to explain the factors we might extract. This is shown somewhat by the development of factor analysis in the field of psychology.

The fact that very few empirical data could be found to conform to the pattern of a general factor and specific factors anticipated by Spearman led to the idea of group factors. There are several variations of this argument.

One theory is that a given set of variables may be reduced only to a number of group factors.[1] Another is that in addition to a general factor, which is not universal, there are several group factors. In psychological terms we might say that there is a general intellectual factor, being measured by all tests; group factors, such as verbal ability, mechanical ability, imagination, and so on, being measured by subsets of the tests; and specific factors, such as addition skill, reading skill, and so on, being measured by each of the component tests.[2]

In recent years, there has been an attempt to develop an objective method that will extract factors in accordance with statistical criteria. The factor that accounts for the largest proportion of total variance in the matrix is extracted first, then the factor that accounts for the next largest proportion of residual variance, and so on. We consider all these developments in this chapter.

Factor analysis begins with a correlation matrix of several variables and attempts to determine if the phenomenon being studied can be expressed in terms of a smaller number of underlying factors. As an example, suppose that a study of organizations has found that decentralization, higher morale,

Table 13-1 Intercorrelations of Organizational Variables

	1	2	3	4
1		.4	.7	0
2			.4	.6
3				.3
4				

1 = decentralization
2 = morale
3 = informalism
4 = styles of management

less formalism in work tasks, and more participatory styles of management are positively related to increased innovation in the organization. We want to know if the four independent variables are really measuring one underlying factor. If we can describe the four variables in terms of one factor, we shall achieve economy in description, which, according to most observers, is a

[1] See William Brown and G. H. Thompson, *The Essentials of Mental Measurement*, Cambridge University Press, New York, 1925.
[2] See L. L. Thurstone, *The Vectors of Mind*, The University of Chicago Press, Chicago, 1935.

proper goal of science.[3] Intercorrelations among the four independent variables are given in Table 13-1.[4]

These correlations are depicted graphically with circles in Figure 13-1. The overlapping areas represent the percentage of variance accounted for by the intercorrelations. There is overlap among decentralization, morale, and informalism, represented by the hatched area. Thus, it appears that these three variables are measuring something in common.[5] There is also overlap among morale, informalism, and styles of management, represented by the darkened area. In addition, variables 3 and 1, variables 2 and 1, variables 4 and 2, and variables 4 and 3 have common overlaps, represented by areas marked with diagonal lines. Finally, each variable has a specific variance

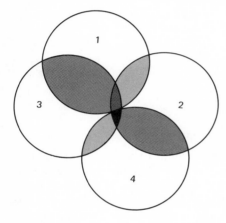

Figure 13-1.

(unshaded area). There are various ways of interpreting this, depending on the theory we are working with (we consider the problem of interpreting factors in more detail below). We may say that three of the four variables (decentralization, morale, and informalism) are measuring one common factor. In addition, there may be four group factors measured by the common variance in the areas marked with diagonal lines. We do not consider the overlap among morale, informalism, and styles of management, because it is

---

[3] Normally we have a large number of independent variables that we want to reduce to a smaller number of explanatory factors. A factor analysis is not usually done on four variables, although it is entirely possible to do so.

[4] The correlations in this example are based on a study done by the author and others. They have been rounded for ease of illustration. See Dennis J. Palumbo "Power and Role Specificity in Organization Theory," *Public Administration Review*, May, 1969.

[5] As we will see in more detail below, the name we may want to give to this underlying factor, if we want to consider it significant at all, depends upon the theory with which we are working.

too small and may have been produced by sampling error. The unshaded area represents specific variance of each variable plus error variance.

These overlaps, which are usually called *loadings* on the factors, can be represented in the form of a table, as in Table 13-2. The values in the cells in this table are the loadings, and they are the correlations of each variable with each of the factors, which are labeled with roman numerals at the head of the table. The numbers in the column labeled h^2 will be explained in a moment.

Table 13-2   Factor Matrix for Table 13-1

| *Variable* | Factor | | | | h^2 |
	I	II	III	*k*	
1	.5	.3	.6	.	.70
2	.5	.5	0	.	.50
3	.8	0	.5	.	.89
.	.	.	.	.	.
.	.	.	.	.	.
.	.	.	.	.	.

The table has not been completed, because it is hypothetical.

Now it is not very meaningful to give names to these factors, that is, roman numerals, because we do not have enough variables in this example. However, in order to give an idea of what might be done in a more realistic model, we suggest names. We might call the general factor the degrees of role specificity, meaning the extent to which each member's role is given relatively exact prescription concerning how he should behave. The group factor measured by the overlap of the variables of decentralization and informalism might be called the structure factor, meaning that it is the structure by which an organization attempts to achieve a larger degree of specificity in roles. We might call the factor measured by the overlap of the variables of morale and styles of management the behavioral factor meaning the supervisory styles by which role specificity is either carried out or thwarted, and so on.

The purpose of pursuing this simplified example is to demonstrate the way factor analysis works rather than to construct a theory of organizational behavior. It demonstrates adequately the purpose of factor analysis and some of its difficulties. It is now time to consider some of these questions a little more rigorously.

FACTOR THEORY

Factor analysis begins with a matrix of linear correlations and tries to determine if these correlations can be expressed in terms of fewer variables.

There are many possible factor solutions, most of which are no longer considered acceptable. We discuss some of these briefly below. For the moment we consider, in general terms, a model that does not postulate that a given matrix yields general and group factors, but only a number of independent factors, and that the number of factors is less than the number of variables the analysis began with. The basic model may be written

$$X_{ij}^2 = a_{i1}X_1{}^2 + a_{i2}X_2{}^2 + \cdots + a_{in}X_n{}^2 + a_iU_i{}^2 + a_ie_i{}^2 \qquad (13\text{-}1)$$

that is, the variance associated with any particular observation X_{ij} is composed of variance common to some of the other variables in the matrix, plus variance unique to the particular variable, plus error variance. The basic problem of factor analysis is to determine the coefficients $a_{i1}, a_{i2}, \ldots, a_{in}$ of the common factors. In variance notation, we may write

$$X_{ij}^2 = \underbrace{\sigma_{x1}^2 + \sigma_{x2}^2 + \cdots + \sigma_{xn}^2}_{\text{Common variance}} + \underbrace{\sigma_{xU}^2}_{\substack{\text{Specific} \\ \text{variance}}} + \underbrace{\sigma_{xe}^2}_{\substack{\text{Error} \\ \text{variance}}} \qquad (13\text{-}2)$$

If we let $h_j{}^2$ represent the common variance of any particular variable, that is, the sum of its common variances on each factor, then we may say that the variance of each variable is $h_j{}^2 + S_j{}^2 + E_j{}^2 = 1.0$, or the total variance of each variable is the sum of the common variance, plus the specific variance, plus the error variance. The communality of a variable is the sum of its independent common loadings. We represent the common loadings of each variable by $h_j{}^2$. Because we represent the individual factor loadings by a_i, we may write the following equation:

$$h_j{}^2 = a_1{}^2 + a_2{}^2 + \cdots + a_n{}^2 \qquad (13\text{-}3)$$

If $h^2 \neq 1$, then $a_1{}^2 + a_2{}^2 + \cdots + a_n{}^2 \neq 1$. But because we said that $h_j{}^2 + S_j{}^2 + E_j{}^2 = 1$, then $1 - h_j{}^2 = S_j{}^2 + E_j{}^2$. This is the uniqueness of a variable plus sampling error.

If we are able to find the communalities of a variable, we can reproduce the correlations, because the correlations of variables can be accounted for by their common factor loadings. Let a_{jk} represent the common factor loadings of two variables. Then

$$r_{jk} = a_{j1}a_{k1} + a_{j2}a_{k2} + \cdots + a_{jr}a_{kr} \qquad (13\text{-}4)$$

Let us illustrate by means of the correlation matrix in Table 13-1. Suppose that the factors shown in Table 13-2 have been found. (We shall consider in a moment how this is done. The factor loadings used in this example are not actual but only hypothetical.) The correlation between variables 1 and 2 is $r_{12} = (.5)(.5) + (.3)(.5) + (.6)(0) = .4$. The correlation between variables 2 and 3 is $r_{23} = (.5)(.8) + (.5)(0) + (0)(.5) = .4$. The correlation between variables 1 and 3 is $r_{13} = (.5)(.8) + (.3)(0) + (.6)(.5) = .7$. The communality h^2 for variable 1 is $h_1{}^2 = (.5)^2 + (.3)^2 + (.6)^2 = .70$, and so on. How do we go from the correlation matrix in Table 13-1 to the factor matrix in Table 13-2?

In order to factor-analyze a correlation matrix, it is necessary first to find values for the principal diagonal of the matrix. Note that the correlation matrix in Table 13-1 does not have coefficients in the principal diagonal or in the lower left-hand half. Of course, if we put the correlations of each variable with itself in the diagonal, the entries will be unity. It is possible to start with these values in the diagonal. For some methods of factor analysis, however, it is better to insert other values in the diagonal.

One of the more difficult problems of factor analysis concerns the question of what values should be inserted in the principal diagonal of the matrix. If the correlation of a variable with itself, that is, unity, is inserted, the resulting factor solution will not be the same as when the communalities are used. To understand why this is so, we must digress a little into some elementary matrix algebra.

One way to phrase the problem of factor analysis is as follows: Given the $n \times n$ correlations in a matrix, can the relationships among the n variables be expressed in terms of fewer than n variables? The number of factors needed to account for the intercorrelations is the *rank* of the matrix. For example, if a 4×4 matrix is of rank four, the four intercorrelations of the matrix cannot be expressed in fewer than four variables. If it is of rank three, the four intercorrelations can be expressed in terms of three factors, and so on. Factor analysis tries to find the fewest number of factors that can express the $n \times n$ correlations. To do so, it is necessary to select diagonal values that minimize the rank of the matrix and thus the number of factors that will be found. The rank of a matrix is determined in part by the diagonal values, because the rank is determined by the order of its highest nonvanishing minors. To explain this, we need to consider some definitions.

A *matrix* is defined as a rectangular array of numbers, exhibited between brackets: $A = \begin{bmatrix} 0 & 1 \\ 5 & 3 \end{bmatrix}$. The number of rows and columns determine the dimensions of the matrix. A matrix with two rows and three columns may be designated as $M_{2 \times 3}$.

A *determinant* can be defined for any square matrix as a real number. For a 2×2 matrix, $A = \begin{bmatrix} a & b \\ c & d \end{bmatrix}$, the determinant is $ad - bc$. For example, for the matrix $A = \begin{bmatrix} 3 & 1 \\ 2 & -5 \end{bmatrix}$, the determinant is $[3 \times (-5) - (1 \times 2)] = -17$. Determinants can be defined for squares matrices of any order in the same way. For example, for a matrix of order three:

$$A = \begin{bmatrix} a_1 & b_1 & c_1 \\ a_2 & b_2 & c_2 \\ a_3 & b_3 & c_3 \end{bmatrix}$$

the determinant is

$$a_1b_2c_3 + a_2b_3c_1 + a_3b_1c_2 - a_3b_2c_1 - b_3c_2a_1 - c_3b_1a_2 \qquad (13\text{-}5)$$

A determinant is usually represented by a rectangular array of numbers, but with vertical bars enclosing the entries. Because the determinant constitutes a function we write $S(A) = \begin{vmatrix} a & b \\ c & d \end{vmatrix}$.

We can expand or reduce a determinant by expanding or reducing its minors. The *minor* of an element in a determinant is the determinant resulting from the deletion of the row and column containing the element. For example, in the determinant

$$|A| = \begin{vmatrix} 1 & 3 & 2 \\ 4 & 0 & 1 \\ -3 & 6 & 1 \end{vmatrix}$$

the minor of element two is

$$|A_{13}| = \begin{vmatrix} 1 & 3 & 2 \\ 4 & 0 & 1 \\ -3 & 6 & 1 \end{vmatrix} = \begin{vmatrix} 4 & 0 \\ -3 & 6 \end{vmatrix}$$

We can also find the minors of any element by rewriting the value of the determinant and factoring. The value of a third-order determinant given in (13-5) can be written

$$a_1b_2c_3 - a_1b_3c_2 + b_1c_2a_3 - b_1c_3a_2 + c_1a_2b_3 - c_1a_3b_2$$

$$= a_1(b_2c_3 - b_3c_2) - b_1(c_3a_2 - c_2a_3) + c_1(a_2b_3 - a_3b_2) \qquad (13\text{-}6)$$

The terms in the parentheses are the minors of a_1, b_1, and c_1 respectively. We may denote these as A_1, B_1, and C_1 and define the determinant of a third-order matrix as $S(A_{3\times3}) = a_1A_1 - b_1B_1 + c_1C_1$; that is, the determinant of the 3×3 matrix is equal to the expansion of the determinant according to the minors of the first row. This value will be the same if we expand according to the minors of any other row or column. Each of the minors (A_1,B_1,C_1) are themselves second-order determinants, and they can also be developed by minors.

Let us give a numerical example and then return to factor theory. Find $S(A)$ for

$$\begin{vmatrix} 2 & 1 & 3 \\ 5 & -1 & 4 \\ 1 & -2 & 3 \end{vmatrix}$$

By the minor for the first column. We have

$$2\begin{vmatrix} -1 & 4 \\ -2 & 3 \end{vmatrix} - 5\begin{vmatrix} 1 & 3 \\ -2 & 3 \end{vmatrix} + 1\begin{vmatrix} 1 & 3 \\ -1 & 4 \end{vmatrix} = -28$$

Now, if we have a 4×4 matrix whose determinant does not vanish, that is, is not equal to zero, it is of rank four, and it cannot be explained by fewer than four variables. If the determinant of such a matrix does vanish, the matrix is of rank three or less, that is, can be explained in terms of three or fewer factors. It is of rank three if one or more of its first minors does not vanish and less than rank three if they do. In the latter case, there are two common factors. If the second-order minors vanish, there is a single common factor.

This can be made intuitively more meaningful by considering Spearman's tetrad difference. Assume the matrix of correlations in Table 13-3. Spearman was able to show that, if these correlations can be accounted for by a single factor, the coefficients in any combination of two columns will be proportional and will vanish when subtracted from each other. For columns 1 and 2 we have

$$\frac{r_{31}}{r_{32}} - \frac{r_{41}}{r_{42}} = r_{31}r_{42} - r_{32}r_{41} = 0$$

and for the hypothetical example we have

$$\frac{.9}{.9} - \frac{.9}{.9} = (.9)(.9) - (.9)(.9) = 0$$

For columns 1 and 3 for the example we have

$$\frac{.8}{.8} - \frac{.9}{.7} = (.8)(.7) - (.8)(.9) = -.16$$

The last difference is not exactly equal to zero, but it is within the limits of sampling error.

Table 13-3

Variable	1	2	3	4
1		.8	.9	.9
2	.8		.8	.7
3	.9	.9		.7
4	.9	.9	.8	

For every four variables, if three tetrad differences are zero (or close to zero), we may conclude that there is one common factor underlying the four variables. These three tetrad differences are: (1) $t_{1234} = r_{12}r_{34} - r_{13}r_{24}$; (2) $t_{1234} = r_{12}r_{34} - r_{14}r_{23}$; (3) $t_{1342} = r_{13}r_{24} - r_{14}r_{23}$, where t stands for the tetrad differences. If these are all zero, the data conform to Spearman's two-factor ideal. Of course, when we have a correlation matrix where the correlations are all very high, and about the same, it is likely that we shall find a single factor underlying all the variables.

When many variables are being investigated, the method of tetrad differences or the method of finding the rank by determinants would be very laborious to use. Other methods of accomplishing much the same purpose have been developed. We consider some of them below. For the moment, we consider a simple method of extracting factors as a means of clarifying factor theory. This method is not practical; it is being used only to illustrate factor theory. Practical methods are considered below.

The Diagonal Method

We noted above that the problem of factor analysis is that values for the principal diagonal of the matrix must be found before factors can be extracted. There are various values that may be used in the principal diagonal, which lead to different solutions. The communalities of each variable h_j^2 can be inserted in the diagonals. This minimizes the rank of the matrix, and the factors extracted reproduce the correlations.

Table 13-4 Correlation Matrix for Diagonal Method

Variable	1	2	3	4	5	6
1	(.8464)	.782	.046	.092	.6992	.0184
2	.782	(.7241)	.0665	.1138	.6540	.0190
3	.046	.0665	(.3626)	.4375	.1585	.034
4	.092	.1138	.4375	(.5313)	.2227	.0534
5	.6992	.6540	.1585	.0534	(.6202)	.0404
6	.0184	.0190	.034	.0404	.0404	(.0933)

Table 13-5 General Factor Matrix

Variable	Factor				h^2
	I	II	III	IV	
1	a_{11}	0	0	0	.8464
2	a_{21}	a_{22}	0	0	.7241
3	a_{31}	a_{32}	a_{33}	0	.3626
4	a_{41}	a_{42}	a_{43}	a_{44}	.5313
5	a_{51}	a_{52}	a_{53}	a_{54}	.6202
6	a_{61}	a_{62}	a_{63}	a_{64}	.0933

In most actual cases, the communalities of the variables are not known and must be approximated. But if they are known, the method of extracting factors is very simple and direct. It is known as the diagonal method and was developed by Thurstone.[6] Let us use, as an example, the correlation matrix in Table 13-4 with communalities (known beforehand) inserted in parentheses in the principal diagonal. The factor matrix we want to find is in Table 13-5. Note that zeros are inserted in the right side of the matrix. This means that one of the variables (in this case variable 1) is selected to involve only one factor. Any of the variables in the matrix may be selected for this position, and this is the principal drawback of the diagonal method. However, as discussed below, it may be used preliminary to a different method of factor analysis. The coefficients a_{11}, a_{21}, a_{22}, ..., a_{64} are called *factor loadings*. They are the correlations of the particular variable with the factor. Hence, if the loading of variable 1 on factor I is .92, this is to be read as the correlation of variable 1 with factor I. Recall that we defined the communalities h_j^2 in Equation (13-3) above as

$$h_j^2 = a_1^2 + a_2^2 + \cdots + a_n^2$$

The communalities are thus the sum of the squares of the factor loadings. In Table 13-5 the value of the loading of variable 1 on factor I must be equal to $\sqrt{h_1^2}$ because variable 1 is not loaded on any other factors. Thus, $a_{11} = \sqrt{h_1^2} = \sqrt{.8464} = .92$. We have entered this value in the cell for a_{11} in Table 13-6. To find the rest of the loadings a_{j1} for factor I, we recall that we

Table 13-6

Variable	\multicolumn{4}{c}{Factor}	h^2
	I	II	III	IV	
1	.92	0	0	0	.8464
2	.85	.04	0	0	.7241
3	.05	.60	a_{33}	0	.3626
4	.10	a_{42}	a_{43}	a_{44}	.5313
5	.76	a_{52}	a_{53}	a_{54}	.6202
6	.02	a_{62}	a_{63}	a_{64}	.0933

defined, in Equation (13-4), the correlation between two variables as the sum of the cross products of their common factor loadings. Thus the correlation of variables 1 and 2 is $r_{12} = a_{11}a_{21} + a_{12}a_{22} + a_{13}a_{23} + a_{14}a_{24}$. But because this equation involves zeros for every cross product except the first, it reduces to $r_{12} = a_{11}a_{21}$; that is, the correlation of variables 1 and 2 is equal to the product of the loading of variable 1 on factor I times the loading of variable 2 on factor I. But we know the values for r_{12} and a_{11}. We may therefore substitute them and solve for a_{21} as follows: $.782 = (.92)(a_{21})$ or $a_{21} = .782/.92 = .85$. We now have the coefficients a_{11} and a_{21}.

[6] See L. L. Thurstone, *op. cit.*

Similarly, the value of a_{31} can be found by

$$r_{13} = a_{11}a_{31}$$
$$(.046) = (.92)(a_{31})$$
$$a_{31} = \frac{.046}{.92} = .05$$

and the loading of variable 4 on factor I can be found in the same way. $a_{41} = r_{14}/a_{11} = .092/.92 = .10$. And $a_{51} = r_{15}/a_{11} = .6992/.92 = .76$. Finally, $a_{61} = .0184/.92 = .02$.

This completes the loadings of each variable on factor I. We now turn to the loadings on factor II. We know by Equation (13-3) that $a_{21}{}^2 + a_{22}{}^2 = h_2{}^2$, and because we know $a_{21}{}^2$ and $h_2{}^2$, we have

$$(.85)^2 + a_{22}{}^2 = .7241$$
$$a_{22}{}^2 = .7241 - (.85)^2$$
$$a_{22} = .04$$

This has been entered into the appropriate cell in Table 13-6.

We turn next to Equation (13-4) to get $r_{23} = a_{21}a_{31} + a_{22}a_{32} + a_{23}a_{33} + a_{24}a_{34}$. Because the cross products beyond the first two involve zeros, this reduces to $r_{23} = a_{21}a_{31} + a_{22}a_{32}$. Note that we know the values of the first four terms in this equation. We may substitute them and solve for a_{32}:

$$.0665 = (.85)(.05) + (.04)(a_{32})$$

or

$$a_{32} = \frac{.0665 - (.85)(.05)}{.04} = .60$$

So as not to get lost, let us collect everything we have done thus far and enter it in Table 13-6. The loadings for factor I are complete and the first two for the second factor are also known now.

The method of finding the rest of the loadings should be fairly clear by now. To find a_{42} and a_{52}, we may use

$$a_{j2} = \frac{r_{2j} - a_{21}a_{j1}}{a_{22}} \qquad (13\text{-}7)$$

Thus

$$a_{42} = \frac{r_{24} - a_{21}a_{41}}{a_{22}}$$

$$= \frac{.1138 - (.85)(.10)}{.04} = .72$$

$$a_{52} = \frac{r_{25} - a_{21}a_{51}}{a_{22}}$$

$$= \frac{.6540 - (.85)(.76)}{.04} = .20$$

$$a_{62} = \frac{r_{26} - a_{21}a_{61}}{a_{22}} = \frac{.0190 - (.85)(.02)}{.04} = .05$$

Then we utilize the definition of h^2 given in Equation (13-4) to obtain a_{33}:

$$(a_{31})^2 + (a_{32})^2 + (a_{33})^2 = h_3{}^2$$
$$(.05)^2 + (.60)^2 + (a_{33})^2 = .3626$$
$$a_{33}{}^2 = .3622 - [(.05)^2 + (.60)^2]$$
$$a_{33} = .01$$

Now that we know a_{33}, we can use the cross-product equation for correlations to find a_{43} and a_{53}: $r_{43} = a_{31}a_{41} + a_{32}a_{42} + a_{33}a_{43} + a_{34}a_{44}$. The last cross product drops out, because it involves a zero. In the remaining equation, we know all the terms except the last:

$$.4375 = (.05)(.10) + (.60)(.72) + (.01)(a_{43})$$

or
$$a_{43} = \frac{.4375 - [(.05)(.10) + (.60)(.72)]}{.01} = .05$$

In general, the equation for loadings after the diagonal loading has been found is

$$a_{kj} = \frac{r_{jk} - (a_{j1}a_{k1} + a_{j2}a_{k2} + \cdots + a_{jj-1}a_{kj-1})}{a_{jj}} \qquad (13\text{-}8)$$

For a_{53}, we have

$$a_{53} = \frac{r_{35} - (a_{31}a_{51} + a_{32}a_{52})}{a_{33}}$$

$$= \frac{.1585 - [(.05)(.76) + (.60)(.20)]}{.01} = .05$$

We leave it to the reader to compute the loadings for a_{63}, a_{44}, a_{45}, a_{46}. The completed factor matrix is presented in Table 13-7.

Table 13-7   Diagonal Factor Solution

Variable	Factor				h^2
	I	II	III	IV	
1	.92	.00	.00	.00	.8464
2	.85	.04	.00	.00	.7241
3	.05	.60	.01	.00	.3626
4	.10	.72	.05	.02	.5313
5	.76	.20	.05	.01	.6202
6	.02	.05	.30	.02	.0933

The reader may suspect that we have gone through this detail simply to try his patience, for we now emphasize that the diagonal method is not a very practical way of solving a correlation matrix. First, the work can become quite tedious if the number of variables is large. Second, we usually do not know the communalities. But the example can serve to illustrate how other solutions differ when we do not know the communalities, and we can use the example above to discuss the end result — the factor matrix — we are trying to obtain.

As we said above, the factor loadings in the cells in Table 13-7 are simply the correlations of the variables with each of the factors. For example, the correlation of variable 1 with factor I is .92, and so on. Geometrically, we cannot represent the factor matrix as a plane, because there are more than two factors extracted. However, let us suppose that we have only two factors for the six variables.[7] If we allow the Y axis to represent factor I and the X axis to represent factor II, we may graph the resulting loadings as in Figure 13-2.

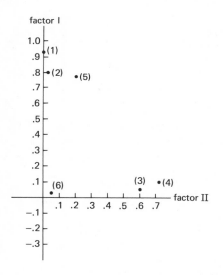

Figure 13-2. Graph of factor loadings.

Note that variables 1, 2, and 5 are highly loaded (correlated) on factor I and that variables 3 and 4 are highly loaded (correlated) on factor II. We may conclude, therefore, that three of the six variables are measuring one underlying factor, two are measuring a second factor, and the remaining

---

[7] For all practical purposes this is so, because the loadings on factors III and IV are small. There is no precise method of determining what size of correlation is needed to be considered a significant loading. This must be done in the context of the research problem.

variable (6) is loaded on a factor by itself. Because the communalities are not equal to 1.00, there appears to be some specific variance plus error variance for each variable. Hence, the model that applies here is $X_1^2 = h_j^2 + S_j^2 + E_j^2$. Unfortunately, there is no precise way of determining how much of $1 - h_j^2$ is specific and how much is error variance.

Estimation of Communality

The diagonal method of extracting factors is not practical because usually we do not know the communalities. When we do not, it is necessary either to estimate the communalities or to use different values in the diagonals. Let us first consider, in general terms, the question of estimating the communalities.

There is no one satisfactory way of estimating the communalities. There are several alternatives. Therefore there is no general solution for factor analysis. We might begin by assuming the correlation matrix to be a given rank.[8] Given this rank, which we call m, we know that there are $n - m$ determinants that must vanish for the matrix to be of this rank.[9] For a four-variable matrix, if the intercorrelations satisfy the following two conditions, then it is of rank one, and the variables can be described in terms of one common factor:

$$r_{12}r_{34} - r_{14}r_{23} = 0$$

$$r_{13}r_{24} - r_{14}r_{23} = 0$$

These are Spearman's tetrad differences, discussed above. If the matrix meets these conditions, we can set the appropriate second-order minors equal to zero and solve for each of the communalities. For the four-variable matrix, the solution for h^2 is given by

$$h_1^2 = \frac{r_{12}r_{13}}{r_{23}} \tag{13-9}$$

$$h_1^2 = \frac{r_{12}r_{14}}{r_{24}}$$

$$h_1^2 = \frac{r_{13}r_{14}}{r_{34}}$$

This can be generalized to a matrix of $n \times n$ dimensions in a straightforward manner. Similarly, the conditions that must be satisfied for a matrix to be rank two have been given.[10] However, the method quickly becomes extremely laborious and is therefore of only theoretical interest at this stage.

---

[8] The rank we assume might be determined by cluster analysis (see below for a discussion of this) or by an estimate based on knowledge of the variables.

[9] For a more formal statement of this, see Harman, p. 70.

[10] See Truman L. Kelley, "Essential Traits of Mental Life," *Harvard Studies in Education*, 26, Harvard University Press, Cambridge, Mass., 1935, p. 58.

There are other theoretical solutions to this problem, but none yet offers a practical method without the use of a computer. The difficulties of this problem have led to a number of arbitrary estimates of communalities, none of which leads to a minimal rank of the correlation matrix. But fortunately for practical research ".... there is much evidence in the literature that for all but very small sets of variables, the resulting factorial solutions are little affected by the use of 'communalities' or unities in the principal diagonal of the correlation matrix" (Harman, pp. 86, 88, 89). Thus, any one of the following arbitrary estimates might be used:

1. The highest correlation of the given variable from among its correlations with all other variables. Note that for the variables in Table 13-4, these values are .782, .782, .437, .437, .699, and .053. These are not greatly different from the true communalities and would yield a factor solution very similar to the one found for this matrix. Note, however, that this arbitrary approximation should not be used for a small number of variables.

2. The average of the correlations of a given variable with all the remaining ones: $\sum_{k=1} r_{jk}/(n-1)$. For the first variable in the matrix in Table 13-4 this is $(.782 + .046 + .092 + .699 + .018)/5 = .3275$. This has the advantage of moderating a high or extreme correlation.

3. Find the factor matrix by a method not requiring estimates of communalities, such as the bifactor method (see below), and use the communalities found by this method for a solution requiring estimates of communalities. This is clarified below when we discuss different types of factor solutions.

4. The squared multiple correlation of each variable with the remaining $n-1$ variables may be used. These are known to be less than the actual communalities, and procedures for computing them for small matrices are not too complicated.

5. Unities may be placed in each of the diagonals. As we said above, these are considered to be as good as placing communalities in the diagonal when we have a large matrix.

The many other methods of estimation are not considered here, because they involve considerations beyond the scope of this book. Moreover, for many practical research problems there are "canned" computer programs of factor analysis that do the work sufficiently well. It is more important to know something about the various types of factor solutions and methods of extracting factors, and it is to this we now turn.

TYPES OF FACTOR SOLUTIONS

Factor analysis does not have a general solution. There are a number of different types of solutions. Some of them have no more than historical interest. Others are being used today but have inadequacies from different points of view. It is necessary to consider some of these in order to get an adequate understanding of factor analysis. (Because the diagonal method was considered above, it is not repeated here.) Note that these are not necessarily practical methods of obtaining factors, but rather types of patterns that may result. Thus, the discussion emphasizes only the general characteristics of the solutions. The practical methods of factor analysis are considered in some detail in the next section.

As a means of illustrating the various types of solutions, we suppose that we have found the following 11 variables to be strongly related (as measured by the Pearson r) to expenditures for local government services and that we want to determine if it is possible to express the relationship in terms of a smaller number of factors. The variables are:

1. Size of community (total population)
2. Number of nonwhite
3. Condition of housing
4. Net migration
5. Median education
6. Median income
7. Number of persons employed in manufacturing
8. Total land area
9. Total local taxes
10. Total intergovernmental revenue
11. Total labor force

Unifactor Solution

An ideal factor solution for these variables would be one in which each of the variables is loaded on only one factor. One possible pattern is shown in Table 13-8. The italicized coefficients are the significant loadings. The others, although not all zero, are not significantly different from zero to be considered. If such a pattern did occur for actual data, it would be fairly easy to interpret. From the similarity among some of the variables loaded on each factor we might conclude that factor I is a demographic factor that refers to crowding or density, factor II relates to social status, and factor III concerns resources. Hence, we might explain the relationship in terms of three factors rather than in terms of 11 separate indicators. Of course, the meaning of the explanation is determined by the theory into which the factors are integrated.

This pattern is considered an ideal that actual data seldom if ever approximate. It might be noted that this type of solution does not require knowledge of the communalities.

Table 13-8

Variable	Factor I	Factor II	Factor III
	I	II	III
1	.86	.00	.03
2	.83	.09	.11
3	.75	.10	.09
4	.80	.02	.01
5	.10	.75	.02
6	.05	.67	.11
7	.00	.86	.03
8	.01	.10	.80
9	.00	.05	.85
10	.08	.06	.70
11	.01	.10	.85

Spearman's Two-factor Solution

We mentioned above that Spearman postulated that intellectual abilities might be explained in terms of two factors: one common to all tests, and the specific factor loadings. The two-factor solution would look like the data shown in Table 13-9. Actually, because each variable is loaded on only one additional factor besides the common factor, the Spearman solution is unifactor, in which only one factor is found. The two-factor solution is not considered satisfactory any more and thus need not be discussed in detail here.

Table 13-9   Illustration of Spearman Two-Factor Theory

Variable	Factor						
	I	II	III	IV	V	. . .	F_n
1	.86	.65	.01	−.05	.00	. . .	.00
2	.80	.02	.95	−.01	.09	. . .	.00
3	.95	.05	.00	.92	.01	. . .	.00
4	.99	.03	−.10	.08	.88	. . .	.00
5	.92	−	−	−	−	. . .	.00
6	.75	−	−	−	−	. . .	.00
7	.68	−	−	−	−	. . .	.00
8	.85	−	−	−	−	. . .	.00
9	.92	−	−	−	−	. . .	.00
10	.95	−	−	−	−	. . .	.00
11	.98	−	−	−	−	. . .	.00

Holzinger's Group-factor Solution

By the 1930s it was found that many actual correlation matrices did not fit the Spearman two-factor idea, because several variables would load on an additional factor besides the general factor. To account for this, Holzinger developed the notion that a correlation matrix might be explained in terms of a general factor, group factors, and specfic factors. The pattern of this solution would look like that shown in Table 13-10. The Holzinger solution helps account for overlap in loadings. In the hypothetical example, variables 1 and 2 are loaded on factor II in addition to being loaded on the general factor (factor I). The former might be considered a group factor. However, outside of psychology, relatively little empirical data seem to fit this pattern exactly. Moreover, its rationale depends on the development of a general theory that can explain the model. Such a theory has not been developed yet. For example, legislative behavior might be explained in terms of a general factor (constituency attitudes) and in terms of group factors (party identification of legislator, personality structure, region represented, and so on). But it is likely that empirical data will not fit such a neat pattern.

Table 13-10   Illustration of Holzinger Solution

Variable	Factor					
	I	II	III	IV	...	n
1	.80	.75	.00	.06	...	
2	.75	.65	.05	.10	...	
3	.86	.00	.90	.00	...	
4	.65	.05	.85	.09	...	
5	.50	.02	.67	.02	...	
6	.90	.00	.08	.75	...	
.						
.						
.						
n						

Principal-factor Solution

The principal-factor solution, frequently called principal-axis, provides a mathematically satisfactory solution in that it extracts first the factor that accounts for the maximum amount of variance, then the one that accounts for the maximum of the residual variance, and so on. The pattern would look like that shown in Table 13-11.

The principal-axis factor pattern can reproduce negative as well as positive coefficients. Thus, in the data in the example, variables 4 and 5 might be considered the oppositive of the factor being measured by variables 1, 2,

Table 13-11 Illustration of Principal-Factor Solution

Variable	Factor I	Factor II	...	n
1	.69	−.35		
2	.89	−.10		
3	.95	−.25		
4	−.99	.15		
5	−.87	.42		
6	.63	−.67		
.	.	.		
.	.	.		
.	.	.		
n				

3, and 6. For example, if the latter were variables such as number of non-white, crowding, and unsound housing and variables 4 and 5 were measures of income and education, they might all be taken to be measures of social status. As income and education go down, crowding and unsound housing increase. We consider the principal-factor method in more detail below, because it is the method most frequently used today and perhaps offers the best method for political science.

Rotation Methods

Once the factor have been found through the principal-factor solution (or one of the other methods), it may be possible to improve the loadings on factors and avoid overlapping of loadings by rotating the axes (what this means in geometric terms is explained below). There are two methods of doing this: orthogonal and oblique. Orthogonal rotation involves rotating the factor axes but keeping them at right angles. Oblique rotation involves rotating the factor axes but permits the angles between the axes to be increased or decreased from 90°. Because the question of rotation is rather important, it is discussed in more detail below.

METHODS OF EXTRACTING FACTORS

Up to this point we have considered some theory of factor analysis, the problem of finding diagonal values, and several of the more important types of factor solutions. Except for the discussion of the diagonal method, which was used to illustrate some of the theory of factor analysis, we have not yet considered how to do a factor analysis. We turn to this now. It might be noted that there are a number of different methods and the labor involved in doing a factor analysis by hand is very great. Most of the important and useful methods can be executed today by "canned" computer programs that are usually easily available. Thus, the discussion of methods that follows does not constitute a practical research guide, because this would take much more space than is warranted by the need for this sort of thing. Instead, we consider only three methods of extracting factors (cluster analysis, centroid, and principal axes) and the two methods of rotation.

Cluster Analysis

Cluster analysis may be used to determine what variables form groups, and we may use this to guess the rank of a matrix to use for estimating communalities. But it also yields similar results as factor analysis when the variables are pure and thus is useful to consider for pedagogical reasons.

There are various methods of cluster analysis, but the method that seems most useful is called Tyron's coefficient of belonging, or the B coefficient. The purpose is simple: we want to see if a set of intercorrelations falls into groups. As an example we use the correlation matrix in Table 13-12.[11]

Table 13-12   Correlation Matrix of Organizational Variables

	1	2	3	4	5	6	7
1		.5	.3	.1	.1	.2	.3
2	.5		.1	.1	.3	.3	.1
3	.3	.1		.7	.2	.3	.4
4	.1	.1	.7		.2	.1	.0
5	.1	.3	.2	.2		.4	.0
6	.2	.3	.3	.1	.4		.6
7	.3	.1	.4	.0	.0	.6	
$\sum_{i=1}^{k} K_i$	1.5	1.4	2.0	1.2	1.2	1.9	1.4

1 = centralization
2 = formalization
3 = specialization
4 = span of control
5 = styles of management
6 = morale
7 = professionalization

[11] The correlations were taken from a study done by the author; to simplify computation, they have been rounded to the nearest tenth. See D. J. Palumbo, et al., *op. cit.*

Because the objective is to find the variables that form groups, we should form the first group from the two variables that correlate highest. We then add to this group the variable that correlates the highest with each of the first two, and then a fourth, and so on, until we reach the point where the newest variable added is no more highly correlated with the variables in the group than with the remaining variables. This means that it does not any more belong to that group than to all the remaining variables. This is what the *B* coefficient measures. It may be defined as 100 times the ratio of the average intercorrelations among the variables of a group to their average correlations with all the remaining variables. In other words, when the average correlations of the variables in the group are no higher than their average correlations with all the remaining variables, *B* will equal 100, and we do not really have a group of variables. Symbolically, *B* is defined as

$$B_j = 100 \left(\frac{G}{n_g}\right)\left(\frac{T}{n_t}\right) \tag{13-10}$$

where G is the sum of correlations in a group, divided by the number of variables in the group n_g, which gives the average correlation, and T is the sum of the correlations of the variables in the group with all the remaining variables, divided by the number of remaining variables.

In actually calculating B, it is useful to set up a table. Table 13-13 gives

Table 13-13 Calculation of B Coefficient

(1) j	(2) R	(3) L	(4) G	(5) M	(6) v	(7) I	(8) r	(9) $\bar{X}_C$	(10) $\bar{X}_R$	(11) B
(3,4)	3.2	.7	.7	1.8	2	1	10	.70	.18	388
(3,4,7)	1.4	.4	1.1	2.4	3	3	12	.36	.20	180*
(3,4,7,6)	1.9	1.0	2.1	2.3	4	6	12	.35	.19	184*
(6,7)	3.3	.6	.6	2.1	2	1	10	.60	.21	285
(6,7,1)	1.5	.5	1.1	2.6	3	3	12	.36	.21	171*
(1,2)	2.9	.5	.5	1.9	2	1	10	.50	.19	263
(1,2,5)	1.2	.4	.9	2.3	3	3	12	.30	.19	157*

Where j = variables in cluster
  R = sum of correlations of variables added to cluster with all other variables
  L = sum of correlations of variables added to cluster with variables already in cluster
  G = sum of correlations among variables in cluster = L + (G from previous line)
  M = sum of correlations of variables in cluster with variables not in cluster = $R - 2(L) + M$ from preceding line
  v = number of variables in cluster
  I = number of intercorrelations in cluster = $v(v - 1)/2$
  r = number of remaining intercorrelations = $v(n - v)$, where n = number of variables
  $\bar{X}_C$ = mean intercorrelation in cluster = G/I
  $\bar{X}_R$ = mean of remaining intercorrelations = M/r
  B = coefficient of belonging = $\bar{X}_C/\bar{X}_R$ 100

all the items necessary for computing B. To illustrate, we calculate the groups for the correlation matrix in Table 13-12. We start with the two variables that have the highest intercorrelation. Table 13-12 shows that these are variables 3 and 4. We enter them in column 1 in Table 13-13. In the second column we enter the sum of the correlations of variables 3 and 4 with all other variables. This is obtained by adding the sums of columns for these two variables from Table 13-12. What is entered in each of the other columns in Table 13-13 is described in the table. The entry in column 5 may be obtained by subtracting from R two times L_1 and then adding M from the previous line. In this case we have $[3.2 - 2(.7)] + 0 = 1.8$. Columns 6, 7, and 8 are self-explanatory. Column 9 is the average correlation of the variables in the cluster obtained by dividing the sum of the correlations of the variables in the cluster (in this case 1.4) by the number of intercorrelations in the cluster. Column 10 is self-explanatory.

As we said above, if the mean correlation of the variables in the cluster is the same as the average correlation of these variables with the variables not in the cluster, B will be 100. In the case of the first row in Table 13-13, it is obvious that the average correlation of the variables in the cluster is almost four times that of their correlation with the remaining variables, and hence we may conclude that the two variables form a cluster. Now we want to see if there are more variables in this cluster, and so we add a third variable. The next variable to be added is the one that correlates highest with the ones in the cluster. Examine Table 13-12 again. Variable 7 has the next highest correlation with variable 3, and variable 5 has the next highest with variable 4. We want to add the variable whose sum of correlations with 3 and 4 is the highest. The sum of the correlations of variable 7 with 3 and 4 is .4. The sum of the correlations of variable 5 with 3 and 4 is also .4. Because for the present data, we have not carried out the correlation coefficients enough places, we have a tie. Generally this does not happen, but if it does, there is no objective criterion for deciding which variable should be added. We must use knowledge of the variables to do this. In the present case, we select variable 7 on the grounds that it is conceptually more likely to be in the same cluster as variables 3 and 4 than variable 5.

The variables now in the cluster are entered on line 2 of column 1 in Table 13-13. The sum of the correlation of the new variable just added with all the other variables is entered in column 2. Column 3 is the sum of the correlations between the variable being entered in the cluster and those already there: $(r_{73} = .4) + (r_{74} = 0) = .4$. Column 4 is the sum of the correlations of all the variables in the cluster: $(r_{34} = .7) + (r_{37} = .4) + (r_{47} = 0) = 1.1$. Note that column 4 can be obtained directly by adding the L of the same row (.4) with the G from the previous row (.7). Column 5 can also be obtained directly from the table in accordance with the directions. Columns 6 to 11 are straightforward and have already been explained. The B coefficient for this new cluster is 180, that is, a drop of 208 points. We may conclude that this is a significant drop.[12] We may therefore reject variable 7. However, let us add one more variable before we do. Consider variable 5. The sum of the correlations of 5 with each of those in the cluster is $(r_{53} = 2.)$ + $(r_{54} = .2) + (r_{57} = 0) = .4$. Note that the correlation of variable 6 with variable 7 is .6. Hence we should also consider variable 6. The sum of its correlations with the variables in the cluster is $(.3) + (.1) + (.6) = 1.0$. Because this is higher than for any other variable, including variable 5, we add it to the cluster. The B coefficient for this cluster is almost the same as the last. However, because the addition of variable 7 reduces B by a large amount, we delete it and also variable 6. We now begin a new cluster by considering the two that have the highest correlation from among the variables not included in the first cluster. An examination of Table 13-12 shows these to be variables 6 and 7. We thus start a new cluster with them. The B coefficient is 285 for these two variables. Next, we refer to Table 13-13 for the variables (not already in a cluster) that correlate highest with 6 and 7. Variable 5 correlates next highest with 6 and variable 1 next highest with 7. The sum of the correlations of each with 6 and 7 is

Variable 5:	$(.4) + (.0) = .4$
Variable 1:	$(.2) + (.3) = .5$

Hence we add variable 1 to the cluster. The B coefficient for this new cluster drops a great deal. Hence we reject variable 1 from this cluster.

So far we have two clusters: variables 3 and 4 and variables 6 and 7. We may ask whether the remaining variables — 1, 2, and 5 — form a cluster. We begin with the two that correlate highest. These are variables 1 and 2. The B coefficient for these variables is 263. When we add the remaining variable 5, the B coefficient drops significantly. Thus, we may conclude that we have three clusters: 3 and 4, 6 and 7, 1 and 2, and one variable that does not seem to belong to any of the clusters, that is, variable 5.

How meaningful is it to say that we have three clusters from among the seven variables with which we began? First, we may conclude that we are not really measuring seven separate variables, but actually only three or four. It appears that variables 1 and 2 are measuring somewhat the same underlying phenomenon. Logically, this appears plausible. One of the ways an organization might try to centralize is to put down in detail the kinds of behavior it expects of its members. This is actually how formalization might be measured. Similarly, it would appear conceptually correct that professionalism and morale should form a cluster. The more professional an organization, the higher the morale. A professional might be defined as one who is highly committed to his job, likes his work, and so on, and these same indices measure morale.

---

[12] There is, unfortunately, no measure of statistical significance here. Judgment must be used.

Without getting into organization theory in detail, the point is that cluster analysis does not provide a method of uniquely determining the minimum number of factors that express a correlation matrix unless the variables are factorily pure. But it can help demonstrate which variables are most highly grouped. We now consider a method that comes closer to a unique solution for the minimum number of factors in a correlation matrix.

Centroid Method

The centroid method is a compromise for the principal-factor method, which is described below. It requires much less computation than the principal-factor method, but it does not provide a unique solution.

As an example, we use correlations for the same seven variables as in Table 13-12, but in this case they have not been rounded (see Table 13-14). We use a step-by-step procedure to compute the first factor:

1. Communalities are usually placed in the principal diagonal. If they are not known, they may be estimated as the highest correlation of the variable with all other variables or by unities. In this example, unities are used (see Table 13-14).

2. Each column is added algebraically, omitting the diagonal. Enter this in the row in Table 13-14 labeled $\sum K$. Do the same for the rows, and check that $\sum K = \sum j$. If any of the column sums are negative, it will be necessary to reflect them. This is described beginning with step 10 below. Because none of the sums in Table 13-13 is negative, we continue.

3. Add the diagonal values to $\sum K$, and enter the sums in the row labeled DK.

4. Add across rows $\sum K$ and DK, and label the sums $\sum\sum K$ and $\sum DK$ respectively.

Table 13-14 Correlation Matrix of Organizational Variables with Estimates of Communalities Placed in the Principal Diagonal

	1	2	3	4	5	6	7	
1	1.000	.489	.291	.095	.068	−.168	−.168	
2	.489	1.000	.102	−.074	.274	.251	−.096	
3	.291	.102	1.000	.664	.220	.290	.368	
4	.095	−.074	.664	1.000	.193	.113	.046	
5	.068	.274	.220	.193	1.000	.368	−.011	
6	−.168	.251	.290	.113	.368	1.000	.579	
7	−.327	−.096	.368	.046	−.011	.579	1.000	
ΣK	.448	.946	1.935	1.037	1.112	1.433	.559	$\Sigma\Sigma K =$ 7.470
DK	1.448	1.946	2.935	2.037	2.112	2.433	1.559	$\Sigma DK =$ 14.470
a_{j1}	.3806	.5115	.7715	.5354	.5552	.6395	.4098	

5. Divide each entry in row DK by $\sqrt{\sum DK}$. Place these directly below row DK in row a_{j1}. These are the factor loadings on the first factor. A check is that $\sum a_{j1} = \sqrt{\sum DK}$ (except for rounding error). The general equation for the first factor loading is thus $a_{j1} = \sum DK_i / \sqrt{\sum DK}$, where $\sum DK_i$ equals the sum of the coefficients in the column for test a and $\sum DK$ equals the sum of all the values of r in the table. Note the similarity of this to the coefficient of belonging.

6. Prepare another table (see Table 13-15) in which the first factor loadings are placed above and to the left of the corresponding column and row numbers for each variable (disregard negative signs).

7. Subtract the product of the column and row factor loading from the value in the cell of the preceding matrix (Table 13-14), and place the result in the corresponding cell of the residual matrix; that is, to find the value of the first row, first column of the residual matrix, we have $(1.0000) -$ $(.3806)(.3806) = .8551$. This is entered in parentheses in Table 13-15. This is continued until all the cells in the residual matrix are completed. (As a check, when this has been done for each of the cell values, the algebraic sum of each column or row should be zero or close to zero, allowing for rounding error.)

8. The communalities are reestimated in Table 13-15. The highest residual correlation of each column, ignoring the sign, is used as the new estimate. These are entered above the value in the parentheses and are italicized.

9. Sum the columns, omitting diagonal values, and enter the sum in row $\sum_{j2}$. If any of these sums are negative (they will always be for the first residual matrix), they must be reflected. This is accomplished in the next several steps.

10. Select the column with the largest negative total for the first reflection. Copy this total directly below in the next row (labeled K_1 because it is the first column that is being reflected), with the sign changed to positive. Place an asterisk at the head of this column and next to its row (column 1, row 1 for the first reflection.)

11. To complete the rest of the row labeled for the column being reflected (K_1 for the first reflection), double the value of the residual correlation of the row being reflected (row 1 for the first reflection) for each column other than the one being reflected, change the sign, and add this amount to the column total. For example, on the first reflection for column 2, we double .2943, $[2(.2943) = .5886]$, change the sign to $-.5886$, and add this to $-.7379$, giving -1.3265. This is entered in the corresponding cell in the row labeled for the column being reflected.

12. The reflection should continue until the new column sums are all positive. Before proceeding to the second reflection, however, check the values obtained. To do so, first total rows $\sum_{j2}$ and K_1. If the computations have been correct the sum of K_1 should be equal to $\sum\sum_{j2} + 4$ (the sum of

Table 13-15 First Residual Matrix

a_{j1} Variable	(.3806) 1*	(.5115) 2*	(.7715) 3	(.5354) 4	(.5552) 5*	(.6395) 6	(.4098) 7	
(.3806) 1*	.4830 (.8551)	.2943	(+) −.0029	(+) −.1088	−.1433	(+) −.4114	(+) −.4830	
(.5115) 2*	.2943	.3479 (.7384)	(+) −.2926	(+) −.3479	−.0100	(+) −.0761	(+) −.3056	
(.7715) 3	(+) −.0029	(+) −.2926	.2926 (.4043)	.2509	(+) −.2083	−.2034	.0518	
(.5354) 4	(+) −.1088	(+) −.3479	.2509	.3479 (.7133)	(+) −.1043	−.2294	−.1734	
(.5552) 5	−.1433	−.0100	(+) −.2083	.1043	.2384 (.6917)	(−) .0129	(+) −.2384	
(.6395) 6	(+) −.4114	(+) −.0761	−.2034	−.2294	(−) .0129	.4114 (.5910)	.3169	
(.4098) 7	(+) −.4830	(+) −.3056	.0518	−.1734	(+) −.2384	.3169	.4830 (.8321)	
Σf_{j2}	−.8551	−.7379	−.4045	−.7129	−.6914	−.5905	−.8317	$\Sigma\Sigma f_{j2} = -4.8240$
K_1	.8551	−1.3265	−.3987	−.4953	−.4048	−.2323	.1343	$\Sigma K_1 = -1.4036$
K_2	1.4437	1.3265	.1865	.2005	−.3848	.3845	.7455	$\Sigma K_2 = 3.9024$
K_5	1.1571	1.3065	.6031	.4091	.3848	.3587	1.2223	$\Sigma K_5 = 5.4416$
DK_2	1.6401	1.6544	.8957	.7570	.6237	.7701	1.7053	$\Sigma DK_2 = 8.0463$
q_{j2}	.5781	.5832	.3157	.2668	.2198	.2714	.6011	

the column being reflected), or $-1.4036 = (-4.8240) + 4(.8551) = -1.4036$. If more reflection is required, select the column with the highest negative total (column 2 in this case), and repeat the process in steps 10 to 12. However, in columns that have been reflected (column 1 in this case) the signs are not reversed before adding the doubled value. The check for the reflection of column 2 is $3.9024 = (-1.4036) + 4(1.3265) = 3.9024$. K_5 still has a negative sign, and thus it is necessary to reflect it. (If a variable is reflected more than once, then, on the odd reflection for this variable, the sign is changed before adding. But the sign is not changed on the even reflection. In the present example, each variable is reflected only once, and thus this rule does not apply.)

13. When the reflections are completed (all totals positive), then reverse the signs of all reflected row values not in reflected columns, and do the same for all reflected column values not in reflected rows. This is indicated in Table 13-15, with the new signs in parentheses.

14. The factor loadings on factor II are obtained in the same way as previously described: (a) Add the reestimated communalities to the last row of adjusted sums (K_5 in this case), label this row DK_2, and sum it $\left(\sum DK_2 = 8.0463\right)$. (b) Obtain the factor loadings by dividing each value in row DK_2 by $\sqrt{\sum DK_2}$. Enter these immediately below row DK_2, and label it a_{j2}. For example, the factor loading of the first variable on factor II is $1.6401/\sqrt{8.0463} = .5781$. Before proceeding, we construct a factor matrix in Table 13-16.

Table 13-16   Factor Matrix

Variable	Factor I	II	III	IV	h^2
1	.3806	−.5781	.4377	−.3244	.7759
2	.5115	−.5832	.1916	.2472	.6996
3	.7715	.3157	.4784	.1256	.9395
4	.5354	.2668	.5184	.3798	.7709
5	.5552	−.2198	.2021	.4216	.5752
6	.6395	.2714	−.5317	.0580	.7687
7	.4098	.6011	−.3774	.3174	.7724

Examination of this table shows that all variables tend to be loaded on the first factor, and variables 1, 2, and 7 tend to be loaded on the second factor. Note that variables 1, 2, and 5 have negative loadings on factor II. The sign for any factor loading is determined by the rule that variables that have been reflected an odd number of times have signs the opposite of their sign on the preceding factor. In the present case, variables 1, 2, and 5 have been reflected once.

Table 13-17 Second Residual Matrix

a_{jp} Variable	(.5781) 1*	(.5832) 2	(.3157) 3	(2.668) 4	(.2198) 5	(.2714) 6*	(.6011) 7*	
(.5781) 1*	*.2704* (.1488)	(+) -.0428	(+) -.1796	(+) -.0454	(+) -.2704	.2545	.1355	
(.5832) 2	(+) -.0428	*.1923* (.0078)	.1085	.1923	-.1382	(+) -.0822	(+) -.0449	
(.3157) 3	(+) -.1796	.1085	*.2891* (.1929)	.1667	.1389	(+) -.2891	(+) -.1380	
(.2668) 4	(+) -.0454	.1923	.1667	*.3338* (.2767)	.0457	(+) -.3018	(+) -.3338	
(.2198) 5	(+) -.2704	-.1382	.1389	.0457	*.2704* (.1901)	(+) -.0726	(-) .1063	
(.2714) 6*	.2545	(+) -.0822	(+) -.2891	(+) -.3018	(+) -.0726	*.3018* (.3377)	.1538	
(.6011) 7*	.1355	(+) -.0449	(+) -.1380	(+) -.3338	(-) .1063	.1538	*.3338* (.1217)	
Σ_{j3}	-.1482	-.0073	-.1926	-.2763	-.1903	-.3374	-.1211	$\Sigma\Sigma_{j3} = -1.2732$
K_6	-.6572	.1571	.3856	.3273	-.0451	.3374	-.4287	$\Sigma K_6 = .0764$
K_1	.6572	.2427	.7448	.4181	.4957	.8464	-.6997	$\Sigma K_1 = 2.7052$
K_7	.9282	.3325	1.0208	1.0857	.2831	1.1540	.6997	$\Sigma K_7 = 5.5042$
DK_3	1.1986	.5248	1.3099	1.4195	.5535	1.4558	1.0335	$\Sigma DK_3 = 7.4956$
a_{j3}	.4377	.1916	.4784	.5184	.2021	.5317	.3774	

Because most of the loadings on the second factor are still high, we should proceed to a third factor. To do so, we obtain a second residual matrix in the same manner as the first; that is, we subtract the product of the second factor loadings from the correlation in the preceding table. The results are shown in Table 13-17. We then return to step 7 above to compute the loadings for the third factor. These have been entered in Table 13-16 from Table 13-17. The row labeled a_{j3} gives the loadings on the third factor.

How many factors should be extracted? Obviously, if the data yield as many factors as there are variables, there is no gain in a factor analysis. It is possible to determine the maximum number of factors that can be uniquely determined by n variables by the formula $r = (2n + 1 - \sqrt{8n + 1})/2$. For example, if we have seven variables, the maximum number of possible factors is $[2(7) + 1 - \sqrt{8(7) + 1}]/2 \doteq 4$. It is generally best to extract at least the maximum number of factors. For the example above, this is four.[13] To do this, a third residual matrix (Table 13-18) has been prepared, and the loadings on the fourth factor have been entered in Table 13-16.

Recall that we said that $h_j^2 + S_j^2 + E_j^2 = 1$, and that $a_1^2 + a_2^2 + \cdots + a_n^2 = h_j^2 = 1$, if there is no specific and error variance. The factor matrix in Table 13-16 shows that there is still some variance not accounted for. For example, the communality of variable 1 is $h_1^2 = (.3806)^2 + (.5781)^2 + (.4377)^2 + (-.3244)^2 = .7759$. Therefore, $1 - .7759 = .2241$ is the amount of specific and error variance for this variable. Table 13-16 indicates that there does not seem to be a clear factor solution to these seven variables. All the variables are loaded on the first factor and most continue to have high loadings on successive factors. We postpone further interpretation of this until the section on rotation.

Principal-factor Method

The centroid method of extracting factors just described was developed by Thurstone as a mathematical compromise for the labor involved in the principal-factor method. The principal-factor method finds first the factor that contributes the most variance, then the factor that contributes the next from the residual variance, and so on. The computations for this method are very laborious and thus, for large matrices, should not be attempted without the aid of a computer.[14] Because the process is very laborious and has been described adequately elsewhere, we do not consider it here (see Hotelling; Fructer). It should be stressed, however, that the factors obtained by the principal-factor method (or by the centroid or other methods) are considered

[13] See Harmon, Chap. 17, for a discussion of statistical methods for testing the number of significant loadings.

[14] See H. Hotelling, "Simplified Calculation of Principal Components," *Psychometrika*, vol. 1, pp. 27–35, 1936.

Table 13-18 Third Residual Matrix

a_{j3} Variable	(.4377) 1*	(.1916) 2	(.4784) 3*	(.5184) 4	(.2021) 5*	(.5217) 6*	(.3774) 7	
(.4377) 1*	*.1819* (.0789)	(+) -.0411	-.0297	(+) -.1815	.1819	.0218	(+) -.0297	
(.1916) 2	(+) -.0411	*.1769* (.1556)	(-) .0168	.0930	(+) -.1769	(+) -.0197	-.0274	
(.4784) 3*	-.0297	(-) .0168	*.0813* (.0603)	(+) -.0813	.0422	.0347	(+) -.0425	
(.5184) 4	(+) -.1815	.0930	(+) -.0813	*.1851* (.0651)	(+) -.0591	(-) .0262	.1382	
(.2021) 5*	.1819	(+) -.1769	.0422	(+) -.0591	*.1826* (.2296)	-.0349	(+) -.1826	
(.5317) 6*	.0218	(+) -.0197	.0347	(-) .0262	-.0349	*.0469* (.0191)	(+) -.0469	
(.3774) 7	(+) -.0297	-.0274	(+) -.0425	.1382	(+) -.1826	(+) -.0469	*.1826* (.1914)	
Σ_{j4}	-.0783	-.1553	-.0598	-.0645	-.2294	-.0188	-.1909	$\Sigma\Sigma_{j4} = -.7970$
K_5	-.4421	.1985	-.1442	.0537	.2294	.0510	.1743	$\Sigma K_5 = .1206$
K_1	.4421	.2807	-.0848	.4167	.5932	.0074	.2337	$\Sigma K_1 = 1.8890$
K_3	.3827	.2471	.0848	.5793	.6776	-.0620	.3187	$\Sigma K_3 = 2.2282$
K_6	.4263	.2865	.1542	.5269	.6078	.0620	.4125	$\Sigma K_6 = 2.4762$
DK_4	.6082	.4634	.2355	.7120	.7904	.1089	.5951	$\Sigma DK_4 = 3.5135$
a_{j4}	.3244	.2472	.1256	.3798	.4216	.0580	.3174	

arbitrary, and rotation is necessary to arrive at more meaningful factors (meaningful from the perspective of the theory being used). Let us consider this question now.

METHODS OF ROTATION

Rotation can best be explained in geometrical terms. We use the first two variables of the matrix in Table 13-16 as an example. Recall now what these variables are so that we can make more meaning out of the purpose of rotation. Figure 13-3 shows clearly that centralization, formalization, morale, and professionalization cluster closely in the upper right-hand quadrant and that specialization and span of control cluster in the lower right-hand quadrant. The solution found by the centroid method indicates that these variables are loaded relatively highly on both factors. But if we rotate the factor axes (to position 2), we maximize the loading of variables 1, 2, 6, and 7 on factor II and minimize their loadings on factor I. At the same time, the loadings of variables 4 and 3 on factor I increase and their loadings on factor II decrease.

The rotation depicted by the straight lines in Figure 13-3 keeps the two

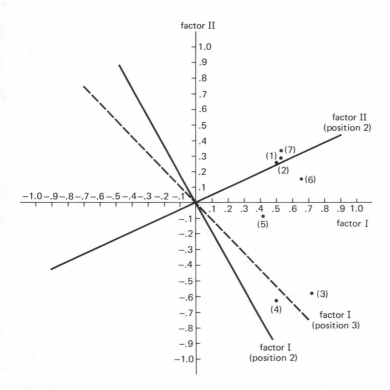

Figure 13-3.

factor axes at right angles and is thus called orthogonal rotation. The rotation seems to have produced a clearer situation in that the loadings of the variables on the respective axes are more factorially pure than with unrotated loadings. It presents a clearer picture of the differences between the sets of variables.

We might also have allowed the angle of the axes of factors I and II to change when we rotated them. For example, we might have decreased the angle of the intersection in the upper right-hand quadrant (see dashed line, position 3, factor I) so as to maximize the loadings of variables 3 and 4 on factor I. This is called oblique rotation, because the axes were allowed to form an acute rather than a right angle. By doing this, we allow the possibility that the factors are correlated rather than independent while making the loadings of variables 3 and 4 on factor I more factorially pure (however, this makes it difficult to compare factors from study to study).

The rotation performed in Figure 13-3 was done visually. The axes were placed where they appeared to minimize the sum of the distances of the loadings from each of the axes. In earlier factor analysis, rotation was carried out in this visual manner. It is apparent that this is not a precise method. In order to overcome this, Thurstone developed five rules of *simple structure* to use as criteria for finding an adequate solution:

1. Each row of the factor matrix should have at least one loading close to zero.

2. For each column, there should be at least as many variables with zero or near zero loadings as there are factors.

3. For every pair of factors there should be several variables with loadings on one factor but not on the other.

4. When four or more factors exist, a large proportion of the variables should have negligible loadings on any pair of factors.

5. For every pair of factors there should be only a small number of variables with appreciable loadings in both columns.

What these principles amount to is a factor pattern close to that shown in Table 13-8. Each variable in this table is a relatively pure measure of its factor. Rotation might also attempt to accomplish the pattern depicted in Holzinger's group-factor theory or some other pattern. There is no objective criterion for determining which pattern is best. Some idea should be in mind prior to the factor analysis, and factor analysis used to see if the pattern assumed does exist. Hence, we might begin by assuming that certain variables will cluster. To use the example in Table 13-12 again, we might assume that the variables of centralization, span of control, and specialization should cluster, because all relate to structural factors. We might also expect styles of management and formalization to cluster as another factor, because they both refer to methods used by supervisors in dealing with subordinates. Finally, we might expect morale and professionalism to cluster, because they both relate to dedication to work. The factor analysis and rotation of these variables done above shows that these assumptions are only partially correct. To approximate this better, we could rotate the axis of factor II a little farther clockwise so as to maximize the loading of variables 2 and 5 on this factor. We should then have a close approximation to the theoretical assumption.

It is clear from the above, however, that rotation and factor analysis are most useful in suggesting ways of revising the measures of variables so as to see whether a simple interpretation is possible. For example, we might measure both morale and professionalism by indices that show how dedicated a person is to his work. The factor analysis demonstrates that these two things are measuring somewhat the same thing. Hence, we might want to combine some aspects of the indices of each variable and use them to measure a new single variable that might be given a new name.[15]

Objective Solutions for Rotation

The methods of rotation and the simple-structure principles described above are entirely subjective. Earlier attempts to make the solutions more objective contained certain arbitrary and subjective elements, although the computations involved were practicable.[16] In the recent decade, however, the use of computers has made it feasible to develop objective techniques, because practical considerations of computing routines can now be somewhat ignored. We discuss several of these here in general terms.[17] The first two assume orthogonal rotations, and the second two oblique.

Quartimax Method. If an ideal unifactor solution is to be approximated, it is necessary to rotate the axes so as to maximize the large factor loadings and minimize the small ones. The objective of the orthogonal rotation in this case would be to maximize the variance of the squared factor loadings. We might define the variance of the squared factor loadings of all r factors for n variables as

$$\sigma_{a^2}^2 = \frac{1}{rn} \sum_{i=1}^{n} \sum_{j=1}^{r} (a_{ij}^2 - \bar{a}^2)^2 \qquad (13\text{-}11)$$

[15] Again, the name to be used for the factor depends entirely on the theoretical framework being used. Factor analysis does not provide the names for us. In the example just discussed, we might apply a broader concept to both professionalism and morale, calling the underlying phenomenon degrees of upward mobility. See Robert Presthus, *The Organization Society*, Random House, Inc., New York, 1965.

[16] See Paul Horst, "A Non-graphical Method for Transforming an Arbitrary Factor Matrix into a Simple Structure Factor Matrix," *Psychometrika*, vol. #2, 6, pp. 77–99, 1941; L. R. Tucker, "A Semi-analytical Method of Factorial Rotation to Simple Structure," *Psychometrika*, vol. #1, 9, pp. 43–68, 1944; L. L. Thurstone, "An Analytical Method for Simple Structure," *Psychometrika*, vol. #3, 19, pp. 173–182, 1954.

[17] Methods of computing rotated factor values are not discussed here. They are generally available in "canned" computer programs.

where
$$\bar{a}^2 = \frac{1}{rn} \sum_{i=1}^{n} \sum_{j=1}^{r} a_{ij}{}^2$$

and where a represents the factor values resulting from the rotation of a. The objective is to maximize the sum (13-11).[18] (This is done through the differential calculus by finding the angle that maximizes the fourth power of the loadings.) Note that we might begin with a centroid or principal-axis solution and then perform an orthogonal rotation to maximize the sum. It might also be emphasized here that the resulting solution does not necessarily make the small factor loadings reach zero. Simple structure is only approximated, not actually achieved (an oblique rotation may have to be used to achieve this, but even here it may not if the empirical data do not fit the unifactor model).

Varimax Method. The varimax method is a modification of the quartimax method. It maximizes the variance of each factor first, and the criterion of maximum simple structure is the maximization of the sum of these simplicities of the individual factors. One form[19] of this criterion is

$$\sigma^2 = \frac{1}{n} \sum_{p=1}^{r} \sum_{i=1}^{n} a^4{}_{ip} - \frac{1}{n^2} \sum_{p=1}^{r} \left(\sum_{i=1}^{n} a^2{}_{ip} \right)^2$$

In a later version of the varimax solution, the variables are weighted equally so that those with larger communalities do not have a greater influence on the rotations. The result is a solution that does not depend on the composition of the variables when they involve the same common factor. This is called the *factorial-invariance* criterion and can be shown to be a better principle than simple structure. It will yield the same solution if only $n-r$ of the original variables are used and they are measures of the same common factor.

Oblimax Method. If we allow the rotation to assume oblique angles rather than restrict it to the orthogonal condition, it is possible to set up criteria similar to the quartimax and varimax methods. One such criterion is to maximize the kurtosis of the "doubled" frequency distribution of rotated factor loadings so as to increase the relative frequencies of the near-zero and large loadings and minimize the intermediate loadings.[20] It can be shown that this is a special case of maximizing the variance of factor loadings given in Equation (13-11). The mathematics required to explain it, however, are beyond the scope of this book. It suffices to note that it cannot be accomplished without the aid of a computer and that it is a way of approximating

[18] See J. O. Neuhaus and C. Wrigley, "The Quartimax Method: An Analytical Approach to Orthogonal Simple Structure," *British Journal of Statistical Psychology*, 7, pp. 81–91, 1954.

[19] See H. Kaiser, "The Varimax Criterion for Analytic Rotation in Factor Analysis," *Psychometrika*, vol. #3, 23, pp. 187–200, 1958.

[20] See C. Pinzha and D. R. Saunders, "Analytic Rotation to Simple Structure, II. Extension to an Oblique Solution," *Research Bulletin*, RB 54–31, Educational Testing Service, Princeton, N. J., 1954.

mathematically the simple-structure criteria postulated by Thurstone. If the data do not have simple structure, the oblimax method will not be of any help in understanding the underlying structure.

Quartimin Method. Another criterion similar to that of Equation (13-11) is to minimize the sum of the cross products of the squared factor loadings.[21] This also yields a mathematically unique estimate of simple structure similar to the quartimax solution, and is thus subject to the same problems in application, although it may come closer to simple structure when the clusters among variables are not so well defined.

Although these four methods provide mathematically precise solutions for the criterion to be maximized, which solution is best depends on the nature of the data being analyzed. Certain solutions are better with complex data than with simple data. In fact, if there are distinct clusters in the data, visual and subjective rotation provides an equally good solution and one that may be better from the theoretical point of view. In addition, one may seriously question whether the simple-structure principles are theoretically sound for political science (or for psychology). A theoretically more interesting line of inquiry concerns the question of whether the linear models of factor analysis are appropriate theoretical vehicles. Of course, factor analysis can conceivably be adapted to curvilinear models as well. But the fundamental assumptions of conventional statistical techniques still remain. Probability models and game theory (discussed in Chapter 14) perhaps offer more fruitful lines of inquiry, at least for theory building in political science.

READINGS AND SELECTED REFERENCES

Cattell, R. *Factor Analysis*, Harper & Row, Publishers, New York, 1952.

Diamond, Solomon. *Information and Error*, Basic Books, Inc., New York, 1959, Chap 12.

Fruchter, Benjamin. *Introduction to Factor Analysis*, D. Van Nostrand, Company, Inc., Princeton, N. J., 1954.

Guilford, J. P. *Psychometric Methods*, McGraw-Hill Book Company, New York, 1936, Chap. 14.

Harman, Harry. *Modern Factor Analysis*, The University of Chicago Press, Chicago, 1960.

Kerlinger, Fred. *Foundations of Behavioral Research*, Holt, Rinehart and Winston, Inc., New York, 1965, Chap. 36.

Thurstone, L. *Multiple Factor Analysis*, The University of Chicago Press, Chicago, 1947.

[21] See J. B. Carroll, "An Analytical Solution for Approximating Simple Structure in Factor Analysis," *Psychometrika*, vol. #1, 18, pp. 23–38, 1953. There is a whole class of methods of oblique transformations which minimize certain expressions to achieve simple structure but which are not considered here. These generally come under the rubric of oblimin methods and have been given special names, depending on which is to be minimized. For example, if we minimize the covariance of squared elements of the rotated factor structure, the solution is called *covarimin*. If we minimize the covariances and the cross products of factor loadings, the solution is called *biquartimin*.

14

Decision Theory; Bayes' Theorem; Game Theory

We noted above that there are two principal schools of thought regarding probability: the objectivist and the subjectivist. The statistics considered so far in this book are based on the objectivist school. The method of testing hypotheses using the objectivist definition of probability is frequently referred to as the traditional or classical approach to decision making. This was described in Chapter 7. In the present chapter we consider some of the newer developments in decision theory.

BAYES' THEOREM

The English minister Thomas Bayes is generally credited with having developed statistical decision theory based on a notion of subjective probabilities. His ideas have not received a great deal of attention until recent years, when they have been revived as a general method for measuring probabilities and making decisions where the classical approach does not help. We begin with an example.

Suppose that there is a wager concerning the next presidential election. Let us say that there are three people involved and that the first bets that the Democratic candidate will receive 40 percent of the total vote, the second that he will receive 60 percent, and the third that he will receive 70 percent. (The wagers are not realistic. Round numbers have been used to make the example easier to follow.) Which of the three is likely to be correct? In answering this question, suppose that you know nothing about the personalities of the men involved, who won the last several elections, what the principal issues are, what the most recent polls say, and so on. Because we have no knowledge about which of the three is likely to be correct, it is reasonable to assign an equal probability to each. Let us represent this as follows:

$$p(H_1) = .33$$
$$p(H_2) = .33$$
$$p(H_3) = .33$$

where H_i represents the wager (or hypothesis) of each person. Let us call these prior probabilities, because they are the probabilities we assign prior to obtaining other information about which of the three is most likely to be correct. We may also consider these our subjective beliefs about the probability of each being correct.

Now, suppose that someone takes a sample of voters and finds that 40 percent of the sample say that they will vote for the Democratic candidate. We now have some additional knowledge. Given this sample evidence, which of the three H_i is most likely to be correct? Common sense would answer that H_1 is. Bayes' theorem gives us a means for measuring the exact probability of its being correct. Or, in more general terms, it gives us a means of revising our prior beliefs, based on sample evidence. To do so, we need two sets of probabilities: first, the conditional probability of each possible population value, given the sample value, and, second, the joint probability of our prior belief and the conditional probabilities.

The conditional probability of each population value can be computed by means of the binomial theorem, as follows:

$$p(4,10;.4) = \binom{10}{4}(.4)^4(.6)^6 = .250$$

$$p(4,10;.6) = \binom{10}{4}(.6)^4(.4)^6 = .111$$

$$p(4,10;.7) = \binom{10}{4}(.7)^4(.3)^6 = .037$$

The joint probabilities of each of the hypotheses, given the sample value, have been computed in Table 14-1. Note that the posterior probabilities are actually the joint probabilities converted to a scale of 0 to 1. They give us the probability of each hypothesis, given the sample evidence. The probability that the first person is correct, given the sample evidence, has been increased from the prior belief of .33 to a new belief of .63. The probabilities of the other hypotheses being correct have been reduced.

Table 14-1 Computation of Posterior Probabilities

P	Prior Probabilities $p(P)$	Conditional Probabilities $p(4,10;P)$	Joint Probabilities col. 2 × col. 3	Posterior Probabilities col. 4/Σ col. 4
.4	.33	.250	.083	.63
.6	.33	.111	.037	.28
.7	.33	.037	.012	.09
Total	. . .	. . .	.132	1.00

Note the difference between what was done in this example and the classical approach to decision making. Rather than compute confidence intervals or make a decision concerning a null hypothesis, in the Bayesian approach we revise our prior belief about an event, based on sample evidence. This revising could be repeated many times if we contined sampling from the same population. If we did so, our belief in the alternative hypotheses would approach closer and closer to the objectivist findings concerning probabilities. Thus, for events that can be repeated many times, the subjectivist and objectivist schools come to the same conclusion. The advantage of the Bayesian approach is that it enables us to compute probabilities for an event that will occur only once, whereas the objectivist approach does not. To get a better understanding of this, let us change the example and describe Bayes' theorem a little more rigorously.

Suppose that we had two separate populations. Population A contains 60 percent Democrats and 40 percent Republicans, and population B contains 30 percent Democrats and 70 percent Republicans. Suppose that we are to choose one or the other of these populations on the basis of selecting numbers from a table of random numbers. If any of the digits 0 to 3 turn up, we take population A, and if any of the digits 4 to 9 turn up, we take population B; that is, the chance of selecting population A is $4/10$, and that of selecting population B is $6/10$. Now suppose that you enter into a guessing game in which you are to guess which population has been selected (without observing what number has been selected). The only hint you are allowed is to take a sample of one from the population before guessing whether it is A or B. You have information about which population it is likely to be before you draw the sample. The probability of selecting population A is $4/10$, and that of selecting B is $6/10$. These are the prior probabilities. Hence, in the absence of other knowledge about which population has been selected, you would do best to guess that it is population B. However, you are allowed to take a sample before guessing. If the sample yields a Democrat, it is more likely that populaiton A has been selected, because it has a larger proportion of Democrats than population B. Thus, you might want to bet that population A has been selected if the sample yields a Democrat. As we said above, to compute the probability that population A or B has been selected, given a Democrat in the sample, we need to compute the joint probability of the prior belief and the sample result. Recall now that in Chapter 3 we defined joint probability as

$$p(A \cap D) = p(A)p(D|A) \qquad (14\text{-}1)$$

$$p(B \cap D) = p(B)p(D|B) \qquad (14\text{-}2)$$

where A = population A; B = population B; D = Democrat.

Now, if a Democrat is selected, he may have come from population A

or *B*. Therefore, the probability that we have selected a Democrat is the sum of the probabilities of drawing a Democrat from either population, or

$$p(D) = p(A \cap D) + p(B \cap D) \qquad (14\text{-}3)$$

First consider the conditional probability that we have population *A* if our sample yields a Democrat. We can define conditional probability as

$$p(A|D) = \frac{p(A \cap D)}{p(D)} \qquad (14\text{-}4)$$

Now we can substitute Equation (14-1) into the numerator of Equation (14-4), and Equation (14-3) into the denominator, getting

$$p(A|D) = \frac{p(A)p(D|A)}{p(A \cap D) + p(B \cap D)}$$

Finally, substituting the definitions of joint probabilities, we have

$$p(A|D) = \frac{p(A)p(D|A)}{p(A)p(D|A) + p(B)p(D|B)} \qquad (14\text{-}5)$$

This is Bayes' theorem. It is the posterior probability of population *A*, given that a Democrat has been selected. For the example, we have

$$p(A|D) = \frac{(.4)(.6)}{(.4)(.6) + (.6)(.3)} = .57$$

The posterior probability that population *B* has been selected if the sample yields a Democrat is

$$p(B|D) = \frac{p(B)p(D|B)}{p(B)p(D|B) + p(A)p(D|A)}$$

$$= \frac{(.6)(.3)}{(.6)(.3) + (.4)(.6)}$$

$$= .43$$

Note that, if we select a Democrat, the probability that we have population *A* is increased from the prior probability of .4 to the posterior probability of .57, and the probability that we have population *B* is decreased from the prior probability of .6 to the posterior probability of .43.

CONTEMPORARY STATISTICAL DECISION THEORY

We now generalize the question of using probabilities (objectively or subjectively arrived at) for making decisions. We said at the beginning of this book that one contemporary definition of statistics is that it is the body of methods that enables us to make decisions in the face of uncertainty. The word "uncertainty" in this context means that two or more events may occur but we are not certain which. These events are sometimes called *states of nature* to emphasize the fact that it will not always be the case that the most unfavorable event will occur.

Although it is uncertain as to what state of nature will take place, it may be possible to set up a criterion for maximizing our ends, provided only that it is possible to measure all possible outcomes in terms that reflect the consequences of each outcome. The following example illustrates these ideas.

Suppose that you are in charge of a campaign for a candidate for some local office in which the principal issue is the state of economic conditions. You have to decide whether to emphasize the personality of your candidate or the issue during the campaign. On the basis of some experience you have had, you believe that, if a recession occurs, the issue should be emphasized throughout the campaign, but if a recession does not occur, the candidate's personality should be emphasized. Suppose that there is a crucial district that will decide the election and that you know that, of the total number of voters in this district, the likely results under the conditions of no recession and recession are as follows:

	Recession	No recession
Emphasize personality	100	300
Emphasize issue	350	150

If the candidate's personality is emphasized and a recession occurs before the election, 100 people in this district will vote for your candidate; but if his personality is emphasized and a recession does not occur, 300 people in this district will vote for your candidate. Similarly, if the issue is emphasized and a recession occurs, you expect 350 people to vote for your candidate, and if a recession does not occur and the issue is emphasized, only 150 people will vote for him. What should you do: emphasize the issue or the candidate's personality?

It is obvious that you will do better emphasizing the candidate's personality if a recession does not occur and the issue if a recession does occur. Your candidate will get the most votes if you emphasize the issue and a recession does occur, that is, 350. But the expected number of votes he will get for each decision alternative depends on the probability of a recession's occurring. If the probability of a recession is very great, the candidate's expected vote will be much higher if the issue is emphasized than if his personality is. To make this specific, suppose that the probability of a recession is 1/3. Given this probability, we can compute the expected vote for each decision alternative. For the alternative of emphasizing the personality, the expected outcome is $(100)(1/3) + (300)(2/3) = 233$. And for the alternative of emphasizing the issue, the expected outcome is $(350)(1/3) + (150)(2/3) = 216$.

Given the probability of a recession of $1/3$, there is no doubt that the candidate's personality should be emphasized, because the expected vote is greater in this case. But if we turn the tables and suppose that the probability of a recession is $2/3$, the expected outcomes in this case are

$$(100)(2/3) + (300)(1/3) = 166$$
$$(350)(2/3) + (150)(1/3) = 283$$

and it would be better to emphasize the issue.

Thus far, everything seems obvious. The critical question is the probability of a recession, and the conclusion we have reached is identical to that which common sense would bring us to. But now let us suppose that we do not know what the probability of a recession is. Given this, we may adopt a criterion for making a decision. For example, we may let the decision depend on whether we are optimistic or not. If we are optimistic (about economic conditions), we emphasize the candidate's personality, because we expect a vote of 300 in this case. Or we may let the decision depend on the flip of a coin. Suppose that we decide which state of nature will occur by the flip of a coin. In this case the probability of a recession is $1/2$. The expected outcomes for each decision alternative are then

$$(100)(1/2) + (300)(1/2) = 200$$
$$(350)(1/2) + (150)(1/2) = 250$$

and it is obvious that the issue should be emphasized in this case (the expected vote is larger for this decision alternative for both states of nature).

A more reasonable decision criterion, however, is to make the decision so as to maximize the expected vote regardless of whether or not a recession occurs. To do so, we may let chance make the decision, as in the last example, but in a slightly different way. We might write "personality" on some slips of paper and "issue" on others, put them in a hat, draw one out, and make the decision in accordance with what appears on the face of the paper. In doing this, we want to fix the odds so as to ensure the same vote regardless of whether or not a recession occurs. We can do this with a little simple algebra. Let p represent the probability that a slip marked "personality" will be drawn, and $1 - p$ the probability that a slip marked "issue" will be drawn. Then, for the event of recession, the expected vote is $100p + 350(1 - p)$; and for the event of no recession, the expected vote is $300p + 150(1 - p)$. Because we want to maximize the vote regardless of which of the two events occurs, we set one equal to the other and solve for p:

$$300p + 150(1 - p) = 100p + 350(1 - p)$$
$$300p + 150 - 150p = 100p + 350 - 350p$$
$$150p + 150 = 350 - 250p$$
$$400p = 200$$
$$p = 1/2$$

This says that we should make it twice as likely that we shall draw a slip marked "issue." The odds are given as the ratio of the numerator to the denominator, or 1:2 in this case. For example, if five slips labeled "personality" and ten labeled "issue" are placed in a hat and if a decision is made in accordance with what is pulled out of the hat on a random draw, the expected long-run vote for the state of nature of no recession is the same as that for the event of recession. These are, respectively,

$$300(1/2) + 150(1/2) = 225$$
$$100(1/2) + 350(1/2) = 225$$

Hence, in the long run, the same result is expected regardless of whether a recession occurs or not. Note that again we are using the assumption of what would occur if the events were repeated many times in order to arrive at a decision for a specific event. In this case, we used the assumption to arrive at a decision criterion that enables us to maximize the expected vote regardless of which state of nature occurs. Later we shall make similar assumptions to arrive at different decision criteria.

This decision process does not guarantee that a correct decision will be made the one time it is used. For example, suppose that the slip pulled out of the hat is marked "issue" (as is likely, because there are twice as many of these), and we decide to emphasize the issue. But suppose that a recession does not occur. In this case the vote would be 150 (not 225), and we should have made an error. Hence, the process simply tells us how to make a decision that will give us a better chance of avoiding an error. But it does not guarantee that an error will be avoided in a particular case.

We can view this decision situation as a game being played between the person directing the campaign and nature. Nature, in this case, is the event of recession or no recession. We have set up criteria for the decision maker to follow so that he will come out with a minimum loss, no matter what nature does. Let us now consider situations where the opponent is not nature but an opposing interest who has the objective of outdoing us. In a game with nature we suppose that our opponent is not trying to outdo us, but is only indifferent. In game theory, we suppose, among other things, that our opponent is trying to win.

GAME THEORY

Terms and Concepts

Game theory is concerned with situations in which there are opposing interests. There may be more than two opposing interests. If there are just two opposing interests, the game is called, not surprisingly, a *two-person game*. Of course, a game may be a two-person game in which one of the players is nature, as in the example above, in which we do not know what course of action nature may take.

A fundamental assumption of game theory is that the opponents are intelligent, rational beings who are trying to outdo each other. If this is not so, game theory, in its present form, does not provide solutions. Similarly, if more than two persons are involved, the problem of what strategies are best has not been solved.

The *payoff* in a game is the amount each person wins. If there are two persons, or interests, say R and C, and R winnings $= C$ winnings, or $R - C = 0$, the game is said to be a *zero-sum game*. Most of the work in game theory has been in zero-sum games. The optimal, or best, strategy can be found for any two-person zero-sum game.

A *strategy* for each person is a complete plan of action. Each player may have more than one strategy. An *optimal strategy* is one that guarantees a player a minimum loss, no matter what his opponent does. A *pure strategy* is one in which a player has one best plan of action. A *mixed strategy* is one in which he must alternate plans of action.

When we say that game theory has worked out solutions to the two-person zero-sum game, we mean that the mathematical solution has been worked out and, as will be seen, is very simple, requiring no more mathematical knowledge than ordinary arithmetic. The conceptual problems of how we measure the payoffs numerically, whether the payoffs are meaningful, and whether this solution is optimal are not a part of game theory. We consider briefly below how the subjective view of probability may be used to solve some of these problems. For the moment we concentrate on the relatively simple problems involved in solved games.

Two-person Zero-sum Games

Any 2×2 matrix with numbers in the cells might be approached by game theory, but for a start we introduce a hypothetical but not totally unrealistic situation. Suppose that there are two candidates for election to the office of mayor of a large city. One is Negro and the other is white. Suppose further that the registered voters in the city are 60 percent white and 40 percent Negro. Finally, suppose that, if the race issue is brought up by both candidates, all the white voters will vote for the white candidate; if race is emphasized by the white candidate but not by the Negro candidate, 90 percent of the white voters, or 50 percent of the total, will vote for the white candidate; if race is emphasized by the Negro candidate but not by the white candidate, 50 percent of all voters will vote for the white candidate; and if race is not emphasized by either candidate, 40 percent of all voters will vote for the white candidate. This is represented in the matrix below, where $C =$ Negro candidate, $R =$ white candidate, and 1 and 2 represent the respective strategies: (1) emphasize race and (2) do not:

$$C$$

	1	2
R 1	60	50
2	50	40

The convention followed in this chapter is that entries in the cells represent payoffs, or winnings, from C to R. Now R (the white candidate) wants to win, but he cannot assume that C (the Negro candidate) will be stupid or careless. He must assume that whichever strategy he chooses, C will make the best countermove available. If he selects strategy 1 (emphasize race), C will select strategy 2, and the election will be a stalemate. If he selects strategy 2, then C will select strategy 2 and win. Hence he will list the worst payoffs that can result for each of his strategies as the row minima. The largest of these minima has been designated with an asterisk:

$$C$$

	6	2	Row minima
R 1	60	50	50*
2	50	40	40
Column maxima	60	50*	

R would obviously prefer strategy 1 in this situation, because this leads to the larger of the two minima that he can expect to win, even if it is a stalemate. C also wants to do the best he can, which, in this case, means a strategy that has the smallest payoff, because the amounts in the boxes represent payment to R. However, he must assume that R will force him into the worst situation, and so he lists the largest payoffs in the column. An asterisk is placed next to the smallest of these maxima. C will naturally prefer the smallest of these,

50. Thus if he uses strategy 2 (do not emphasize race), his losses cannot exceed 50. If he uses strategy 1, then R is sure to win.

Note what happens. Each candidate has arrived at a strategy which ensures him a minimum loss and which guarantees the same payoff, 50. This is a situation called saddlepoint, or a *strictly determined game*; that is, whenever the larger of the row minima is equal to the smaller of the column maxima, the game has a saddlepoint in the sense that any other strategies will produce a worse situation. In this case, neither candidate wins (they both get exactly 50 percent of the vote), and this is thus a *fair game.*[1]

Let us suppose that this is what happens and that a runoff election is called, except that new developments occurring between the two elections produce the following payoff matrix:

		C		Row minima
		1	2	
R	1	30	60	30
	2	50	40	40*
Column maxima		50*	60	

Now R can expect that the most he can get is 40 percent of the vote by following strategy 2, whereas C can be sure of keeping his share down to 50 percent by using strategy 1. There is no saddlepoint here. The game is *nonstrictly determined*. What strategies should be followed? If R follows strategy 1, he can be sure that C will follow his strategy 1, forcing him down to 30 percent, whereas he can get 40 percent of the vote by following strategy 2. Thus, in either case, he seems destined to lose. This would seem, therefore, to be an unfair game. Could he do better if he alternated his strategies? Let us compute mixed strategies and see.

In computing an optimum mixed strategy, we want to find a solution for the game that is the same as the solution for pure strategies, that is, a solution that guarantees each player a minimum no matter what his opponent does. Suppose that we let the optimal mix be determined by the flip of a coin. Suppose that R will follow strategy 1 if the flip turns up a head and strategy 2 if it turns up a tail. When he uses this strategy against C's strategy 1, he will win 30 percent half of the time and 50 percent half of the time, or his expected long-term winnings will be $(30 + 50)/2 = 40$ percent, and against C's strategy 2, his long-term winnings will be $(60 + 40)/2 = 50$ percent. Thus, if he alternates randomly between strategies 1 and 2, he can ensure that his long-term winnings will be 45 percent instead of 40 percent that he would get by just following strategy 2. But, of course, you might

---

[1] Normally, a fair game is one in which the value of the game is zero, because this represents a situation in which neither person wins. In the present example, because we are using percentages, a fair game has a value of 50 percent.

say that he cannot win any way, which is true. This is, therefore, an unfair game, in contrast to the earlier one. *R* should therefore refuse to play this game or do something that will change the payoffs in the boxes.

Let us leave the example, however, and consider some simple matrices, so that we can find an easy method of finding the optimal mix of strategies in nonstrictly determined games; that is, instead of alternating between two strategies so that in the end the actor plays each of his strategies an equal number of times, he may want to play one strategy more frequently because it will ensure a larger payoff. Consider the following matrix:

C

		1	2	Row minima
	1	10	8	8*
R				
	2	5	7	5
Column maxima		10	8*	

What is the best strategy for *R* to use in this case? *R* should use strategy 1, because this is a strictly determined game with a saddlepoint. The row minimum (8) for *R* 1 is equal to the column maximum (8) for *C* 2. Thus, in this case we seek no further for strategies. Mixed strategies apply only to nonstrictly determined games.

Now consider the following matrix:

C

		1	2	Row minima
	1	6	12	6
R				
	2	14	10	10*
Column maxima		14	12*	

There is no saddlepoint, and so a mixed strategy is called for. The way to find the optimum mixed strategy to be used is as follows. Consider *C*. Subtract the numbers in the second row from those in the first, giving -8 and $+2$. One of the numbers will always be negative, and negative signs will be ignored. These numbers are reversed to find the optimal mix. Thus, *C* should mix his strategies in the ratio of 2:8 or 1:4; that is, he should alternate between the two, playing his strategy 2 four times for every time he plays strategy 1. The optimum strategy for *R* is found in the same manner except that everything is turned on its side: column 2 is subtracted from column 1, and the results are reversed (signs ignored), giving $\frac{4}{6}$ or $\frac{2}{3}$. The optimum mix for *R* is that he should play his second strategy three times for every two times he plays his first strategy. Note the inversion of the numbers to determine the optimum mixed strategy. The subtraction resulted in the absolute ratio of $\frac{6}{4}$ but the actual ratio of *R*'s strategy 1 to his strategy 2 is $\frac{4}{6}$. *R* will have to use some chance device other than flipping a coin to get a practical result of 2:3, because a coin produces the ratio of 1:1. One way to do so is with a table of random numbers. For example, to get a 2:3 ratio, the table of random numbers can be entered at any point. If the first digit selected is a 0 or 1, play strategy 1; if it is a 2, 3, or 4 play strategy 2; if it is any other digit (5, 6, 7, 8, 9), skip to the next number. This procedure works for any set of odds. For 5:4, for example, the table of random numbers would be entered (at any place, without looking). If the first digit found is a 0, 1, 2, 3, or 4, play strategy 1; if it is a 5, 6, 7, or 8, play strategy 2; if it is a 9, skip to the next number; and so on.

Returning to the example, if *C* uses his mixed strategy in the ratio of 1:4 and *R* uses his in the ratio of 2:3, what is the expected value of the game? It is the payoff that results when the best mix is used against either pure strategy of the opponent. The expected loss for *C*, who should use a mixed strategy in the ratio of 1:4, against *R* 1 is $(1 \times 6 + 4 \times 12)/(1 + 4) = 10.8$. Similarly for *R*, whose mix should be 2:3, the expected gain against *C* 1 is $(2 \times 6 + 3 \times 14)/(2 + 3) = 10.8$. Note that the value of the game is the same no matter whose mix is being considered. This is also a way of checking to see if the correct mix or optimal strategy has been found, for if the wrong odds are used, the expected value of the game will be different when computed for each player.

Note that in a nonstrictly determined game, it does not matter which strategy the opponent is using if one of the players is using his optimum mixed strategy. For example, in the matrix

		C	
		1	2
R	1	6	12
	2	10	8

if *C* uses his optimum mixed strategy, it does not matter which strategy *R* plays, because the expected value of the game against either *R* strategy is

$$\frac{1 \times 6 + 1 \times 12}{1 + 1} = 9$$

$$\frac{1 \times 10 + 1 \times 8}{1 + 1} = 9$$

Hence, once *C* plays his optimum mixed strategy, it does not matter what

R does, and the reverse is also true. Note also that, if the stakes of the game are doubled, the optimum mix for each player is unaffected, but the value of the game is doubled:

$$C$$

	1	2
R 1	12	24
2	20	16

Value of game:

$$\frac{1 \times 12 + 1 \times 24}{1 + 1} = 18$$

In general, the strategies are unaffected if a constant is added, subtracted, or multiplied into the payoffs, but the value is changed by the amount of the constant. This fact is useful in helping to find solutions to games that may have negative values in the cells. For example, consider the following matrix:

$$C$$

	1	2	Row minima
R 1	−5	10	−5
2	40	−10	−10
Column maxima	40	10	

To find the optimum mixed strategy of this game requires subtracting positive from negative members and negative from positive. To avoid this, we may simply add a constant to make all values in the cells positive. For example, we may add the constant 20, producing the matrix

$$C$$

	1	2
R 1	15	30
2	60	10

Because there is no saddlepoint, this is a nonstrictly determined game, and the optimum mixed strategies are, for *C*, 20:45, or 4:9; for *R*, 50:15, or 10:3. The value of the game found by the converted matrix is $(4 \times 15 + 9 \times 30)/(4 + 9) - 20 \doteq 5.38$ and, by using the original matrix, $[4 \times (-5) + 9 \times 10]/(4 + 9) \doteq 5.38$.

2 × n Games

What we have considered so far can be generalized with relatively little difficulty in order to cover games in which one of the players has more than two strategies available. Consider the military question in which C has a chance of using either conventional military forces or guerilla forces to seize a piece of land belonging to R, who might defend it by using tanks, troops, or air power. If C uses strategy 1 (conventional forces) and R defends by his strategy 1 (tanks), the cost to C will be 200 men. If C uses strategy 2 (guerilla troops) and R uses his strategy 1, the cost to C will be 400 men. Similarly, the respective payoffs from C to R when R uses strategy 2 (troops) against C's 1 and 2 are 500 and 300; and when R uses strategy 3 (air power), they are 600 and 500. All this is entered in the following matrix:

		C		Row minima
		1	2	
	1	200	400	200
R	2	500	300	300
	3	600	500	500*
Column maxima		600	500*	

What strategies should be used? The game has a saddlepoint and is thus strictly determined. C should use guerilla forces, and R should use air power to defend. The value of the game is a payment of 500 men to R. Note that, if C uses strategy 1, then R is certain to use his strategy 3, and the cost to C will be 600 men. Also, we must assume that, if R defended with his strategy 1, then C would certainly use his strategy 1, and his cost would be 200 men. But C cannot get away with this small loss, for if he does use strategy 1, then R will use strategy 3, forcing the cost up to 600 men.

It appears, therefore, that in a 2 × n game, if there is a saddlepoint, the optimal strategies are those which converge on it. This is technically called *minimax strategy*, and we consider alternative and less conservative strategies below. For the moment, consider the more extended matrix:

		C		Row minima
		1	2	
	1	3	6	3
R	2	5	4	4
	3	4	7	4
	4	6	5	5*
Column maxima		6*	7	

There is no saddlepoint in this game. To solve this game, we must first try to reduce it to a 2 × 2 game. This can be done by trying to reduce *R*'s strategies. If one of the strategies is superior to another, it is dominant, and the inferior one may be eliminated. For example, consider *R* 1 and *R* 2. *R* 1 against *C* 1 results in a payoff of 3, and *R* 2 against *C* 1 results in a payoff of 5. Thus *R* should prefer 2 over 1 against *C* 1. But against *C* 2, *R* 1 is better. Thus *R* 1 is not superior to *R* 2 on a box-by-box basis. Consider *R* 1 and *R* 3 next. The payoffs for *R* 1 against either *C* 1 or *C* 2 are smaller than the losses for *R* 3. Thus, *R* 3 is superior to *R* 1 (from *R*'s perspective), and *R* 1 should be dropped. Next, compare *R* 2 and *R* 3. Neither is superior on a box-by-box basis. Compare *R* 2 and *R* 4. *R* 4 yields a larger payoff against each of *C*'s strategies and is dominant. Thus we may drop *R* 2. The game has now been reduced to

		C	
		1	2
R	3	4	7
	4	6	5

The solution to this reduced 2 × 2 game is a mixed strategy, and it is also a solution to the larger matrix. The solution is, for *C*, 1:1; for *R*, 0:0:1:3 — that is, *R* should never play strategies 1 and 2, and he should play strategy 4 three times for every time he plays *R* 3. The value of the game, as computed against *C* 1, is $[(0 \times 3) + (0 \times 5) + (1 \times 4) + (3 \times 6)]/(1 + 3) = 22/4 = 5.5$.

The same thing is done if *C* has more than one strategy. Consider

		C				Row minima
		1	2	3	4	
R	1	3	9	2	5	2*
	2	6	10	1	8	1
Column maxima		6	10	2*	8	

It has a saddlepoint. The optimum strategies are *R* 1 and *C* 3. The value of the game is 2. Now consider

		C				Row minima
		1	2	3	4	
R	1	−4	1	3	6	−4
	2	9	0	8	−3	−3*
Column maxima		9	1*	8	6	

There is no saddlepoint. To reduce it to a 2×2 matrix, we consider C's strategies. In the case, because the entries in the box represent payments to R, we want to find the dominated C strategies and retain them. Consider $C\,1$ and $C\,2$. For $C\,1$ against $R\,1$, C wins 4, whereas he loses 1 with $C\,2$ against $R\,1$. Against $R\,2$, $C\,1$ loses 9, and $C\,2$ wins him 0. Thus neither $C\,1$ nor $C\,2$ is dominant from C's perspective on a box-by-box basis. Consider $C\,1$ and $C\,3$. Neither is dominant from C's perspective, because against $R\,1$ he does better with $C\,1$ than $C\,3$, but against $R\,2$, he does better with $C\,3$. Next consider $C\,1$ and $C\,4$. $C\,1$ is better against $R\,1$, but $C\,4$ is better against $R\,2$. It turns out on a strategy-by-strategy comparison that $C\,3$ is dominated by $C\,2$, and thus $C\,3$ can be eliminated, leaving

		C		
		1	2	4
R	1	-4	1	6
	2	9	0	-3

Now we must find the 2×2 game from this 2×3 matrix whose solution yields a minimum game value. This is also the solution to the matrix we began with. We start by considering

		C	
		1	2
R	1	-4	1
	2	9	0

The optimum strategies are, for R, $9:5$; for C, $1:13$. The value of the game is $[9 \times (-4) + 5 \times 9]/(9 + 5) = 9/14$. The same R strategy against $C\,4$ yields a value of $[9 \times 6 + 5 \times (-3)]/(9 + 5) = 39/14$. Hence, the best strategy for the original matrix is, for R, $9:5$; for C, $1:13:0:0$. For larger games, finding a solution by this method can be tedious. There are short methods that are not considered here[2]

3×3 Games

Three-by-three games are essentially the same as $2 \times n$ games. If there is a saddlepoint, the game is solved, and the work is easy. Consider

---

[2] See J. D. Williams, *The Complete Strategyst*, McGraw-Hill Book Company, New York, 1954.

$$C$$

		1	2	3	Row minima
	1	1	8	2	1*
R	2	-5	10	3	-5
	3	-2	5	14	-2
Column maxima		1*	8	2	

This game has a saddlepoint. Its solution is, for C, 1:0:0; for R, 1:0:0. And its value is 1; that is, C should play his strategy 1 and R should do the same. Consider next

$$C$$

		1	2	3
	1	6	3	5
R	2	7	8	4
	3	5	6	2

It has no saddlepoint, and we can try to reduce it by looking for dominant strategies. Consider R. His winnings are larger for strategy 2 than for strategy 3 for each of C's strategies. Thus we can eliminate R 3. Next consider C. He wants to keep the numbers as small as possible. C 3 is smaller than C 1 against each of the two strategies that R might use, that is, R 1 and R 2. Thus we can eliminate C 1, giving a 2 × 2 game:

$$C$$

		1	2
	1	3	5
R			
	2	8	4

The reader should now be able to find the solution to this game, which is also the solution to the larger game. He should find that the value of the game is 4.66+.

If a 3 × 3 game does not have a saddlepoint and none of the strategies is dominant, another method must be used to solve the game. Let us consider a long method for solving such games.[3]

Consider the following game:

$$C$$

		1	2	3	Row minima
	1	2	5	1	1
R	2	10	3	12	3
	3	4	12	6	4*
Column maxima		10*	12	12	

[3] For a general iterative method for solving games, see J. D. Williams, *ibid.*

There is no saddlepoint and no dominance in this game, and so we use a long method to find its solution. First, we must find the optimum mix for *C*. To do so, subtract each row from the preceding row to get

-8	2	-11
6	-9	6

To get the optimum mix for *C* 1, we focus on the numbers in the last two columns:

2	-11
-9	6

The optimum mix for *C* 1 is the difference between the diagonal products: $2 \times 6 - (-11) \times (-9) = 12 - 99 = -87$. For *C* 2, we focus on

-8	-11
6	6

and get $-8 \times 6 - (-11) \times 6 = -48 - (-66) = 18$. Finally, for *C* 3, we focus on

-8	2
6	-9

and get $(-8) \times (-9) - 2 \times 6 = 72 - 12 = 60$. Hence, *C*'s best mixed strategy is $87:18:60$. We can find *R*'s best mixed strategy in the same way:

-3	4
7	-9
-8	6

The mix for *R* 1 is

7	-9
-8	6

and $7 \times 6 - (-9) \times (-8) = 42 - 72 = -30$. For *R* 2,

-3	4
-8	6

and $(-3) \times 6 - 4 \times (-8) = -18 - (-32) = 14$. For *R* 3,

-3	4
7	-9

and $(-3) \times (-9) - 4 \times 7 = 27 - 28 = -1$. *R*'s optimum mix is thus $72:14:1$. The value of the game is

$$C$$

		1	2	3	Row odds
	1	2	5	1	30
R	2	10	3	12	14
	3	4	12	6	1
Column odds		87	18	60	

$$\frac{30 \times 2 + 14 \times 10 + 1 \times 4}{30 + 14 + 1} = 4.53$$

The Problem of Measuring Payoffs

Now we show that, even if the payoff matrix is wrong, as long as the error is randomly distributed, it does not greatly affect the solution of the game. Let us alter the matrix by introducing measurement error that is random and see what happens. We add $+2$ or -1 to each of the cells on a random basis, decided by the toss of a coin. If the coin turns up heads, 2 is added; if it turns up tails, 1 is subtracted. The only restriction is that the values 1 and 12 will not be changed so as to retain the original scale. The new matrix, with measurement error included, is

$$C$$

		1	2	3	Row minima
	1	1	4	1	1
R	2	12	2	12	2
	3	3	12	8	3
Column maxima		12	12	12	

Now consider the best mixed strategies. For R we begin with

-3	3
10	-10
-9	4

The R odds are:

R 1: $10 \times 4 - (-10) \times (-9) = -50$

R 2: $-3 \times 4 - 3 \times (-9) = 15$

R 3: $-3 \times (-10) - 3 \times 10 = 0$

or 50:15:0. The value of the game for R is $(50 \times 1 + 15 \times 12 + 0 \times 3)/(50 + 15 + 0) \doteq 3.54$.

Thus, if *R* plays this game, the erroneous game, and follows the best mixed strategy, he wins a maximum of 3.54 as opposed to the 4.53 he would have won in the true game. The conclusion is that measurement error leads to incorrect solutions, but if it is randomly distributed, the results will not be drastically changed.

It also is possible to find solutions to games even if exact numerical values to outcomes cannot be given. For example, assume the following situation:

		C 1 War	*C* 2 Peace
R 1 Build nuclear force		End of world	Continue as is
R 2 Disarm		Become Communist	Best of all worlds

where *C* is Russia and *R* is the United States and the items in the boxes represent the consequences of each policy. Now, it seems silly to give numerical values to these consequences, because the magnitudes may be all wrong. But we may be able to rank the consequences in order of desirability, such as the following matrix:

		C 1	2	Row minima
R 1		4th	2nd	4th
R 2		3rd	1st	3rd*
Column maxima		3rd*	1st	

Now, the game has a saddlepoint, which dictates the optimum strategy: *C* 1, *R* 2. We might interpret it this way. If *R* selects strategy 1, then *C* must select strategy 1, and the payoff to *R* is that which he desires least. If *R* selects strategy 2, then *C* will counter with strategy 1, which is the best that *R* can hope for under the circumstances.

Of course, not all 2 × 2 games have a saddlepoint. Assuming the same game, it might be argued that the possible consequences are

A — Excellent
B — Good
C — Fair
D — Poor
F — Fail

and the matrix should be

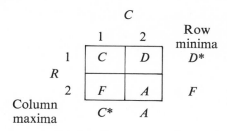

that is, becoming Communist is worse than the end of the world, and continuing things as they are is also. The maximum of the row minima is poor, and the minimum of the column maxima is fair. The value of the game to R is thus between poor and fail. The odds for R's mixed strategies can be computed in the same manner as with numbers; that is, for R 1, it is excellent minus fail, and for R 2, it is fair minus poor. To obtain a value for this, consider the line segment

$$\overset{\cdot}{F} \qquad \overset{\cdot}{D} \qquad\qquad \overset{\cdot}{C} \quad \overset{\cdot}{B} \qquad \overset{\cdot}{A}$$

It is apparent that subtracting F from A yields a larger amount than subtracting D from C, and thus R should favor his first strategy or play it more than his second.

There is little doubt that this is not a completely satisfactory discussion of games that do not have exact payoffs. But it might be said that thinking about the question in terms of ordinal values helps clarify what is involved. But what about mixed strategies here? It does not seem helpful to say that R should play strategy 1 more often than strategy 2, because this seems to be a game that is played only once. After all, how many times can there be an end of the world? The strict answer is that game theory, as currently developed, applies to games that are played only once. For if the game is repeated many times, the frequencies of choice of strategies by the opponent can contain information about the pattern of his strategy. In this case, the objective view of probability can be used to compute expected payoffs. Or, alternatively, the subjective view of probability can be used to test hypotheses about his strategy if the game is not repeated many times. These subjective probabilities can be modified by a sampling process if we know something about the relationship between the characteristics of the players and the strategies they may follow.[4]

Nonzero-sum Games

The assumption throughout this chapter so far is that each player is rational in the sense that he wants to maximize his gains or minimize his losses. If

[4] See Morton Kaplan, *System and Process in International Politics*, John Wiley & Sons, Inc., New York, 1957, for a theory built on this idea.

one of the players does not do so, the solutions found are not correct in that the player following the best strategy can improve his payoff by capitalizing on his opponents failure to play his best strategy. For example, in the game

		C 1	C 2	Row minima
R	1	10	80	10*
	2	2	120	2
Column maxima		10*	120	

the best strategy for C is C 1, and for R it is R 1, which is the saddlepoint. The value of the game is the payment of 10 to R. But if C is not trying to minimize his losses or lacks complete information, he may play his C 2, in which case R, if he is trying to maximize, should shift to R 2 and win a payoff of 120. Furthermore, suppose that we also relax the assumption that the winnings of R equal the losses of C. Under these conditions we can construct more realistic games. Let us consider some of these.

There are many different possible solutions for nonzero-sum games, and none is unique. We consider four solutions:

1. Maximin
2. Equilibrium
3. Cooperative
4. Threat

The conditions of the game are the same as before: two players, R and C, each attempting to maximize his self-interest and each having alternative strategies. However, the payoffs are represented as double entries in the matrix, symbolized as a_{ij} = amount received by R and b_{ij} = amount received by C. As an example, let us consider the following game. R and C represent two aspirants for the Republican party presidential nomination, and they are faced with two up-coming state primary elections they can enter. C is strong in one state, R is strong in the other, and the payoffs are delegates to the Republican nominating convention. These are entered in the matrix

		C enters	
		State 1	State 2
R enters	State 1	(15,10)	(22,20)
	State 2	(11,13)	(10,15)

where the first entry represents the payoff for R and the second the payoff for C. For example, if they both enter the primary in state 1, R will get 15 delegates from the state and C will get 10. If R enters the primary in state 1 and *C* enters the primary in state 2, then *R* will get 22 delegates (from state 1) and *C* will get 20 delegates (from state 2). We exclude the possibility of both candidates entering both states.

Given these conditions, what strategies should each follow? It is obvious that *R* should enter state 1 and *C* state 2, because for *R*, state 1 is preferred no matter what *C* does (it is a dominant strategy), and for *C*, state 2 is preferred no matter what *R* does.

However, let us suppose that the survey that produced the payoff numbers was wrong and that a second and better survey shows that the expected delegate distributions are

	C enters	
	State 1	State 2
R enters State 1	(15,30)	(22,20)
State 2	(11,13)	(30,15)

If we suppose that each candidate wants to maximize his expectations and figures that his opponent will behave so as to keep his winnings to a minimum, each person will follow a maximin principle. For *R*, the payoff matrix is

$$A = \begin{array}{|c|c|} \hline 15 & 22 \\ \hline 11 & 30 \\ \hline \end{array}$$

which is a strictly determined game (with a saddlepoint), and *R*'s best strategy is to enter state 1. For *C*, the payoff matrix is

$$B = \begin{array}{|c|c|} \hline 30 & 20 \\ \hline 13 & 15 \\ \hline \end{array}$$

and because it is a strictly determined game, the optimal strategy for *C* is to enter state 2. Thus, the maximin strategy produces the payoff vector (22,20), which is the same solution produced in the last game, which had dominant strategies. But there is an important difference. In the first game, neither *R* nor *C* could improve his situation no matter what his opponent did. In the second game, if *R* believes or knows that *C* will not change his strategy, he should change to enter state 2, where he can increase his delegates to 30 and diminish *C*'s to 15. The payoff vector is (30,15). Similarly, for *C*, if he knows that *R* will stick to the maximin solution, he should change to enter state 1, producing the payoffs (15,30). Because in this game there are alternative solutions, the maximin solution (22,20) is not stable. The two alternative solutions, (30,15) and (15,30), are equilibrium-point solutions. (Note that the first game had one equilibrium-point solution, and it was the same as the maximin solution.)

We thus have three different solutions to the second game: (22,20), (30,15), and (15,30). Game theory cannot tell us which is preferable. A consideration of exogenous factors, such as traditions, institutions, information, and the like, would be needed to find the solution.

Let us consider a game that has mixed strategies as the optimal solution and introduce two additional solutions. Assume the same kind of game, but the expected number of delegates are now as in the matrix

		C enters	
		State 1	State 2
R enters	State 1	(10,10)	(20,30)
	State 2	(30,20)	(15,15)

The maximin solution to this game is one of mixed strategies. For *R*, the odds are 3:2. For *C*, the odds also are 3:2. The value of the game is 18, as the reader can easily verify. But, of course, if *R* believes that *C* will play his optimum mixed strategy, he will shift to state 2 for the equilibrium point (30,20). Similarly, for *C*, the equilibrium point is (20,30). These produce better results than the value for the maximin solution.

Finally, let us suppose that *R* and *C* are not such hateful enemies after all, and because there are other candidates for the Republican party nomination, there may be some incentive for them to get together. Suppose that they do and agree to the following strategies (a reprehensible act, of course, in American politics, but it is done nevertheless): *R* will enter state 1 and *C* will enter state 2. The solution is (20,30), and it is a cooperative solution. It provides more delegates for both candidates than the maximin solution, but the solution for *R* is not so good as his equilibrium point (30,20). Hence, we may suppose that, before *R* agrees to this, he will demand a side payment from *C*. For example, if *C* agrees to a side payment of 5, the cooperative solution is (25,25). Now, although this is better for both candidates, there is no unique solution to how much the side payments are likely to be. One possible solution might be as follows. Suppose that *C* is more powerful than *R* in the sense that he has more delegates and that there are more states whose primaries he might enter. *R* might threaten *C*, suggesting that unless he agrees to a side payment of 2, which produces the solution (28,22), he will enter all subsequent primaries and ruin *C*'s chances. If *C* believes this threat, the threat solution is (28,22).

None of these solutions can be generalized for all possible nonzero-sum games. The issues and problems involved in *n*-person, nonzero-sum games are vast. For example, an important question concerns the number of coalitions that might be formed and the number of persons or interests that would be involved in each. Riker has proposed a size principle concerning this, suggesting that the size of a winning coalition will always be minimal, to ensure the largest possible payoff to each participant in the coalition.[5] But whether the assumptions concerning the information and knowledge necessary to make this principle empirically useful are too restrictive is difficult to determine. The issues involved are quite complex. It appears, however, that no one theory is going to meet all demands. A number of different solutions will be optimal under varying game rules.[6] The most important contribution of game theory thus far resides in the new concepts and perspective it has provided. Even if it never gives us practical guides for action, by calling attention to some of the different kinds of assumptions we must make to find solutions to games, it has increased our knowledge of politics.

SIMULATION

Before decision theory such as that described above can be used in research, it is necessary to find realistic values for the payoff matrices. A possible solution to this problem in certain circumstances is simulation.

When we simulate a phenomenon, we build a model of it and study how the model behaves under different conditions. These models may be actual small-scale physical models, such as an airplane in a wind tunnel, or symbolic models. In the latter case, we allow symbols to represent properties of the system we are studying and allow them to take on different values so that we may study what is likely to happen to the system under different circumstances.[7] In order to describe this, we use a hypothetical example that is meant to demonstrate the principle of simulation, and not an actual application. It would take too much space to describe an actual application.

---

[5] See William Riker, *The Theory of Political Coalitions*, Yale University Press, New Haven, Conn., 1962.

[6] See Evar Nering, "Coalition Bargaining in *n*-Person Games," in M. Dresher, L. S. Shapley, and A. W. Tucker (ed.), *Advances in Game Theory*, Princeton University Press, Princeton, N. J., 1964.

[7] Some writers use the word "simulation" to refer to laboratory experiments in which a situation in real life is replicated. See Morris Zelditch and William Evan, "Simulated Bureaucracies: A Methodological Analysis," in Harold Guetzkow (ed.), *Simulation in Social Science*, Prentice-Hall, Inc., Englewood Cliffs, N. J., 1962, pp. 48–60. For example, the experiment by Simon and Guetzkow described above concerning communications structures and problem solving might be called a simulation of what might occur under conditions outside the laboratory. Or war games (or business games) might attempt to simulate actual conditions. Here we refer to them as experiments and reserve the word "simulation" for symbolic representation of events (by letters and numbers) and for small-scale physical models of real things.

Therefore, the assumptions made in the example may seem a little unrealistic.

Assume that senator *A* has exactly one hour before a bill of his is to come to a vote on the Senate floor. The vote is going to be very close, and he would like to improve the chances of passage by using the hour to get more votes behind him. He would like to confer with two key members, senators *X* and *Y* in particular, during this hour, because each of these senators can control 11 and 18 votes respectively. Senator *A* believes that he has 40 votes without the support of these two key senators, and therefore if he can get either one to support him, he is assured of winning. Both senators *X* and *Y* have indicated that they are wavering. Senator *A* also knows that senator *Y* will be harder to extract support from, although he controls more votes. He must therefore decide whether or not to visit both senators *X* and *Y* before he has to be on the floor, or only one or the other, and if so, which of the two he should visit. He must visit them in person and cannot telephone. Hence, time is a crucial variable. Suppose that he decides to visit senator *Y* first, because he controls more votes, and then go on to senator *X*. He does not know exactly how long it will take to convince either senator. But he knows that, if he spends enough time he can convince either one, and he knows approximately how long it will take to convince either senator. He also knows the approximate time it will take to travel to each senator's office and the time it will take to return to the Senate floor. Finally, we know that there is a possibility that he will be stopped on the way to senator *Y*'s office, and this could cause him some delay.

What chance does senator *A* have of seeing both senators *X* and *Y* and convincing them to support his bill? Should he try to see both senators or only one? We can answer these questions by simulation if the following facts are known:

1. The probable travel time required from senator *A*'s office to senator *Y*'s ranges from 5 to 12 minutes in the following distribution:

Minutes	Frequency	Probability (relative F)	Cumulative probability	Random number assignment
5	1	.05	.05	00–04
6	2	.10	.15	05–14
7	5	.25	.40	15–39
8	6	.30	.70	40–69
9	3	.15	.85	70–84
10	2	.10	.95	85–94
11	0	0		none
12	1	.05	1.00	95–99
Total	20	1.00		

2. Probable time delays enroute to senator Y's office:

Minutes	Frequency	Probability (relative F)	Cumulative probability	Random number assignment
0	10	.50	.50	00–49
1	3	.15	.65	50–64
2	2	.10	.75	65–74
3	1	.05	.80	75–79
4	0	0		none
5	1	.05	.85	80–84
6	1	.05	.90	85–89
7	0	0		none
8	2	.10	1.00	90–99
Total	20	1.00		

3. Time required to convince senator Y:

Minutes	Frequency	Probability (relative F)	Cumulative probability	Random number assignment
10	2	.10	.10	00–09
11	7	.35	.45	10–44
12	6	.30	.75	45–74
13	2	.10	.85	75–84
14	0	0		none
15	0	0		none
16	1	.05	.90	85–89
17	0	0		none
18	0	0		none
19	1	.05	.95	90–94
20	0	0		none
21	0	0		none
22	0	0		none
23	1	.05	1.00	95–99
Total	20	1.00		

4. Travel time to senator X's office:

Minutes	Frequency	Probability (relative F)	Cumulative probability	Random number assignment
3	1	.05	.05	00–04
4	4	.20	.25	05–24
5	5	.25	.50	25–49
6	7	.35	.85	50–84
7	2	.10	.95	85–94
8	1	.05	1.00	95–99
Total	20	1.00		

5. Time required to convince senator X:

Minutes	Frequency	Probability (relative F)	Cumulative probability	Random number assignment
15	2	.10	.10	00–09
16	9	.45	.55	10–54
17	5	.25	.80	55–79
18	2	.10	.90	80–89
19	0	0		none
20	1	.05	.95	90–94
21	1	.05	1.00	95–99
Total	20	1.00		

6. Return time to Senate floor:

Minutes	Frequency	Probability (relative F)	Cumulative probability	Random number assignment
5	2	.10	.10	00–09
6	8	.40	.50	10–49
7	6	.30	.80	50–79
8	1	.05	.85	80–84
9	2	.10	.95	85–94
10	1	.05	1.00	95–99
Total	20	1.00		

Each of these distributions represents how long each activity, 1 to 6, is expected to take. So that the reader understands the frequency distribution for each activity, in an actual problem they would be based on past experience, preferably from an actual record, but they might also be subjective estimates. Each event is something that varies. We do not know the exact time for each, but only the likely times. Our question is whether senator A can see both senators X and Y, convince them both to support him, and still return to the Senate floor in time. Let us now simulate the sequence one time.

We allow the time it will take to perform each activity to be decided by a chance device. Consider the first activity, the trip to senator Y's office. The most probable time for this is 8 minutes. However, any particular trip can be as short as 5 minutes or as long as 12 minutes, possibly depending on whether the elevator is there when senator A arrives. We could set up a chance device for determining the time it will take on any one trip by taking cards marked with the respective minutes in proportion to the frequency of each possible time. For activity 1, we would take one card marked 5 minutes, two marked 6 minutes, five marked 7 minutes, and so on. These would be thoroughly shuffled, and the one selected would be the time for this activity. Alternatively, we can generate the times by a process known as *Monte Carlo simulation*. To do so, we assign two-digit numbers in accordance with the cumulative probabilities of each time. This accomplishes the same thing as the card situation described above. We draw two-digit numbers from the table of random numbers to determine the length of time for each activity. This process is continued until the entire sequence is completed. The following results were obtained for one complete simulation of the example:

Activity	Time (min)	Time used up, (min)
1. Time to senator Y's office	8	8
2. Delay time enroute	2	10
3. Time required to convince senator Y	19	29
4. Time to senator X's office	5	34
5. Delay time enroute	0	34
6. Time required to convince senator X	16	50
7. Time to return to senate	6	56
8. Delay time enroute	2	58

In this one run, senator A was able to accomplish the entire sequence within the alloted time. We might decide on the basis of this one run that this is the procedure the senator should use. However, this is only one of a number of outcomes that might occur. Hence, we should run the sequence many times (several hundred). We should see that, on some of the trials, senator A would make it but not on others. The ratio of the successes to the total trials is the probability of success and might be used in making a decision. For example, if it turned out that the probability was small, then it might be decided to concentrate on just one senator. This would have to be considered in the context of the number of votes needed to win, and so on. This example is not realistic, because we normally should not resort to such an elaborate procedure for making decisions of this kind. But the principle can be and has been applied to situations like flood control, designing bus depots, and computing optimal service facilities.[8]

PROBABILISTIC MODELS

The statistics considered prior to this chapter can be called *deterministic*, because they assume that a definite value is associated with the dependent variable for every value of the independent variables. In this chapter we have considered some probabilistic models that assume that indeterminacy is inherent in events. These models postulate that the world we must deal with is probabilistic rather than deterministic. A particular event can take on a number of different values, and the most we may try to do is to compute

[8] See Harold Guetzkow (ed.), *Simulation in Social Science*, Prentice-Hall, Inc., Englewood Cliffs, N. J., 1962, and Harold Guetzkow et al., *Simulation in International Relations: Developments for Research and Teaching*, Prentice-Hall, Inc., Englewood Cliffs, N. J., 1963.

the probability of its taking on any of these values. We cannot predict exactly which values will occur.

Let us briefly consider this a little more formally. We consider the random processes known as *random walk* and *Markov chains*.

Random walk is used in probability theory to describe the movement of an event where the direction is determined by a stochastic process, such as the outcome of a Bernoulli trial (a Bernoulli trial is an experiment having two possible outcomes, such as the flip of a coin). As an illustration, consider a bill moving through Congress. There are a number of stages it might go through, but for simplicity here we suppose that it can be assigned to the rules committee, from there to a subject-matter committee, and from there to the floor of the House. At each stage it can be killed or returned to the prior stage. Suppose that the decision as to whether it will move forward or not is made by the flip of a coin. We can generate a random walk by allowing the bill to move forward if the coin turns up heads, but killing it if the coin turns up tails. What is the probability of the bill's passing? If a tail comes up, the bill is killed. This is called a random walk with an absorbing barrier. Once a tail comes up, the (stochastic) process ends. It is simple in this case to compute the probability of a bill's passing. Of course, we know that the probability of any particular bill's being passed from one stage to the next is not .5. We may generate a random walk for this example using a binomial distribution under various assumptions in order to make it more realistic. We do not consider this in detail here, because the random walk does not provide a good model of empirical events. Instead, let us consider a Markov process, because it is more realistic.

In contrast to the random walk, in the Markov process it is assumed that the outcome of an event depends on preceding events. We certainly know this to be true of processes such as a bill passing through Congress, elections, and the like. As a simple example of a Markov process, suppose that there are four parties in a local community. A survey shows the probability that any voter will switch from one party to another is depicted in Table 14-2. The table shows that a person who voted Republican in the past has a probability that he will vote Republican in the next election of .7. The probability

Table 14-2 Transition Probabilities for Voting in Community X

	To			
From	1	2	3	4
Republican	.70	.04	.10	.16
Democrat	.02	.80	.08	.10
Independent	.10	.15	.60	.15
Liberal	.05	.25	.30	.40

that he will switch to Democratic is .04, and so on. This is called a *matrix of transitional probabilities* or a *stochastic matrix*. The diagonal values measure party loyalty. The Liberals are the most fickle (by these hypothetical data). Suppose that these figures have been computed from data for the past fifteen years so that it is not likely that they will change drastically from election to election. Suppose also that the percentage of the total vote going to each of the four parties for the past several elections is as follows:

Party	Percentage of vote
Republican	.40
Democrat	.45
Independent	.10
Liberal	.05
Total	1.00

We can now use the transitional probabilities to predict the probable outcome of the next election and the long-run tendency. For every 100 Republicans, we expect that 70 will vote Republican in the next election, 4 will switch to Democratic, 10 to independent, and 16 to Liberal. Hence, the expected percentage of vote for Republicans for the next election is $(.70)(.40) + (.02)(.45) + (.10)(.10) + (.05)(.05) = .3015$. The expected percentage for each of the other parties can also be computed in the same manner. The long-run rank of each party in this community can be computed by multiplying the transitional probabilities times the respective party percentages after each election and summing. The expected percentage of the total vote going to each party after the next several elections has been computed in Table 14-3. This also is shown in the form of a graph in Figure 14-1. After the

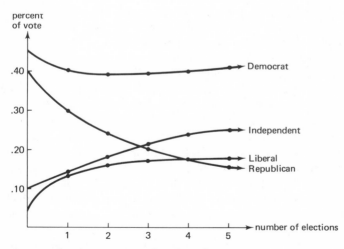

Figure 14-1. Long-run ranks of parties.

fourth election there is a convergence to the equilibrium state of the system. The Republicans, who started out in second rank, end up in last rank. As can be seen easily, additional computations will not change the relative positions of the parties. The only thing that can induce a different ranking is a change in the transitional probabilities. The mathematics of Markov processes do not tell us how to compute these transitional probabilities. However, they make it possible to take a different view of political behavior than conventional regression analysis.

Table 14-3 Expected Shift in Votes among Parties, Assuming Constant Transitional Probabilities

	Percentage of vote after election				
Party	*First*	*Second*	*Third*	*Fourth*	*Fifth*
Republican	.3015	.2414	.204	.18	.16
Democrat	.4035	.3935	.394	.40	.41
Independent	.1510	.1956	.226	.24	.25
Liberal	.1440	.1688	.176	.18	.18
Total	1.0000	1.0000	1.000	1.00	1.00

Markov chains make it possible to view behavior as being partly indeterminate in a different way than regression analysis. In conventional regression analysis it is assumed that a probability distribution of the dependent variable is generated for fixed values of the independent variable. The probability distributions are assumed to be normal and the expected value is the conditional mean $E(Y_{ig}|X) = \bar{Y}_g$. The difference between the expected and the actual value is assumed to be due to sampling error; therefore these models are generally called stochastic. They are stochastic in the sense that we do not know which Y value the sample will turn up. But the models also are determinate in the sense that once the regression coefficients $a, b_1 \ldots, b_n$, are determined, the models yield the same prediction each time they are applied. Hence, we may say that the regression models assume that the "laws" relating the behavior of the variables do not change. Markov chains, on the other hand, yield different results each time they are applied if it is assumed that the transition probabilities change. In such a model, the "laws" relating the behavior of the variables are assumed to change. If it is assumed that some of the change is due to random shocks, such as a scandal, or other unpredictable issue, these models are probabilistic in the sense that they assume some indeterminacy in the behavior being studied above and beyond sampling error. This latter seems more realistic, and hence Markov chains do afford better models of some aspects of political behavior.

The important thing to note about probabilistic models is that they do not assume that there is a single, unique prediction for the dependent variable as is the case with determinate models such as those employed in conventional regression analysis. With a determinate model, such as the simple linear regression equation, the prediction of Y is not expected to be accurate. The model is determinate only in the sense that it gives the same prediction each time for various values of the independent variable. A stochastic model, on the other hand, implies a possible sequence of outcomes and a probability function generating the sequence. Hence, it does not produce a single, unique result. Causal inferences are not possible with such models. But it is likely that the populations that we study (political behavior) contain indeterminate behavior, and thus probabilistic models such as Markov chains may be more appropriate. More likely, the best models, yet to be developed, probably will contain a mixture of determinate and indeterminate functions.

READINGS AND SELECTED REFERENCES

Ackoff, Russell, et al. *Scientific Method*, John Wiley & Sons, Inc., New York, 1962, Chap. 9.

Coplin, William D. *Simulation in the Study of Politics*, Markham Publishing Co., Chicago, 1968.

Jeffrey, R. C., *The Logic of Decision*, McGraw-Hill Book Company, New York, 1965.

Kemeny, John G., J. Laurie Snell, and Gerald Thompson. *Introduction to Finite Mathematics*, Prentice-Hall, Inc., Englewood Cliffs, N. J., 1966, Ch. VI.

Luce, Duncan R., and Howard Raiffa. *Games and Decisions: Introduction and Critical Survey*, John Wiley & Sons, Inc., New York, 1957.

McKinsey, J. C. C. *Introduction to the Theory of Games*, McGraw-Hill Book Company, New York, 1952.

Martin, Francis G. *Computer Modeling and Simulation*, John Wiley & Sons, Inc., New York, 1968.

Rapoport, Anatol. *Fights, Games and Debates*, The University of Michigan Press, Ann Arbor, Mich., 1960.

Spurr, William A., and Charles P. Bonini. *Statistical Analysis for Business Decisions*, Richard D. Irwin, Inc., Homewood, Ill., 1967, Chaps. 15–17.

Von Neumann, John, and Oskar Morgenstern. *Theory of Games and Economic Behavior*, Princeton University Press, Princeton, N. J., 1944.

Williams, J. D., *The Complete Strategyst*, McGraw-Hill Book Company, New York 1954.

15

Sample Design and Elements of Sampling Theory

At several points we have mentioned that the methods of inference considered in this book suppose that the data have been collected by random sampling; otherwise, the methods can be used only in a descriptive manner. Whenever we use the sampling distributions and methods of inference described above (normal, t, F, binomial, Poisson, and x^2), the sample data must be collected by random sampling. Otherwise, a test of statistical significance is useless. Of course, it is possible to make inferences if random sampling has not been used. However, it would not be possible to measure the standard error as we did above. In this chapter we consider sample designs that are nonrandom. We also consider random sampling in more detail and discuss alternative methods of trying to achieve a random sample.

The reason for sampling is clear. We sample because it is not possible to study the entire population, that is, the population is infinite, or because it is more practical to study a sample, that is, the population is finite but very large. Let us consider these two kinds of populations before turning to the question of sample design.

A population may be a group of people, houses, boxes of soap, legislators, and so on. The specific nature of the population depends on what the unit of analysis is that we are studying. If we are studying voting behavior in a presidential election, the population is all those who are eligible to vote.[1] The unit of analysis is a person. We may be interested in several characteristics of this unit, such as religious affiliation, age, income, and so on. This is a finite population. It may not be possible to obtain a list of this population, and we should have to resort to a special sample design in order to draw a probability sample from this population. We consider this question below.

Now consider the kinds of questions in which political science is usually interested. We generally study things such as community decision making, role perception of legislators, United States Supreme Court decisions relations among nations of the world, differences in party systems, and so on.

---

[1] The practical population in this case may be all those who are registered to vote, because all those who are eligible may not be allowed to vote because they have not registered or have moved, and so on.

The unit of analysis in these cases is not people but cities, counties, legislatures, government agencies, nations, and so on. Such populations may be finite or infinite. If we want to generalize, for example, to all cities in the United States, the population is finite. But more often we want to generalize to infinite sets of elements in theoretical models. Usually, we want to generalize about legislative behavior rather than a particular set of legislators. Or we want to generalize about international politics rather than the politics of specific nations. The populations in these cases are infinite and analytical. We consider this question later in this chapter. For the moment, we restrict ourselves to a discussion of the methods of sampling appropriate to finite populations.

The methods of sampling that might be used for finite populations can be classified under two broad headings: probability and nonprobability samples. A probability sample is one in which every member of the population has a known probability of being selected. A nonprobability sample is one in which the probability of any member's being included in the sample is not known. It is not possible to measure the standard error of the estimate with nonprobability samples. The statistics discussed in this book all assume probability samples if we intend to infer something about a population parameter. However, nonprobability sampling has its place in research, and we first discuss some of the more frequently used forms of nonprobability samples.

The principal form of nonprobability sample is the *judgment sample*. This is exactly what is implied by the word "judgment." Someone's judgment is used about which units should be included in the sample. Usually, past research or experience is used in making the determination. Judgment samples may be used in pretests of questionnaires or in pilot studies, because the investigator wants to be sure that certain units will be included. A form of judgment sampling is used by pollsters and the television networks in predicting the outcome of presidential elections. CBS calls its system Voter Profile Analysis. In conjunction with Louis Harris and Associates, it maintains socioeconomic and demographic data on the voting precincts of each state and the relationship of these factors to past voting behavior. A small number of precincts (32 to 60) are selected from each state as representative of the voting population of the entire state. These are combined into a model for each state and stored in a computer. On election night, the results for these particular precincts are called in and compared with past results. On the basis of the comparison, projections are made for the entire state and the election results announced hours before the total vote is counted. Although the procedure is not a probability sample, it has been used for several years and is reported to have an error of less than 1 percent.

Another form of judgment sample is the *quota sample*. This is also used frequently in opinion polling because of its economy. It may sometimes be combined with probability sampling. For example, suppose that we are interested in the opinions of vocational guidance counselors in the high schools of a state. We do not know exactly how many there are, and there is no list available. We might select a sample of high schools at random and instruct our interviewers to interview a given quota of vocational guidance counselors in each school. The cost of this method is lower than obtaining a probability sample, and larger samples can therefore be obtained with less money. However, there is no way to measure the probable error in such samples.

Another form of nonprobability sampling is the *hit-or-miss* technique. This is used in street-corner interviews. The interviewer stands on a street corner and stops passers-by, asking their opinions about various matters. If the things being studied are randomly distributed in the population the sample is drawn from, such as a well-shuffled deck of cards, this procedure will produce a random sample. Sciences such as medicine, archeology, and astronomy often rely on this method. But if the assumption about the characteristic being randomly distributed in the population is wrong, there is no way of measuring the probable error. In the example about the man on the street, if the interviewer is standing on the corner at about noon, he is likely to choose a large number of white-collar and professional people.

In contrast to nonprobability sampling, probability sampling is systematic and based on mechanical procedures. It is not haphazard. For example, if I wanted a probability sample of 10 students out of a class of 50, in general I should not be able to obtain one by writing each of their names on a slip of paper, placing the slips in a hat, mixing them, and then pulling out 10 at random because of the possibility of mechanical imperfections in the procedure. Each slip of paper may not be exactly the same size and weight, they might stick together, and they may not be thoroughly shuffled. Thus, the physical operation used to produce a probability sample is usually a table of random numbers that, with proper use, will produce a sample that is very close to the desired probability model.

The basic method of producing a probability sample is through simple random sampling. This produces a sample in which each member of the population has an equal chance of being selected. The other methods of probability sampling, such as stratified, systematic, and cluster sampling, are essentially modifications of simple random sampling. Let us consider each of them in turn.

SIMPLE RANDOM SAMPLING

Simple random sampling is a sampling procedure that gives each member of the population an equal chance of being selected. The standard and best physical method of ensuring this result is the use of a table of random numbers, such as that reproduced in the Appendix. The operational procedure is simple. Each element in the population is listed and given a number, from 0 to *N*. The table of random numbers is entered at any point. Each number that appears in the table is read in order (and the order can be to read either up, down, or sideways; it does not matter, so long as the direction once started is always the same). Whenever the number that appears in the table of random numbers corresponds to the number of an element on the list, that element is selected into the sample. This process is continued until the desired sample size is reached. If we are sampling without replacement, which is usually the case, and the same number appears before the sample is completed, it is skipped. As an illustration, suppose that we have a population of 12 people and want a simple random sample of 2. Each person is assigned a number. We can start with 0 and number to 11 or start with 1 and number to 12:

01	John	06	Carrie	11	Susan
02	Dick	07	Larry	12	Jean
03	Harry	08	Betty		
04	Mary	09	Ann		
05	Jane	10	Linda		

Next, the table of random numbers is entered, starting on any page, with any row or column, by opening to a page and placing your finger on the page without looking. The first two-digit number near where our finger was placed is 87, which does not correspond to any number on the list; so we ignore it. The second two-digit number is 02, and Dick is therefore the first member selected. The next three two-digit numbers are 91, 35, and 21 respectively, and they all are skipped. The next one encountered is 05, and the next member of the sample is Jane, for a sample consisting of Dick and Jane.

Of course, this is all very simple, because the process of simple random sampling is very simple. In actuality, we do not sample from a population of 12, and the example is not realistic. So let us now consider something a bit more realistic.

Suppose that the population consists of the 10,232 full-time undergraduates of a university. We want a simple random sample of 400. Each student is numbered (hopefully beforehand by the administration) from 1 to 10,232. We enter the table of random numbers the same way as before, but we read five-digit numbers now. Suppose that the first five-digit number encountered is 54,253. We ignore this and go on to the next. It is 10,097. Thus the student who is numbered 10,097 on the list becomes the first member of the sample. We continue this process until 400 students are selected. If the number 10,097 (or any other number previously selected) appears a second time before we select a sample of 400, it is ignored. This gives us a sample without replacement. This process accomplishes the purpose of giving us a sample in which every member has the same chance of being included. What is this chance? It is n/N. In the example above, it is 400/10,232. Let us explain why this is so.

Suppose that we have five cards numbered 1 to 5. We select samples of size 3 from this population (without replacement). What is the probability that the number 2 will be included in the sample? As we said above it is n/N, or 3/5. If the number 2 is to be a member of the sample, it will be selected either on the first, second, or third draw. The probability that it will be selected on the first draw is 1/5. The probability that it will be selected on the second draw is the probability that it will not be selected on the first draw (4/5) times the probability that it will be selected on the second draw given that it was not selected on the first (1/4). Finally, the probability that it will be selected on the third draw is the probability that it will not be selected on the first draw (4/5), times the probability that it will not be selected on the second draw given that it was not selected on the first draw (3/4), times the probability that it will be selected on the third draw given that it was not selected on the first or second draw (1/3). Hence, the probability that the number 2 is a member of the sample is

$$p(\text{1st} \cup \text{2nd} \cup \text{3rd}) = 1/5 + (4/5 \times 1/4) + (4/5 \times 3/4 \times 1/3) = 3/5$$

The probability that any other number in the population will be included in the sample is also 3/5. In general, the probability of any member's being included in the sample is n/N, which is the definition of simple random sampling.

Simple random sampling is an unambiguous concept. But it is often difficult to obtain such a sample, because a list of the population is not available in the form we want it in. A ready-made list may be presented to us in which the persons are listed alphabetically, or according to some criterion not related to the research design, such as names listed by the department in which the person works, or the list may not be complete. The remedies vary with each case. If the list is not complete, and the missing elements are few, they can safely be ignored (unless the study is specifically focusing on new arrivals, and they are the ones left off the list). If the list is divided into groups uneven in size and they cannot (economically) be recast into a single list, then separate simple random samples might be drawn from each list, either in the proportion of the number of elements in each cluster or by weighting the individual selected from the cluster in accordance with the cluster size. This gets us into stratified and cluster sampling, which we now consider.

STRATIFIED SAMPLING

It may be that the population a sample is to be drawn from consists of a wide variety of subgroups. For example, suppose that the population is all counties of the United States. This population consists of counties in metropolitan areas, counties in rural areas, counties that are coterminous with cities, counties in the South, North, and so on. For certain kinds of estimates, it will be better to place the counties into categories before sampling. Each category used is called a stratum. When we sample randomly from each stratum, we have a stratified random sample.

We might draw into our sample the same number from each stratum. But if the total number in each stratum is different, the end result is *disproportionate* stratified random sampling. In this case, a mean computed for the population based on the means of each stratum would have to be weighted in accordance with the number in the stratum.

When weighting is necessary, the population mean can be defined as $\bar{Y} = W_1\bar{Y}_1 + W_2\bar{Y}_2 + \cdots + W_n\bar{Y}_n = \sum_{i=1}^{n} W_i\bar{Y}_i$, where $\bar{Y}$ is the estimated mean based on strata; W_n is the weight of each stratum, computed as $W_i = N_i/N$, where N_i is the number of elements in the ith stratum and N is the total number in the population; and $\bar{Y}_n$ is the mean for each stratum. Let us illustrate this.

The following notation is used:

N = number of units in the population

n = number of elements in the entire sample

f = sampling fraction, or proportion of each stratum to be included in each sample; it is equal to n/N

N_i = number of units in the ith stratum

n_i = number of units selected into the sample from the ith stratum

$\bar{Y}_i$ = mean of the N_i units in the ith stratum

$\bar{y}_i$ = mean of the n_i units in the sample selected from the ith stratum

S_i^2 = variance of the N_i units in the ith stratum

s_i^2 = variance of the n_i units in the sample selected from the ith stratum

$\bar{Y}$ = mean of the population

$\bar{y}$ = mean of the sample

Note that the following equations hold:

$$N = \sum N_i = N_1 + N_2 + \cdots + N_i$$

$$\bar{Y}_i = \frac{\sum_{i=1}^{N_i} N_i}{N_i}$$

$$n = \sum n_i = n_1 + n_2 + \cdots + n_i$$

$$\bar{y}_i = \frac{\sum_{i=1}^{n_i} n_i}{n_i}$$

$$S_i{}^2 = \frac{\sum\limits_{i=1}^{N_i} (Y_i - \bar{Y}_i)^2}{N_i}$$

$$s_i{}^2 = \frac{\sum\limits_{i=1}^{n_i} (y_i - \bar{y}_i)^2}{n_i}$$

As an example, suppose that we have a population of $1{,}000(N = 1{,}000)$ that is broken down into three strata, $A = 200$, $B = 300$, $C = 500$. A disproportionate stratified random sample of 30 is drawn from each stratum. Suppose that the means of the three samples are 8, 10, and 9 respectively. Because the sample comprises more than 5 percent of the population, a finite population correction is required. The sample mean in this case is

	Stratum			
	A	*B*	*C*	Total
N_i	200	300	500	1,000
n_i	30	30	30	90
$\bar{y}_i$	8	10	9	
s_i	2	4	2	

$$\bar{y} = \sum_{i=1}^{n_i} W_i \bar{y}_i \left(\sqrt{1 - \frac{n}{N}} \right)$$

$$= \left[\frac{200}{1{,}000}(8) + \frac{300}{1{,}000}(10) + \frac{500}{1{,}000}(9) \right]\left[\sqrt{1 - \frac{90}{1{,}000}} \right] = 8.2$$

If a *proportionate* stratified sample is desired, a sampling fraction of $n/N = 100/1{,}000 = 1/10$ is used for each stratum. For stratum A it would be $1/10\,(200) = 20$, and so on. It can be seen that $n_i/n = N_i/N$ for proportionate stratified sampling. This is said to yield a self-weighting sample, and the population mean $\bar{Y}$ can be estimated directly with the means of the strata $\bar{y}_i$.

The standard error of the mean computed by strata $\bar{y}_s$ is given by

$$S_{\bar{y}_s} = \sqrt{\sum W_i{}^2 S_{\bar{y}_i}{}^2}$$

where $S_{\bar{y}_i} = s_i/\sqrt{n_i}$. The finite population correction would be added if the sample is less than 5 percent of the population. The standard error of the mean in each stratum in the example above is

$$S_{\bar{y}_1} = \frac{2}{\sqrt{30}} \sqrt{1 - \frac{30}{200}} = .336$$

$$S_{\bar{y}_2} = \frac{4}{\sqrt{30}} \sqrt{1 - \frac{30}{300}} = .694$$

$$S_{\bar{y}_3} = \frac{2}{\sqrt{30}} \sqrt{1 - \frac{30}{500}} = .354$$

The standard error of the sample mean is

$$S_{\bar{y}_s} = \sqrt{\sum W_i^2 S_{\bar{y}_i}^2} = \sqrt{\left(\frac{2}{10}\right)^2 (.34)^2 + \left(\frac{3}{10}\right)^2 (.69)^2 + \left(\frac{5}{10}\right)^2 (.35)^2} = .279$$

Because one of the main purposes of stratified sampling is to improve the estimate of $\bar{Y}$ (by reducing $S_{\bar{y}_s}^2$), it is efficient to sample disproportionately from each stratum. This is especially true if the data are highly skewed. We might take a larger proportion into the sample from strata that have larger variances, and a smaller proportion from strata with smaller variances. An optimum allocation for disproportionate stratified sampling is

$$n_i = n \frac{N_i s_i}{\sum N_i s_i}$$

where the symbols are as defined above. To illustrate this, consider the data in the example above again. We took a sample of 30 from each of the stratum. But an optimum sample for each stratum would be

$$n_1 = 90 \frac{200(2)}{2,600} = 14$$

$$n_2 = 90 \frac{300(4)}{2,600} = 42$$

$$n_3 = 90 \frac{500(2)}{2,600} = 35$$

Samples of these sizes would reduce the standard error of the mean from that calculated with samples of 30 from each stratum. If cost is a factor and varies with each stratum, to get an optimum allocation, considering cost, we should multiply both the numerator and denominator in this equation by the square root of the estimated cost of sampling one item in each stratum.

SYSTEMATIC SAMPLING

Systematic sampling may be used as a substitute for simple random sampling when the latter is not practicable. It consists of selecting every kth individual after the original individual is selected at random. The rate is determined by how large a sample is desired. For example, if the universe consists of 10,000 units and a sample of 400 is desired, the sampling rate is $10,000/400 = 25$. We should therefore select every twenty-fifth person. The first individual to be selected would be selected at random from among the first 25 individuals on the list (by using a table of random numbers). Then every twenty-fifth person following this would be selected into the sample. If the first person selected is the fifteenth, the next will be the fortieth, the next the sixty-fifth, and so on until the list is complete.

Systematic sampling is much easier than simple random sampling. When individuals untrained in sampling (interviewers) are to do the sampling in the field, it is much easier to instruct them to select every kth person from a list they obtain than to follow a process of using a table of random numbers. Their work can also be more easily checked when they use systematic sampling.

Because, for systematic sampling, each unit has a probability of being selected equal to $1/k$, the sample mean is an unbiased estimate of the population mean. However, the variance offers some problems. Because we are, in effect, dividing the units into strata each of k size, the variance of the sample may not be a good estimate of population variance. There may be a systematic pattern in the data, occurring at every kth unit, and this will be picked up. For example, if dwelling units on blocks were selected in this manner, and the first selected were a corner house, every kth unit might also be a corner house. Similarly with an alphabetical list of names, there might be systematic variance associated with the listing. In these cases, if the list can be thoroughly shuffled first, these problems may be alleviated, and the subsequent selection treated as a simple random sample. If not, then the variance of the estimated mean can be computed by

$$S_{\bar{y}}^2 = \frac{1-f}{2n(n-1)} \sum_{i=1}^{n} (y_i - y_{i+1})^2$$

where f is the sampling rate. As an example, suppose that we want to estimate the mean years of schooling with a sample of 10 selected from a population of 1,000. The sample results are 12, 10, 8, 7, 4, 6, 5, 3, 5, 10. We suppose that, unknown to the sampler, the list was based on a factor related to the schooling of the individual. This appears in the systematic descending order in the sample. The sample mean is $\bar{X} = 70/10 = 7$. This is the estimate of the population mean. The variance of the mean is $S_y^2 = \dfrac{.99}{20(9)} (56) = .308$, where $56 = (12 - 10)^2 + (10 - 8)^2 + \cdots + (5 - 10)^2$.

In general, if the sampler suspects that there is an increasing or decreasing trend, or periodic fluctuation in the data, or if it is not possible to mix the list randomly, systematic sampling should be avoided. For example, the list in the example above might have been constructed by some other factor related to the variable we are studying. It could be a list of employees made on the basis of position held in the organization, and this is related to education. There are procedures for avoiding these problems, such as paired selections and replicated sampling (see Kish and Cochran). In general, simple random sampling is preferred except where it is extremely difficult to implement. If systematic sampling is used, every effort should be made to ensure that the list from which the sample is selected does not contain a periodic tendency.

CLUSTER SAMPLING

In practical terms the kind of sampling problem faced in survey research is frequently as follows. The research aims at studying the attitudes of voters (or party officials) in the various election districts of a city (or town). There is no single list available containing all the voters, and it is too expensive to make one. However, a list (or map) of the election districts is readily available. As a practical solution, election districts can be selected at random (first-stage cluster sampling) from this larger list, and within each of the districts, blocks may be selected at random (second-stage cluster sampling) and all the persons in these blocks interviewed.[2] The first-stage sample of clusters is called the primary sampling units, and the second-stage sample is called the secondary sampling units. This sampling method is also called area probability sampling or just area sampling. The cost and ease of area sampling make it much preferable to simple random sampling for large surveys, although it is not so precise.

The important aspect of cluster sampling is the selection of units, each of which contains more than one individual member of the population. Suppose that we want a random sample of 400 voters in a city that contains an estimated 200,000 voters. There is no list of these voters available. We can obtain the sample by sampling city blocks. Suppose that the city has 5,000 blocks. This means there is an average of 40 voters to a block. To get a sample of 400, we can take 400/200,000 of the blocks, or a sample of about 10 blocks. If the blocks do have an average of 40 voters each, the sample size will be $10 \times 40 = 400$, as desired. The probability of selection of each of the voters is 1/500, or the same as the probability of the selection of the block he lives in. The actual sample size will depend on which blocks are selected. If one block has a large apartment house, it may have 100 voters. If this is not offset by a block with fewer voters, the sample will be larger than 400.

The same process holds for sampling from wider populations, such as states. Suppose that we are interested in the attitudes of all elected local officials in a fairly populous state. There is no list of these officials available, but they are estimated to number about 10,000. We want a sample of 400. There is a list of all election districts for local office, which number 800. We thus want to sample $(400/10,000)800 = 32$ election districts. There is an average of 12 1/2 elected officials per district. Hence we can expect the sample to contain $32 \times 12 \ 1/2 = 400$ elected officials. The clusters may be selected by the use of a table of random numbers or by systematic sampling.

---

[2] Simple random sampling could be used within each block selected, in which case we should have a three-stage cluster sample. But it will frequently be less expensive to interview the entire block.

Because the particular districts selected may be homogeneous in certain respects, for example, some may be from upper middle-class suburbs, but others from central city poor districts, the variance of the estimated mean will be much higher, and the chance of error greater. But this may be offset by the practicability of the method of sampling and its lower cost.

If the number of elements E in each cluster c are the same, that is, we have equal clusters, then the probability of selection of any of the N population members is equal to $c/C = n/N = f$, where C equals the number of clusters in the population and the other symbols are as defined above. For the last example, we have $32/800 = 400/10,000 = 1/25$. $\bar{X}_c$ is an estimate of μ with the same properties as the mean of a simple random sample. It is computed by

$$\bar{X}_c = \frac{\sum N_i \bar{X}_i}{N_i}$$

where N_i is the number of secondary units in the ith primary unit, that is, blocks in an election district, and $\bar{X}_i$ is the mean of the sampled secondary units in the ith primary unit, that is, mean numbers of voters in a block. $\bar{X}_c$ is a slightly biased estimate of the population mean, but the bias is not large if a large sample is used. The variance of the mean is

$$S_{\bar{X}_c}^2 = \sqrt{1 - \frac{n}{N}} \frac{S_c^2}{c}$$

where

$$S_c^2 = \frac{1}{c-1} \sum (\bar{Y}_c - \bar{Y})^2$$

c is the number in the cluster, $\bar{Y}_c$ is the cluster mean, and $\bar{Y}$ is the population mean. The finite population correction can be ignored, as before, for small n relative to N.

Consider the following simple example. Suppose that we want to estimate the total number of elected officials who devote more than 10 hours during a week to a particular question, say, air pollution, in a population of 5 districts. We select a simple random sample of 2 districts from the universe. Suppose that the universe values are as follows:

District	Number of officials who devote 10 hours or more to the question
1	3
2	1
3	5
4	3
5	3
Total	15

Recall that the number of different possible samples of 2 from this universe is $_NC_n = \binom{5}{2} = 10$. These are the following combinations [(1,2) (1,3) (1,4) (1,5) (2,3) (2,4) (2,5) (3,4) (3,5) (4,5)]. If the random sample we select is districts 1 and 2, the number in the sample is 4. The mean per district is 2. But the universe mean is $15/5 = 3$. The estimated total based on the sample total of 4 for this selection can be obtained by multiplying the sample mean per district (2) times the total number of districts (5). For this sample, we underestimate the population value. If the random sample we select is districts 2 and 3, the estimated total based on these districts is $3 \times 5 = 15$, or exactly the population total. The reader should compute the estimated total based on any of the other possible samples. For example, for the sample (1,5) it is also $3 \times 5 = 15$. Each of the 10 samples is equally likely to be drawn. The probability of getting any particular sample is thus $1/10$. Each of the districts has the same probability of being selected. Because each one appears 4 times in the 10 possible samples, the probability of selecting any one district is $4/10$. The standard deviation of all possible estimates of $\bar{X}$ is

$$S_{\bar{X}}^2 = \sqrt{1 - \frac{n}{N}\frac{S_c^2}{c}}$$

$$= \sqrt{1 - \frac{2}{5}\left[\frac{[1/(2-1)](2-3)^2+(4-3)^2+(3-3)^2+(3-3)^2+(3-3)^2+(2-3)^2+(2-3)^2}{2}\right.}$$

$$\left. + \frac{(4-3)^2+(4-3)^2+(3-3)^2}{2}\right]$$

$$= 2.33$$

Let us use a different example now to demonstrate some of the problems that can occur in cluster sampling. We suppose that each district has five members whose incomes are:

District	Member	Income of elected officials
1	1	$10,000
	2	20,000
	3	15,000
	4	12,000
	5	6,000
2	1	20,000
	2	30,000
	3	25,000
	4	15,000
	5	18,000

District	Member	Income of elected officials
3	1	$6,000
	2	4,000
	3	10,000
	4	8,000
	5	7,000
4	1	12,000
	2	11,000
	3	6,000
	4	20,000
	5	25,000
5	1	6,000
	2	7,000
	3	14,000
	4	5,000
	5	4,000

The population mean μ is $\mu = \sum X_i/N = 316,000/25 = \$12,640$. We suppose that we do not know this. We select two clusters into the sample by the use of a table of random numbers (without replacement): clusters (1,4). The mean income $\bar{X}$ for this sample is $\bar{X} = \sum X_i/n = 137,000/10 = \$13,700$. This is an unbiased estimate of μ. If the number of clusters in the sample is large, the sampling distribution of $\bar{X}$ can be approximated with the normal distribution and used to compute confidence intervals for μ.

The sample estimate of the population mean was almost $1,000 higher in this case. We could have done worse. The distribution of incomes within each cluster is not random. Districts 2 and 4 have relatively high means; districts 3 and 5 have low means. If the distributions were random within each, cluster sampling would yield population estimates that are as precise as simple random sampling. Because this is rarely so, estimates obtained from cluster sampling are generally less precise than those obtained by random sampling. The example above readily shows why this is so. If we had been unlucky enough to draw clusters 2 and 4 into the sample, the estimate of μ would have been too high and, on the average, much higher than samples obtained by simple random sampling.

So far in our discussion of cluster sampling we have assumed that the number in each cluster is approximately the same. In practical sampling situations, this is not generally so. This creates problems concerning estimates of population values such as means and proportions. The best practical solution is to try to reduce the variation in size by various selection techniques, such as stratifying according to size, creating artificial clusters of approximately equal size, subsampling with probability proportional to cluster size, and paired selection (see Kish, pp. 184–246). Let us consider a few of these techniques briefly here.

Suppose that we want a sample of 2,000 students in a particular university and all we have is a list of classes and the number of students in each. There are 5,000 classes, which range from 8 to 250 in size, with an average number of 40 in a class. A sample of approximately 2,000 students can be selected by taking 50 classes. To do so, we can form 25 strata of 100 classes each, and select at random 2 classes from each stratum to yield a total of 50 classes. This method of paired selection will result in a sample that is about 2,000, and the sampling fraction is 1/50. The mean and variance of the sample can then be computed as described above.

We can reduce the variation in sample size by stratification according to size of cluster. We can then select from each stratum in a manner to ensure a uniform probability of sampling fraction, selecting more from the middle range and less from the extremes. Alternatively, we can select a large number of clusters and subsample an equal number of elements from within each of the selected clusters. Neither results in a sample in which each member of the population has an equal probability of being selected. The members of the larger clusters have less chance of being selected (inversely proportional to the size of the cluster) than those of the smaller clusters. This can be compensated for by weighting each member selected proportionate to the number in the cluster, but this is not considered efficient.

There are a multitude of particular designs by which it is possible to achieve the objective of giving every member of the population an equal chance of being selected through cluster sampling. Once this objective is achieved, the problem of estimating population parameters is straightforward. Because of the number of possible techniques and because many of them are applicable to specific situations, we do not consider them here.

NONSAMPLING ERROR

The theory of sampling, strictly speaking, is concerned with the error introduced by the sampling procedure. In an ideal design, this error is minimized for an individual sample. In principle, it is zero when we consider average results in the long run. The error in estimates, such as for the mean, refer to what is expected in the long run if a particular set of procedures is followed. But even if this error is minimized, there are other sources of error besides sampling error. Measurement error in one. Usually two kinds of measurement error are distinguished: (*a*) systematic error, called *bias*, which affects the validity of the measurement, and (*b*) variable error, which is frequently assumed to be random and is related to reliability. Another pair of terms is sometimes used to refer to these kinds of error: accuracy and precision. A measurement free of systematic error is *accurate*. A measurement free of variable error, including sampling error, is *precise*.

A general and simplified model of these errors can be written $y_i = Y_i + \sum_g B_g + \sum_e V_e$, which says that a particular observation contains the true population parameter value plus systematic bias B_g plus variable error V_e. Both B_g and V_e have two sources: sampling and nonsampling. Variable sampling error has been considered in many places in this book for various statistics. Systematic sampling error can be produced by using the wrong sample design or using, for example, the sample variance as an estimate of the population variance without a correction. Let us now consider the nonsampling aspects of each of these.

It is possible to get a measure of nonsampling variable error by measuring differences of replications of units within the sample. The coefficient of reliability is one method. The simplest form is the split-half technique, using a correction factor. As an example, suppose that we have measured the morale of organizations on a 10-item scale. We divide the scale in half by taking the odd-numbered items in one half and the even-numbered ones in the other. The scores of the first half are correlated with those of the second half. This gives an estimate of the reliability of the measurement, but because the scale has been cut in half, we must correct the correlation by the Spearman-Brown prophecy equation.[3] This is $r_i = 2r_{12}/(1 + r_{12})$, where r_i equals the index of reliability and r_{12} is the correlation of the two halves. For example, if $r_{12} = .70$, then $r_i = [2(.70)]/(1 + .70) = .82$.

The measurement of nonsampling bias is possible by checking the results of one measurement against those of an external and more accurate measurement of the same thing. The more accurate measurement is called the *criterion*. For example, we may operationally measure the amount of centralization in an organization by salary ratios. The salary ratio of the top 2 percent of the people in the organization against the bottom 40 percent might be compared. We may say that, the higher the ratio, the more centralized the agency.[4] To check the validity of this measurement we can use another measurement of the same phenomenon. Suppose that we use perceptions of individuals in the organizations. To measure the net bias in the first, we first convert each to standard-score form. The difference between the two is a measure of the systematic bias and variable error. If there is no variable error (the measuring instrument is reliable), the difference can be attributed to systematic error (provided that we are dealing with the population and not a sample).

We might represent the different sources of error by the simple model:

$$X_{\text{true}} = X_i \pm [X_i - E(X_i)] \pm [E(X_i) - X] \pm [X - M_i] \pm [M_i - X_{\text{true}}]$$

[3] See J. P. Guilford, *Psychometric Methods*, McGraw-Hill Book Company, New York, 1936, pp. 411 f.

[4] See Thomas L. Whistler, "Measuring Centralization of Control in Business Organizations," in Robert Tannenbaum, Riving R. Weschler, and Fred Massarik, *Leadership and Organization: A Behavioral Science Approach*, McGraw-Hill Book Company, 1961, New York, 1961, pp. 333–346.

where X_{true} represents the true population value, that is, mean, median, correlation, difference of means, proportions, and so on. The first component X_i is the sample value. The second is the variable sampling error, which can be taken to be the difference between the actual sample value and the average of the sampling distribution. The third component $[E(X_i) - X]$ is the constant error. The fourth is the variable nonsampling error, which may be the amount of error produced by measurement, and, if randomly distributed, $\sum M_i = 0$. $[M_i - X_{true}]$ represents a constant nonsampling error that is also zero if the measurement instrument is perfectly valid and there are no errors in using it, such as, transcribing error, clerical error, misplaced forms, and so on. As a hypothetical example, we might have

$$100 = 90 \pm 20 \pm 8 \pm 0 \pm 2$$
$$= 90 + 20 - 8 - 0 - 2$$
$$= 100$$

Thus, the observation in the sample is 90, and the difference between this and the expected average value of the sampling distribution is 20. This means that there is another source of error besides variable sampling error. This has been represented as consisting of constant sampling error and constant nonsampling error. Of course, this does not at all represent what the relative importance or contribution each of these sources is likely to make in the long run. In some cases (the sample is large and the population has a small variance), the variable sampling error may be negligible, and the variable nonsampling error (a very unreliable instrument) may be large.

It is also important to note that the different errors will have a different effect on diverse statistics. For example, the mean $\bar{X}$ will be affected by constant nonsampling error, such as nonresponses, but the correlation coefficient may be unaffected if the constant error is the same for both characteristics being studied. The latter derives from the fact that the correlation coefficient is a pure number, as described in Chapter 9. Similarly, a constant bias on means will not affect a difference of means if both are affected in the same way. Hence a relatively large measurement error is tolerable if it affects all variables in the same way and we are measuring relationships. For example, suppose that we are studying the relationship between centralization of organizations and conservatism. We have measured centralization on salary ratios and conservatism on an attitude scale containing 40 items. Suppose that the salary measurement consistently underestimates centralization for each organization by a factor of $1/2X_i$, so that an organization that has an observed score of 4 is really 6, and so on. Suppose also that the conservatism scale consistently overestimates conservatism for each observation by a factor of 2. Hypothetical scores for true and measured values of each variable with the respective correlations are shown in Table 15-1.

Table 15-1   Correlations for True Population Values and Measured Values Containing Systematic Measurement Error

X (centralization)				Y (conservatism)			
$X_{true} = x_i + \frac{1}{2}x_i$	$(X_{true})^2$	x_i	x_i^2	$Y_{true} = y_i - \frac{1}{2}y_i$	$(Y_{true})^2$	y_i	y_i^2
3	9	2	4	2	4	4	16
6	36	4	16	3	9	6	36
9	81	6	36	2	4	4	16
12	144	8	64	4	16	8	64
15	225	10	100	6	36	12	144
45	495	30	220	10	69	34	276

$$\Sigma (X_{true})(Y_{true}) = 180$$
$$\Sigma x_i y_i = 240$$

$$r = \frac{5(240) - (30)(34)}{\sqrt{5(220) - (30)^2} \ \sqrt{5(276) - (34)^2}} = \frac{180}{\sqrt{200} \ \sqrt{224}} \doteq .85$$

$$r_{true} = \frac{5(180) - (45)(17)}{\sqrt{5(495) - (45)^2} \ \sqrt{5(69) - (17)^2}} = \frac{135}{\sqrt{450} \ \sqrt{50}} \doteq 8.5$$

Although the measurements for each variable are not accurate, the correlations for the observations containing measurement error are exactly the same as the true population correlation. In this example, the sample n is equal to the population N, and therefore there is no sampling error. Also, we suppose that there is no variable nonsampling error. It is apparent that, if we introduced sampling error into this example, the sample r would not equal the population ρ. However, confidence intervals for r can tell us what we can expect ρ to be. Generally speaking, we can suppose that random, variable, nonsampling errors will equal zero in the long run, because the errors of underestimation may be canceled by other errors of overestimation. Hence, for correlation analysis, we do not need to worry very much about systematic measurement error or variable error. The same is not true for estimating particular parameters, such as μ.

NONRESPONSE

In survey research, among the sources of error, nonresponse error is perhaps the most pervasive and common and thus requires singling out (there are others, such as noncoverage, clerical errors, and so on, but, for most of these, there is little remedy other than common sense and experience). Nonresponse may be defined as those observations which are not made because of refusal to answer, lost forms, not-at-homes, and so on. The rate of response is easy to compute. It is $R = 1 - (n - r)/n$. For example, if the original sample size is 400 of which 300 are actually obtained, the response rate is $1 - (400 - 300)/400 = .75$. Nonresponse, in this case, is .25 or 25 percent.

Nonresponse can introduce a substantial bias into a study. The amount of and kind of bias is related to the kind of nonresponse. There are various ways to classify nonresponse. Each of the categories affects the sample results in a different way:

1. *Not-at-homes.*   If the study involves the attitudes of the head of the household and is conducted at a time when he is most likely to be home (evening), it is possible that certain types of respondents are more likely to be at home than others (those who work during the day, married people, perhaps middle aged or older people rather than younger, and so on).

2. *Refusals.*   Certain types of respondents are more likely to cooperate than others, such as those with more education, higher income, and so on. Also, this may vary with the type of question being asked.

3. *Incapacity.*   This may be due to illness, language barriers, or illiteracy. This can introduce a bias that may be somewhat different from the last category.

4. *Not found.*   This includes those who moved and those who are inaccessible (cannot make appointment).

5. *Lost forms.*   A small number may occur because of misplacement of the completed questionnaire or interview schedules.

These categories apply to entire schedules as well as to parts of an interview or questionnaire schedule, or single questions. There are really no general and precise ways to correct for the problems created by nonresponse of this kind. There are some rough corrections that might be made, which we now consider.

Overall Response Rate

The proportion of nonrespondents will vary widely, depending on a large number of factors, including the nature of the population being studied, the type of instrument being used (interview schedule, self-administered questionnaire, mail questionnaire), the skill of those administering the forms, the number of call-backs that can be made, and the kinds of questions being asked. Only general rules of thumb and experience help in this area. A poorly administered and designed mail survey might yield a response rate of only 10 percent, but a carefully designed interview survey might achieve a 95 percent response rate. This is not to say that a mail survey inevitably has a low response rate. There are examples of mail surveys that have achieved a

90 percent response or better. However, they dealt principally with specialized or professional populations. It goes almost without saying that a survey with a 10 percent response rate is as good as no survey at all.

Corrections for Missing Respondents

Most attempts to estimate the effect of nonresponse depends on information about the nonrespondents. This may be estimated by extrapolating characteristics from the nonrespondents found on call-backs, or from knowledge from previous studies. Substituting new respondents is not a solution, because they will more likely resemble the respondents already obtained than the nonrespondents.

The amount of bias introduced by nonresponse can be very large. Hansen and his colleagues (Hansen et al., 1953) report on a 1937 federal government census of unemployment that had 11,000,000 respondents and a 67 percent response rate. When the results were checked with a 2 percent sample survey, it was shown that the errors of the large survey were many times as large as the sampling error of the small sample, and this was due to nonresponse bias. Unfortunately, there are no precise methods for correcting for this.[5] The best solution is sufficient call-backs to ensure at least a 95 percent response rate. But, short of this, it may be possible to make some corrections. If we know some of the characteristics of the nonrespondents, it may be possible to make satisfactory adjustments. Hansen and his colleagues describe a 1948 federal government survey of retail stores in which nonresponses accounted for 13.8 percent of sales volume: 6 percent of these were refusals or would file the monthly report only occasionally, 5.5 percent resulted from late filing and illness, and 2.3 percent were refusals for the area sample. *Imputations* were made for these based on knowledge about the size of the store's payroll, as shown in the social security records, or from other sources.

In political science, many analogous examples come to mind. A survey of voters to estimate the proportions that belong to one party or another (or even to discover something about attitudes) that has a nonresponse rate of 10 percent could be corrected if something is known about the income, education, or occupation of the nonrespondents. Suppose that the 10 percent amounts to 200 voters and that, of these, 70 percent have an income of about $15,000 a year. We might predict that $.70 \times 200 \times .90 = 126$ are Republicans if we know that in general 90 percent of people in this income level are Republicans.

[5] A special sampling method developed by Politz for handling nonrespondents is not discussed here. It has been critized on the basis that it is too difficult to execute. See A. Politz and W. Simmons, "An Attempt to Get the Not-at-homes into the Sample without Callbacks," *Journal of the American Statistical Association*, vol. 44, pp. 9–31, 1949. See also Z. W. Birnbaum and M. G. Sirken, "Bias Due to Non-availability in Sampling Surveys," *Journal of the American Statistical Association*, vol. 45, pp. 98–110; 1950, and J. A. Clausen and R. N. Ford, "Controlling Bias in Mail Questionnaires," *Journal of the American Statistical Association*, vol. 42, pp. 497–511, 1947.

But there is no way of computing the possible error in this figure. Only ingenuity, judgment, and some gambling (on the conservative side) can help in this situation. This method of imputation is perhaps the best way to correct for nonresponse, but it is safe only so long as the nonresponse rate is low.

Corrections for Missing Items

This presents less of a problem than does a completely missing respondent, because other information is available on the respondent to use as a basis for imputing what his missing response may be. This is particularly useful for attitudinal scales that contain a number of items. We assume a scale of 10 items (measuring a variable such as conservatism) as an example. Suppose that each item has a choice range of 1 (strongly disagree) to 6 (strongly agree) and that higher scores represent more conservatism. Assume also 5 respondents with the following pattern of responses:

Respondent	\multicolumn{10}{c}{Items}									
	1	2	3	4	5	6	7	8	9	10
1	6	4	—	5	3	5	2	—	6	1
2	5	4	3	3	6	5	1	6	3	—
3	2	2	5	3	1	2	3	4	1	3
4	6	—	—	4	—	—	5	—	—	—
5	—	5	6	—	4	3	—	6	—	6

Respondents 1 and 2 and perhaps 5 can be given imputed scores for the missing items based on the items they did respond to. The usual procedure is to give each as a score of each missing item, the mean of the items the person did respond to. Thus, for respondent 1, we have $32/8 = 4$, or a total score of $6 + 4 + 4 + 5 + 3 + 5 + 2 + 4 + 6 + 1 = 40$. It would be dangerous to do the same for respondent 4, and respondent 5 is a borderline case. In the latter case, one could not go too far wrong in imputing scores for the missing items. In the former case, less error might be introduced by dropping this respondent, although there is no empirical evidence available to prove this.

The same line of reasoning can be used to impute scores to missing respondents for particular variables. Suppose that the scores for the five respondents above (with respondent 4 missing) are to be correlated with another variable. Let us assume this to be the Y variable and the respective scores as follows:

Respondent	X	Y
1	40	20
2	40	30
3	26	40
4	(39)	10
5	50	5

We might handle respondent 4's score on X by giving him the average of the other four respondents (entered in parentheses). The correlation between X and Y given these values is approximately $-.85$. The correlation between X and Y with respondent 4 dropped is approximately $-.94$. The effect of dropping respondent 4 is to increase the correlation by $.09$. It may be true that the correlation with the imputed value is closer to the population value, because respondent 4 undoubtedly is very close to 39 on the X variable; that is, he is likely to be in the vicinity of 39, because we cannot assume that, given the way he responded to items 1, 4, and 7, he would have an average value as low as 1 or 2 for the remaining items. Supposing that the lowest reasonable average value for the remaining items is 3, his total score on X would be 36, and the correlation would be very close to $-.85$. This method does not generally lead a researcher too far astray. But it should also be noted that, if the number of respondents who omit a large number of response items is high and the sample is small, the resulting correlation could vary rather widely under different assumptions. The only answer, in this case, is that the questionnaire or interview schedule is bad, and the whole thing should be done again.

SAMPLING FROM ANALYTICAL UNIVERSE

So far in this chapter we have considered sampling from a finite population. When we sample from an infinite population, it is not possible to obtain a simple random sample, because there is no process by which we can assure that each member of the population has the same chance of being included in the sample. We cannot use a table of random numbers, because we cannot list and number the members of the population.

Hence, it would seem that conventional probability theory cannot help us generalize to infinite universes, because we have no way of developing confidence limits or of computing the standard error. One possible conclusion in that the process of generalizing to analytical, or infinite, universes is analytical, for which probability theory, in its classical sense, can offer no help. But it is also true that in probability theory we make assumptions about what will happen if we repeat an experiment an infinite number of times, and we regard a finite set of trials as a sample from the infinite population of all trials.

One way to approach this question is as follows: Suppose that we study all legislators (national, state, and local) in the United States and we have measures of each member of this population. This population itself might be considered a random sample from a "super" population of all possible finite populations that could have been produced at the instant of observation under the conditions obtaining. This line of reasoning has proved fruitful in experimental research. Suppose that we are able to set up an experiment in which all variables are controlled and we manipulate the independent variable (say that it is the structure of communications). We measure a correlation r between this independent variable and the time it takes to solve problems. Strictly speaking, the value of r applies to the N groups studied. But we might test r to see if it is significantly different from zero, using N as the sample size and the null hypothesis as $\rho = 0$. What is the universe to which the decision applies? It is defined as all the possible results that would be obtained by repeating the same experiment an infinite number of times.[6] We could thus say that we should expect the same r, within the limits of the standard error, for another N groups. We cannot say that the r for the next group will be exactly equal to the r found in the first experiment. This is because of the element of chance. Suppose that the r found in the experiment is .7, that is, the less structured the communications, the shorter the time required to solve problems. In an applied situation, we could say that, if communications are left unstructured, the time required to solve problems will be shorter, provided that the other factors associated with the dependent variable are present in the same way in the case at hand as they were in the experiment. This, of course, will rarely be so, and hence a particular prediction may be quite off. We may attribute some of this to sampling error, some to measurement error, and so on. We can measure the sampling error by the standard error of the estimate, as described above.

The use of probability theory in statistical inference can be justified in two ways. The phenomena being studied might be considered to be a process in which certain of the components are determined randomly, and these are assumed to be random both in the universe and in the sample. On the other hand, the individual observations may be selected at random from a universe in which the quantities are nonrandom (or completely causally ordered). In this case, certain quantities in the sample are random but not in the universe. If we adopt the first point of view, the stochastic model that is used [such as $Y' = a + b(x) + e$] will be assumed to represent the process which generates the quantities studied. This point of view is necessary if we are not able to sample at random from a population, and even more necessary when our interest is not simply to evaluate characteristics of the population from which random samples are drawn, but more importantly to discover laws that have more general validity. In the latter case, our generalization is not about some quantity in the population, but about more permanent characteristics of the laws in the theories directing our inquiry.

The same reasoning may be applied to nonexperimental research design. Consider again the example concerning the relationship between social status and role perception for legislators. Suppose that we have a positive ρ for the population. The "super" population to which we can generalize is the popu-

---

[6] G. Snedecor, *Statistical Methods Applied to Experiments in Agriculture and Biology*, 5th ed., Iowa State College Press, Ames, Iowa, 1956.

lation of all possible results we could obtain by repeating the study under identical conditions an infinite number of times. Of course, the requirement that it be done under identical conditions does not seem to be so practical as in an experimental design. But, in principle, we can conceive of the problem this way and hence test statistical significance. Another way to handle this question is to conceive of the population as a sample in time. Time-series analysis would be an appropriate mode of analysis, and we should try to build a dynamic factor into our theory. This was done in Chapter 14 in the Markov models which assume changes in transition probabilities. But the question of statistical significance should not get out of hand. In many cases the more important question relates to the strength and nature of the relationship rather than statistical significance. As we said before, statistical significance is not coterminous with theoretical significance. This is not to say that sampling questions can be ignored, for we can be quite misled by bad samples. It means only that theory is at least as important as statistics in making meaning out of empirical data.

READINGS AND SELECTED REFERENCES

Blalock, Hubert, Jr. *Social Statistics*, McGraw-Hill Book Company, New York, 1960, Chap. 22.

Cochran, William. *Sampling Techniques*, 2d ed., John Wiley & Sons, Inc., New York, 1963.

Deming, W. E. *Some Theories of Sampling*, John Wiley & Sons, Inc., New York, 1950.

Hansen, Morris, William Hurwitz, and William Madow. *Sample Survey Methods and Theory*, 2 vols., John Wiley & Sons, Inc., New York, 1953.

Kish, Leslie. *Survey Sampling*, John Wiley & Sons, Inc., New York, 1965.

McCarthy, Philip J. *Introduction to Statistical Reasoning*, McGraw-Hill Book Company, New York, 1957, Chaps. 5, 6, 10.

Spurr, William A., and Charles P. Bonini. *Statistical Analysis for Business Decisions*, Richard D. Irwin, Inc., Homewood, Ill., 1967, Chap. 14.

Appendix

Table 1 Areas under the Normal Curve*

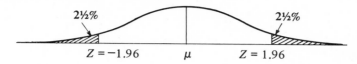

Z	.00	.01	.02	.03	.04	.05	.06	.07	.08	.09
0.0	0000	0040	0080	0120	0159	0199	0239	0279	0319	0359
0.1	0398	0438	0478	0517	0557	0596	0636	0675	0714	0753
0.2	0793	0832	0871	0910	0948	0987	1026	1064	1103	1141
0.3	1179	1217	1255	1293	1331	1368	1406	1443	1480	1517
0.4	1554	1591	1628	1664	1700	1736	1772	1808	1844	1879
0.5	1915	1950	1985	2019	2054	2088	2123	2157	2190	2224
0.6	2257	2291	2324	2357	2389	2422	2454	2486	2518	2549
0.7	2580	2612	2642	2673	2704	2734	2764	2794	2823	2852
0.8	2881	2910	2939	2967	2995	3023	3051	3078	3106	3133
0.9	3159	3186	3212	3238	3264	3289	3315	3340	3365	3389
1.0	3413	3438	3461	3485	3508	3531	3554	3577	3599	3621
1.1	3643	3665	3686	3718	3729	3749	3770	3790	3810	3830
1.2	3849	3869	3888	3907	3925	3944	3962	3980	3997	4015
1.3	4032	4049	4066	4083	4099	4115	4131	4147	4162	4177
1.4	4192	4207	4222	4236	4251	4265	4279	4292	4306	4319
1.5	4332	4345	4357	4370	4382	4394	4406	4418	4430	4441
1.6	4452	4463	4474	4485	4495	4505	4515	4525	4535	4545
1.7	4554	4564	4573	4582	4591	4599	4608	4616	4625	4633
1.8	4641	4649	4656	4664	4671	4678	4686	4693	4699	4706
1.9	4713	4719	4726	4732	4738	4744	4750	4758	4762	4767
2.0	4773	4778	4783	4788	4793	4798	4803	4808	4812	4817
2.1	4821	4826	4830	4834	4838	4842	4846	4850	4854	4857
2.2	4861	4865	4868	4871	4875	4878	4881	4884	4887	4890
2.3	4893	4896	4898	4901	4904	4906	4909	4911	4913	4916
2.4	4918	4920	4922	4925	4927	4929	4931	4932	4934	4936
2.5	4938	4940	4941	4943	4945	4946	4948	4949	4951	4952
2.6	4953	4955	4956	4957	4959	4960	4961	4962	4963	4964
2.7	4965	4966	4967	4968	4969	4970	4971	4972	4973	4974
2.8	4974	4975	4976	4977	4977	4978	4979	4980	4980	4981
2.9	4981	4982	4983	4984	4984	4984	4985	4985	4986	4986
3.0	4986.5	4987	4987	4988	4988	4988	4989	4989	4989	4990
3.1	4990.0	4991	4991	4991	4992	4992	4992	4992	4993	4993
3.2	4993.129									
3.3	4995.166									
3.4	4996.631									
3.5	4997.674									
3.6	4998.409									
3.7	4998.922									
3.8	4999.277									
3.9	4999.519									
4.0	4999.683									
4.5	4999.966									
5.0	4999.997133									

*From Harold O. Rugg, *Statistical Methods Applied to Education* (Boston, Houghton Mifflin Co., 1917), Appendix table III, pp. 389–390, with the permission of the publisher.

Table 2   Distribution of t^*

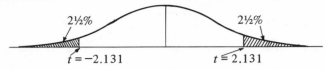

$$2\frac{1}{2}\% \qquad\qquad 2\frac{1}{2}\%$$

$$t = -2.131 \qquad\qquad t = 2.131$$

$$df = 15, \quad p = .05$$

df	.9	.8	.7	.6	.5	.4	.3	.2	.1	.05	.02	.01	.001
								Probability					
1	.158	.325	.510	.727	1.000	1.376	1.963	3.078	6.314	12.706	31.821	63.657	636.619
2	.142	.289	.445	.617	.816	1.061	1.386	1.886	2.920	4.303	6.965	9.925	31.598
3	.137	.277	.424	.584	.765	.978	1.250	1.638	2.353	3.182	4.541	5.841	12.924
4	.134	.271	.414	.569	.741	.941	1.190	1.533	2.132	2.776	3.747	4.604	8.610
5	.132	.267	.408	.559	.727	.920	1.156	1.476	2.015	2.571	3.365	4.032	6.869
6	.131	.265	.404	.553	.718	.906	1.134	1.440	1.943	2.447	3.143	3.707	5.959
7	.130	.263	.402	.549	.711	.896	1.119	1.415	1.895	2.365	2.998	3.499	5.408
8	.130	.262	.399	.546	.706	.889	1.108	1.397	1.860	2.306	2.896	3.355	5.041
9	.129	.261	.398	.543	.703	.883	1.100	1.383	1.833	2.262	2.821	3.250	4.781
10	.129	.260	.397	.542	.700	.879	1.093	1.372	1.812	2.228	2.764	3.169	4.587
11	.129	.260	.396	.540	.697	.876	1.088	1.363	1.796	2.201	2.718	3.106	4.437
12	.128	.259	.395	.539	.695	.873	1.083	1.356	1.782	2.179	2.681	3.055	4.318
13	.128	.259	.394	.538	.694	.870	1.079	1.350	1.771	2.160	2.650	3.012	4.221
14	.128	.258	.393	.537	.692	.868	1.076	1.345	1.761	2.145	2.624	2.977	4.140
15	.128	.258	.393	.536	.691	.866	1.074	1.341	1.753	2.131	2.602	2.947	4.073
16	.128	.258	.392	.535	.690	.865	1.071	1.337	1.746	2.120	2.583	2.921	4.015
17	.128	.257	.392	.534	.689	.863	1.069	1.333	1.740	2.110	2.567	2.898	3.965
18	.127	.257	.392	.534	.688	.862	1.067	1.330	1.734	2.101	2.552	2.878	3.922
19	.127	.257	.391	.533	.688	.861	1.066	1.328	1.729	2.093	2.539	2.801	3.883
20	.127	.257	.391	.533	.687	.860	1.064	1.325	1.725	2.086	2.528	2.845	3.850
21	.127	.257	.391	.532	.686	.859	1.063	1.323	1.721	2.080	2.518	2.831	3.819
22	.127	.256	.390	.532	.686	.858	1.061	1.321	1.717	2.074	2.508	2.819	3.792
23	.127	.256	.390	.532	.685	.858	1.060	1.319	1.714	2.069	2.500	2.807	3.767
24	.127	.256	.390	.531	.685	.857	1.059	1.318	1.711	2.064	2.492	2.797	3.745
25	.127	.256	.390	.531	.684	.856	1.058	1.316	1.708	2.060	2.485	2.787	3.725
26	.127	.256	.390	.531	.684	.856	1.058	1.315	1.706	2.056	2.479	2.779	3.707
27	.127	.256	.389	.531	.684	.855	1.057	1.314	1.703	2.052	2.473	2.771	3.690
28	.127	.256	.389	.530	.683	.855	1.056	1.313	1.701	2.048	2.467	2.763	3.674
29	.127	.256	.389	.530	.683	.854	1.055	1.311	1.699	2.045	2.462	2.756	3.659
30	.127	.256	.389	.530	.683	.854	1.055	1.310	1.697	2.042	2.457	2.750	3.646
40	.126	.255	.388	.529	.681	.851	1.050	1.303	1.684	2.021	2.423	2.704	3.551
60	.126	.254	.387	.527	.679	.848	1.046	1.296	1.671	2.000	2.390	2.660	3.460
120	.126	.254	.386	.526	.677	.845	1.041	1.289	1.658	1.980	2.358	2.617	3.373
∞	.126	.253	.385	.524	.674	.842	1.036	1.282	1.645	1.960	2,326	2.576	3.291

*From Table III of R. A. Fisher and Frank Yates, *Statistical Tables for Biological, Agricultural and Medical Research* (1953 edition), published by Oliver & Boyd, Ltd., Edinburgh, and by permission of the authors and publishers.

Table 3 Distribution of χ^2

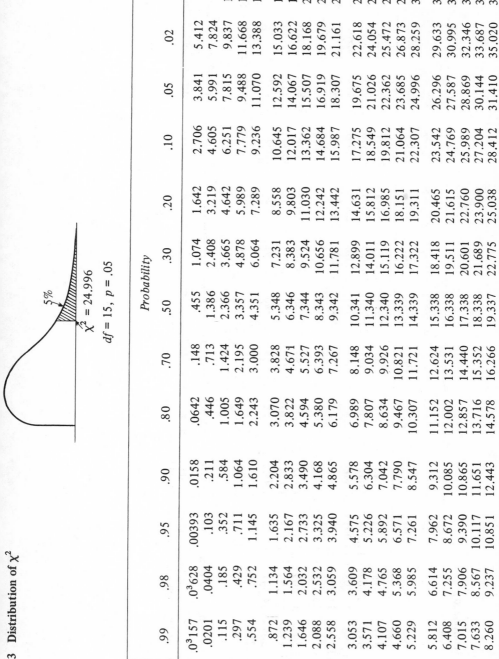

5%

$\chi^2 = 24.996$

$df = 15, \ p = .05$

df	.99	.98	.95	.90	.80	.70	.50	.30	.20	.10	.05	.02	.01	.001
							Probability							
1	.0³157	.0³628	.00393	.0158	.0642	.148	.455	1.074	1.642	2.706	3.841	5.412	6.635	10.827
2	.0201	.0404	.103	.211	.446	.713	1.386	2.408	3.219	4.605	5.991	7.824	9.210	13.815
3	.115	.185	.352	.584	1.005	1.424	2.366	3.665	4.642	6.251	7.815	9.837	11.345	16.266
4	.297	.429	.711	1.064	1.649	2.195	3.357	4.878	5.989	7.779	9.488	11.668	13.277	18.467
5	.554	.752	1.145	1.610	2.243	3.000	4.351	6.064	7.289	9.236	11.070	13.388	15.086	20.515
6	.872	1.134	1.635	2.204	3.070	3.828	5.348	7.231	8.558	10.645	12.592	15.033	16.812	22.457
7	1.239	1.564	2.167	2.833	3.822	4.671	6.346	8.383	9.803	12.017	14.067	16.622	18.475	24.322
8	1.646	2.032	2.733	3.490	4.594	5.527	7.344	9.524	11.030	13.362	15.507	18.168	20.090	26.125
9	2.088	2.532	3.325	4.168	5.380	6.393	8.343	10.656	12.242	14.684	16.919	19.679	21.666	27.877
10	2.558	3.059	3.940	4.865	6.179	7.267	9.342	11.781	13.442	15.987	18.307	21.161	23.209	29.588
11	3.053	3.609	4.575	5.578	6.989	8.148	10.341	12.899	14.631	17.275	19.675	22.618	24.725	31.264
12	3.571	4.178	5.226	6.304	7.807	9.034	11.340	14.011	15.812	18.549	21.026	24.054	26.217	32.909
13	4.107	4.765	5.892	7.042	8.634	9.926	12.340	15.119	16.985	19.812	22.362	25.472	27.688	34.528
14	4.660	5.368	6.571	7.790	9.467	10.821	13.339	16.222	18.151	21.064	23.685	26.873	29.141	36.123
15	5.229	5.985	7.261	8.547	10.307	11.721	14.339	17.322	19.311	22.307	24.996	28.259	30.578	37.697
16	5.812	6.614	7.962	9.312	11.152	12.624	15.338	18.418	20.465	23.542	26.296	29.633	32.000	39.252
17	6.408	7.255	8.672	10.085	12.002	13.531	16.338	19.511	21.615	24.769	27.587	30.995	33.409	40.790
18	7.015	7.906	9.390	10.865	12.857	14.440	17.338	20.601	22.760	25.989	28.869	32.346	34.805	42.312
19	7.633	8.567	10.117	11.651	13.716	15.352	18.338	21.689	23.900	27.204	30.144	33.687	36.191	43.820
20	8.260	9.237	10.851	12.443	14.578	16.266	19.337	22.775	25.038	28.412	31.410	35.020	37.566	45.315

Table 3 (continued)

df						Probability								
	.99	.98	.95	.90	.80	.70	.50	.30	.20	.10	.05	.02	.01	.001
21	8.897	9.915	11.591	13.240	15.445	17.182	20.337	23.858	26.171	29.615	32.671	36.343	38.932	46.797
22	9.542	10.600	12.338	14.041	16.314	18.101	21.337	24.939	27.301	30.813	33.924	37.659	40.289	48.268
23	10.196	11.293	13.091	14.848	17.187	19.021	22.337	26.018	28.4	32.007	35.172	38.968	41.638	49.728
24	10.856	11.992	13.848	15.659	18.062	19.943	23.337	27.096	29.55	33.196	36.415	40.270	42.980	51.179
25	11.524	12.697	14.611	16.473	18.940	20.867	24.337	28.172	30.675	34.382	37.652	41.566	44.314	52.620
26	12.198	13.409	15.379	17.292	19.820	21.792	25.336	29.246	31.795	35.563	38.885	42.856	45.642	54.052
27	12.879	14.125	16.151	18.114	20.703	22.719	26.336	30.319	32.912	36.741	40.113	44.140	46.963	55.476
28	13.565	14.847	16.928	18.939	21.588	23.647	27.336	31.391	34.027	37.916	41.337	45.419	48.278	56.893
29	14.256	15.574	17.708	19.768	22.475	24.577	28.336	32.461	35.139	39.087	42.557	46.693	49.588	58.302
30	14.953	16.306	18.493	20.599	23.364	25.508	29.336	33.530	36.250	40.256	43.773	47.962	50.892	59.703

For larger values of n, the expression $\sqrt{2\chi^2} - \sqrt{2n-1}$ may be used as a normal deviate with unit variance, remembering that the probability for χ^2 corresponds with that of a single tail of the normal curve.

From Table IV of R. A. Fisher and Frank Yates, *Statistical Tables for Biological, Agricultural and Medical Research* (1953 edition), published by Oliver & Boyd, Ltd., Edinburgh, and by permission of the authors and publishers.

College Printing Office

Table 4 Distribution of F
($p \leq 0.05$)

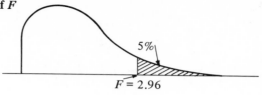

5%

$F = 2.96$

$n_1 = 12, \ n_2 = 12$

n_2 \ n_1	1	2	3	4	5	6	8	12	24	∞
1	161.4	199.5	215.7	224.6	230.2	234.0	238.9	243.9	249.0	254.3
2	18.51	19.00	19.16	19.25	19.30	19.33	19.37	19.41	19.45	19.50
3	10.13	9.55	9.28	9.12	9.01	8.94	8.84	8.74	8.64	8.53
4	7.71	6.94	6.59	6.39	6.26	6.16	6.04	5.91	5.77	5.63
5	6.61	5.79	5.41	5.19	5.05	4.95	4.82	4.68	4.53	4.36
6	5.99	5.14	4.76	4.53	4.39	4.28	4.15	4.00	3.84	3.67
7	5.59	4.74	4.35	4.12	3.97	3.87	3.73	3.57	3.41	3.23
8	5.32	4.46	4.07	3.84	3.69	3.58	3.44	3.28	3.12	2.93
9	5.12	4.26	3.86	3.63	3.48	3.37	3.23	3.07	2.90	2.71
10	4.96	4.10	3.71	3.48	3.33	3.22	3.07	2.91	2.74	2.54
11	4.84	3.98	3.59	3.36	3.20	3.09	2.95	2.79	2.61	2.40
12	4.75	3.88	3.49	3.26	3.11	3.00	2.85	2.69	2.50	2.30
13	4.67	3.80	3.41	3.18	3.02	2.92	2.77	2.60	2.42	2.21
14	4.60	3.74	3.34	3.11	2.96	2.85	2.70	2.53	2.35	2.13
15	4.54	3.68	3.29	3.06	2.90	2.79	2.64	2.48	2.29	2.07
16	4.49	3.63	3.24	3.01	2.85	2.74	2.59	2.42	2.24	2.01
17	4.45	3.59	3.20	2.96	2.81	2.70	2.55	2.38	2.19	1.96
18	4.41	3.55	3.16	2.93	2.77	2.66	2.51	2.34	2.15	1.92
19	4.38	3.52	3.13	2.90	2.74	2.63	2.48	2.31	2.11	1.88
20	4.35	3.49	3.10	2.87	2.71	2.60	2.45	2.28	2.08	1.84
21	4.32	3.47	3.07	2.84	2.68	2.57	2.42	2.25	2.05	1.81
22	4.30	3.44	3.05	2.82	2.66	2.55	2.40	2.23	2.03	1.78
23	4.28	3.42	3.03	2.80	2.64	2.53	2.38	2.20	2.00	1.76
24	4.26	3.40	3.01	2.78	2.62	2.51	2.36	2.18	1.98	1.73
25	4.24	3.38	2.99	2.76	2.60	2.49	2.34	2.16	1.96	1.71
26	4.22	3.37	2.98	2.74	2.59	2.47	2.32	2.15	1.95	1.69
27	4.21	3.35	2.96	2.73	2.57	2.46	2.30	2.13	1.93	1.67
28	4.20	3.34	2.95	2.71	2.56	2.44	2.29	2.12	1.91	1.65
29	4.18	3.33	2.93	2.70	2.54	2.43	2.28	2.10	1.90	1.64
30	4.17	3.32	2.92	2.69	2.53	2.42	2.27	2.09	1.89	1.62
40	4.08	3.23	2.84	2.61	2.45	2.34	2.18	2.00	1.79	1.51
60	4.00	3.15	2.76	2.52	2.37	2.25	2.10	1.92	1.70	1.39
120	3.92	3.07	2.68	2.45	2.29	2.17	2.02	1.83	1.61	1.25
∞	3.84	2.99	2.60	2.37	2.21	2.10	1.94	1.75	1.52	1.00

Lower 5 per cent. points are found by interchange of n_1 and n_2, i.e. n_1 must always correspond with the greater mean square.

Abridged from Table V of R. A. Fisher and Frank Yates, *Statistical Tables for Biological, Agricultural and Medical Research* (1953 edition), published by Oliver & Boyd, Ltd., Edinburgh, and by permission of the authors and publishers.

Table 4 Distribution of F
$(p \leq 0.01)$

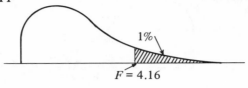

1%

$F = 4.16$

$n_1 = 12,\ n_2 = 12$

n_2 \ n_1	1	2	3	4	5	6	8	12	24	∞
1	4052	4999	5403	5625	5764	5859	5982	6106	6234	6366
2	98.50	99.00	99.17	99.25	99.30	99.33	99.37	99.42	99.46	99.50
3	34.12	30.82	29.46	28.71	28.24	27.91	27.49	27.05	26.60	26.12
4	21.20	18.00	16.69	15.98	15.52	15.21	14.80	14.37	13.93	13.46
5	16.26	13.27	12.06	11.39	10.97	10.67	10.29	9.89	9.47	9.02
6	13.74	10.92	9.78	9.15	8.75	8.47	8.10	7.72	7.31	6.88
7	12.25	9.55	8.45	7.85	7.46	7.19	6.84	6.47	6.07	5.65
8	11.26	8.65	7.59	7.01	6.63	6.37	6.03	5.67	5.28	4.86
9	10.56	8.02	6.99	6.42	6.06	5.80	5.47	5.11	4.73	4.31
10	10.04	7.56	6.55	5.99	5.64	5.39	5.06	4.71	4.33	3.91
11	9.65	7.20	6.22	5.67	5.32	5.07	4.74	4.40	4.02	3.60
12	9.33	6.93	5.95	5.41	5.06	4.82	4.50	4.16	3.78	3.36
13	9.07	6.70	5.74	5.20	4.86	4.62	4.30	3.96	3.59	3.16
14	8.86	6.51	5.56	5.03	4.69	4.46	4.14	3.80	3.43	3.00
15	8.68	6.36	5.42	4.89	4.56	4.32	4.00	3.67	3.29	2.87
16	8.53	6.23	5.29	4.77	4.44	4.20	3.89	3.55	3.18	2.75
17	8.40	6.11	5.18	4.67	4.34	4.10	3.79	3.45	3.08	2.65
18	8.28	6.01	5.09	4.58	4.25	4.01	3.71	3.37	3.00	2.57
19	8.18	5.93	5.01	4.50	4.17	3.94	3.63	3.30	2.92	2.49
20	8.10	5.85	4.94	4.43	4.10	3.87	3.56	3.23	2.86	2.42
21	8.02	5.78	4.87	4.37	4.04	3.81	3.51	3.17	2.80	2.36
22	7.94	5.72	4.82	4.31	3.99	3.76	3.45	3.12	2.75	2.31
23	7.88	5.66	4.76	4.26	3.94	3.71	3.41	3.07	2.70	2.26
24	7.82	5.61	4.72	4.22	3.90	3.67	3.36	3.03	2.66	2.21
25	7.77	5.57	4.68	4.18	3.86	3.63	3.32	2.99	2.62	2.17
26	7.72	5.53	4.64	4.14	3.82	3.59	3.29	2.96	2.58	2.13
27	7.68	5.49	4.60	4.11	3.78	3.56	3.26	2.93	2.55	2.10
28	7.64	5.45	4.57	4.07	3.75	3.53	3.23	2.90	2.52	2.06
29	7.60	5.42	4.54	4.04	3.73	3.50	3.20	2.87	2.49	2.03
30	7.56	5.39	4.51	4.02	3.70	3.47	3.17	2.84	2.47	2.01
40	7.31	5.18	4.31	3.83	3.51	3.29	2.99	2.66	2.29	1.80
60	7.08	4.98	4.13	3.65	3.34	3.12	2.82	2.50	2.12	1.60
120	6.85	4.79	3.95	3.48	3.17	2.96	2.66	2.34	1.95	1.38
∞	6.64	4.60	3.78	3.32	3.02	2.80	2.51	2.18	1.79	1.00

Lower 1 per cent. points are found by interchange of n_1 and n_2, i.e. n_1 must always correspond with the greater mean square.

Table 4   Distribution of F
$(p \leq 0.001)$

.1%

$F = 7.00$

$n_1 = 12, \ n_2 = 12$

n_2 \ n_1	1	2	3	4	5	6	8	12	24	∞
1	405,284	500,000	540,379	562,500	576,405	585,937	598,144	610,667	623,497	636,619
2	998.5	999.0	999.2	999.2	999.3	999.3	999.4	999.4	999.5	999.5
3	167.0	148.5	141.1	137.1	134.6	132.8	130.6	128.3	125.9	123.5
4	74.14	61.25	56.18	55.34	51.71	50.53	49.00	47.41	45.77	44.05
5	47.18	37.12	33.20	31.09	29.75	28.84	27.64	26.42	25.14	23.78
6	35.51	27.00	23.70	21.92	20.81	20.03	19.03	17.99	16.89	14.75
7	29.25	21.69	18.77	17.19	16.21	15.52	14.63	13.71	12.73	11.69
8	25.42	18.49	15.83	14.39	13.49	12.86	12.04	11.19	10.30	9.34
9	22.86	16.39	13.90	12.56	11.71	11.13	10.37	9.57	8.72	7.81
10	21.04	14.91	12.55	11.28	10.48	9.92	9.20	8.45	7.64	6.76
11	19.69	13.81	11.56	10.35	9.58	9.05	8.35	7.63	6.85	6.00
12	18.64	12.97	10.80	9.63	8.89	8.38	7.71	7.00	6.25	5.42
13	17.81	12.31	10.21	9.07	8.35	7.86	7.21	6.52	5.78	4.97
14	17.14	11.78	9.73	8.62	7.92	7.43	6.80	6.13	5.41	4.60
15	16.59	11.34	9.34	8.25	7.57	7.09	6.47	5.81	5.10	4.31
16	16.12	10.97	9.00	7.94	7.27	6.81	6.19	5.55	4.85	4.06
17	15.72	10.66	8.73	7.68	7.02	6.56	5.96	5.32	4.63	3.85
18	15.38	10.39	8.49	7.46	6.81	6.35	5.76	5.13	4.45	3.67
19	15.08	10.16	8.28	7.26	6.62	6.18	5.59	4.97	4.29	3.52
20	14.82	9.95	8.10	7.10	6.46	6.02	5.44	4.82	4.15	3.38
21	14.59	9.77	7.94	6.95	6.32	5.88	5.31	4.70	4.03	3.26
22	14.38	9.61	7.80	6.81	6.19	5.76	5.19	4.58	3.92	3.15
23	14.19	9.47	7.67	6.69	6.08	5.65	5.09	4.48	3.82	3.05
24	14.03	9.34	7.55	6.59	5.98	5.55	4.99	4.39	3.74	2.97
25	13.88	9.22	7.45	6.49	5.88	5.46	4.91	4.31	3.66	2.89
26	13.74	9.12	7.36	6.41	5.80	5.38	4.83	4.24	3.59	2.82
27	13.61	9.02	7.27	6.33	5.73	5.31	4.76	4.17	3.52	2.75
28	13.50	8.93	7.19	6.25	5.66	5.24	4.69	4.11	3.46	2.70
29	13.39	8.85	7.12	6.19	5.59	5.18	4.64	4.05	3.41	2.64
30	13.29	8.77	7.05	6.12	5.53	5.12	4.58	4.00	3.36	2.59
40	12.61	8.25	6.60	5.70	5.13	4.73	4.21	3.64	3.01	2.23
60	11.97	7.76	6.17	5.31	4.76	4.37	3.87	3.31	2.69	1.90
120	11.38	7.32	5.79	4.95	4.42	4.04	3.55	3.02	2.40	1.54
∞	10.83	6.91	5.42	4.62	4.10	3.74	3.27	2.74	2.13	1.00

Lower 0.1 per cent. points are found by interchange of n_1 and n_2, i.e. n_1 must always correspond with the greater mean square.

Table 5   Distribution of Z for Given Values of r*

r	.000	.001	.002	.003	.004	.005	.006	.007	.008	.009
.000	.0000	.0010	.0020	.0030	.0040	.0050	.0060	.0070	.0080	.0090
.010	.0100	.0110	.0120	.0130	.0140	.0150	.0160	.0170	.0180	.0190
.020	.0200	.0210	.0220	.0230	.0240	.0250	.0260	.0270	.0280	.0290
.030	.0300	.0310	.0320	.0330	.0340	.0350	.0360	.0370	.0380	.0390
.040	.0400	.0410	.0420	.0430	.0440	.0450	.0460	.0470	.0480	.0490
.050	.0501	.0511	.0521	.0531	.0541	.0551	.0561	.0571	.0581	.0591
.060	.0601	.0611	.0621	.0631	.0641	.0651	.0661	.0671	.0681	.0691
.070	.0701	.0711	.0721	.0731	.0741	.0751	.0761	.0771	.0782	.0792
.080	.0802	.0812	.0822	.0832	.0842	.0852	.0862	.0872	.0882	.0892
.090	.0902	.0912	.0922	.0933	.0943	.0953	.0963	.0973	.0983	.0993
.100	.1003	.1013	.1024	.1034	.1044	.1054	.1064	.1074	.1084	.1094
.110	.1105	.1115	.1125	.1135	.1145	.1155	.1165	.1175	.1185	.1195
.120	.1206	.1216	.1226	.1236	.1246	.1257	.1267	.1277	.1287	.1297
.130	.1308	.1318	.1328	.1338	.1348	.1358	.1368	.1379	.1389	.1399
.140	.1409	.1419	.1430	.1440	.1450	.1460	.1470	.1481	.1491	.1501
.150	.1511	.1522	.1532	.1542	.1552	.1563	.1573	.1583	.1593	.1604
.160	.1614	.1624	.1634	.1644	.1655	.1665	.1676	.1686	.1696	.1706
.170	.1717	.1727	.1737	.1748	.1758	.1768	.1779	.1789	.1799	.1810
.180	.1820	.1830	.1841	.1851	.1861	.1872	.1882	.1892	.1903	.1913
.190	.1923	.1934	.1944	.1954	.1965	.1975	.1986	.1996	.2007	.2017
.200	.2027	.2038	.2048	.2059	.2069	.2079	.2090	.2100	.2111	.2121
.210	.2132	.2142	.2153	.2163	.2174	.2184	.2194	.2205	.2215	.2226
.220	.2237	.2247	.2258	.2268	.2279	.2289	.2300	.2310	.2321	.2331
.230	.2342	.2353	.2363	.2374	.2384	.2395	.2405	.2416	.2427	.2437
.240	.2448	.2458	.2469	.2480	.2490	.2501	.2511	.2522	.2533	.2543
.250	.2554	.2565	.2575	.2586	.2597	.2608	.2618	.2629	.2640	.2650
.260	.2661	.2672	.2682	.2693	.2704	.2715	.2726	.2736	.2747	.2758
.270	.2769	.2779	.2790	.2801	.2812	.2823	.2833	.2844	.2855	.2866
.280	.2877	.2888	.2898	.2909	.2920	.2931	.2942	.2953	.2964	.2975
.290	.2986	.2997	.3008	.3019	.3029	.3040	.3051	.3062	.3073	.3084
.300	.3095	.3106	.3117	.3128	.3139	.3150	.3161	.3172	.3183	.3195
.310	.3206	.3217	.3228	.3239	.3250	.3261	.3272	.3283	.3294	.3305
.320	.3317	.3328	.3339	.3350	.3361	.3372	.3384	.3395	.3406	.3417
.330	.3428	.3439	.3451	.3462	.3473	.3484	.3496	.3507	.3518	.3530
.340	.3541	.3552	.3564	.3575	.3586	.3597	.3609	.3620	.3632	.3643
.350	.3654	.3666	.3677	.3689	.3700	.3712	.3723	.3734	.3746	.3757
.360	.3769	.3780	.3792	.3803	.3815	.3826	.3838	.3850	.3861	.3873
.370	.3884	.3896	.3907	.3919	.3931	.3942	.3954	.3966	.3977	.3989
.380	.4001	.4012	.4024	.4036	.4047	.4059	.4071	.4083	.4094	.4106
.390	.4118	.4130	.4142	.4153	.4165	.4177	.4189	.4201	.4213	.4225
.400	.4236	.4248	.4260	.4272	.4284	.4296	.4308	.4320	.4332	.4344
.410	.4356	.4368	.4380	.4392	.4404	.4416	.4429	.4441	.4453	.4465
.420	.4477	.4489	.4501	.4513	.4526	.4538	.4550	.4562	.4574	.4587
.430	.4599	.4611	.4623	.4636	.4648	.4660	.4673	.4685	.4697	.4710
.440	.4722	.4735	.4747	.4760	.4772	.4784	.4797	.4809	.4822	.4835
.450	.4847	.4860	.4872	.4885	.4897	.4910	.4923	.4935	.4948	.4961
.460	.4973	.4986	.4999	.5011	.5024	.5037	.5049	.5062	.5075	.5088
.470	.5101	.5114	.5126	.5139	.5152	.5165	.5178	.5191	.5204	.5217
.480	.5230	.5243	.5256	.5279	.5282	.5295	.5308	.5321	.5334	.5347
.490	.5361	.5374	.5387	.5400	.5413	.5427	.5440	.5453	.5466	.5480

*From *Statistical Tables and Problems* by A. E. Waugh, Copyright, 1952, by McGraw-Hill, Inc., used by permission of McGraw-Hill Book Company.

Table 5 (continued)

r	.000	.001	.002	.003	.004	.005	.006	.007	.008	.009
.500	.5493	.5506	.5520	.5533	.5547	.5560	.5573	.5587	.5600	.5614
.510	.5627	.5641	.5654	.5668	.5681	.5695	.5709	.5722	.5736	.5750
.520	.5763	.5777	.5791	.5805	.5818	.5832	.5846	.5860	.5874	.5888
.530	.5901	.5915	.5929	.5943	.5957	.5971	.5985	.5999	.6013	.6027
.540	.6042	.6056	.6070	.6084	.6098	.6112	.6127	.6141	.6155	.6170
.550	.6184	.6198	.6213	.6227	.6241	.6256	.6270	.6285	.6299	.6314
.560	.6328	.6343	.6358	.6372	.6387	.6401	.6416	.6431	.6446	.6460
.570	.6475	.6490	.6505	.6520	.6535	.6550	.6565	.6579	.6594	.6610
.580	.6625	.6640	.6655	.6670	.6685	.6700	.6715	.6731	.6746	.6761
.590	.6777	.6792	.6807	.6823	.6838	.6854	.6869	.6885	.6900	.6916
.600	.6931	.6947	.6963	.6978	.6994	.7010	.7026	.7042	.7057	.7073
.610	.7089	.7105	.7121	.7137	.7153	.7169	.7185	.7201	.7218	.7234
.620	.7250	.7266	.7283	.7299	.7315	.7332	.7348	.7364	.7381	.7398
.630	.7414	.7431	.7447	.7464	.7481	.7497	.7514	.7531	.7548	.7565
.640	.7582	.7599	.7616	.7633	.7650	.7667	.7684	.7701	.7718	.7736
.650	.7753	.7770	.7788	.7805	.7823	.7840	.7858	.7875	.7893	.7910
.660	.7928	.7946	.7964	.7981	.7999	.8017	.8035	.8053	.8071	.8089
.670	.8107	.8126	.8144	.8162	.8180	.8199	.8217	.8236	.8254	.8273
.680	.8291	.8310	.8328	.8347	.8366	.8385	.8404	.8423	.8442	.8461
.690	.8480	.8499	.8518	.8537	.8556	.8576	.8595	.8614	.8634	.8653
.700	.8673	.8693	.8712	.8732	.8752	.8772	.8792	.8812	.8832	.8852
.710	.8872	.8892	.8912	.8933	.8953	.8973	.8994	.9014	.9035	.9056
.720	.9076	.9097	.9118	.9139	.9160	.9181	.9202	.9223	.9245	.9266
.730	.9287	.9309	.9330	.9352	.9373	.9395	.9417	.9439	.9461	.9483
.740	.9505	.9527	.9549	.9571	.9594	.9616	.9639	.9661	.9684	.9707
.750	.9730	.9752	.9775	.9799	.9822	.9845	.9868	.9892	.9915	.9939
.760	.9962	.9986	1.0010	1.0034	1.0058	1.0082	1.0106	1.0130	1.0154	1.0179
.770	1.0203	1.0228	1.0253	1.0277	1.0302	1.0327	1.0352	1.0378	1.0403	1.0428
.780	1.0454	1.0479	1.0505	1.0531	1.0557	1.0583	1.0609	1.0635	1.0661	1.0688
.790	1.0714	1.0741	1.0768	1.0795	1.0822	1.0849	1.0876	1.0903	1.0931	1.0958
.800	1.0986	1.1014	1.1041	1.1070	1.1098	1.1127	1.1155	1.1184	1.1212	1.1241
.810	1.1270	1.1299	1.1329	1.1358	1.1388	1.1417	1.1447	1.1477	1.1507	1.1538
.820	1.1568	1.1599	1.1630	1.1660	1.1692	1.1723	1.1754	1.1786	1.1817	1.1849
.830	1.1870	1.1913	1.1946	1.1979	1.2011	1.2044	1.2077	1.2111	1.2144	1.2178
.840	1.2212	1.2246	1.2280	1.2315	1.2349	1.2384	1.2419	1.2454	1.2490	1.2526
.850	1.2561	1.2598	1.2634	1.2670	1.2708	1.2744	1.2782	1.2819	1.2857	1.2895
.860	1.2934	1.2972	1.3011	1.3050	1.3089	1.3129	1.3168	1.3209	1.3249	1.3290
.870	1.3331	1.3372	1.3414	1.3456	1.3498	1.3540	1.3583	1.3626	1.3670	1.3714
.880	1.3758	1.3802	1.3847	1.3892	1.3938	1.3984	1.4030	1.4077	1.4124	1.4171
.890	1.4219	1.4268	1.4316	1.4366	1.4415	1.4465	1.4516	1.4566	1.4618	1.4670
.900	1.4722	1.4775	1.4828	1.4883	1.4937	1.4992	1.5047	1.5103	1.5160	1.5217
.910	1.5275	1.5334	1.5393	1.5453	1.5513	1.5574	1.5636	1.5698	1.5762	1.5825
.920	1.5890	1.5956	1.6022	1.6089	1.6157	1.6226	1.6296	1.6366	1.6438	1.6510
.930	1.6584	1.6659	1.6734	1.6811	1.6888	1.6967	1.7047	1.7129	1.7211	1.7295
.940	1.7380	1.7467	1.7555	1.7645	1.7736	1.7828	1.7923	1.8019	1.8117	1.8216
.950	1.8318	1.8421	1.8527	1.8635	1.8745	1.8857	1.8972	1.9090	1.9210	1.9333
.960	1.9459	1.9588	1.9721	1.9857	1.9996	2.0140	2.0287	2.0439	2.0595	2.0756
.970	2.0923	2.1095	2.1273	2.1457	2.1649	2.1847	2.2054	2.2269	2.2494	2.2729
.980	2.2976	2.3223	2.3507	2.3796	2.4101	2.4426	2.4774	2.5147	2.5550	2.5988
.990	2.6467	2.6996	2.7587	2.8257	2.9031	2.9945	3.1063	3.2504	3.4534	3.8002

r	z
.9999	4.95172
.99999	6.10303

Table 6   Random Numbers I*

03 47 43 73 86	36 96 47 36 61	46 98 63 71 62	33 26 16 80 45	60 11 14 10 95
97 74 24 67 62	42 81 14 57 20	42 53 32 37 32	27 07 36 07 51	24 51 79 89 73
16 76 62 27 66	56 50 26 71 07	32 90 79 78 53	13 55 38 58 59	88 97 54 14 10
12 56 85 99 26	96 96 68 27 31	05 03 72 93 15	57 12 10 14 21	88 26 49 81 76
55 59 56 35 64	38 54 82 46 22	31 62 43 09 90	06 18 44 32 53	23 83 01 30 30
16 22 77 94 39	49 54 43 54 82	17 37 93 23 78	87 35 20 96 43	84 26 34 91 64
84 42 17 53 31	57 24 55 06 88	77 04 74 47 67	21 76 33 50 25	83 92 12 06 76
63 01 63 78 59	16 95 55 67 19	98 10 50 71 75	12 86 73 58 07	44 39 52 38 79
33 21 12 34 29	78 64 56 07 82	52 42 07 44 38	15 51 00 13 42	99 66 02 79 54
57 60 86 32 44	09 47 27 96 54	49 17 46 09 62	90 52 84 77 27	08 02 73 43 28
18 18 07 92 46	44 17 16 58 09	79 83 86 19 62	06 76 50 03 10	55 23 64 05 05
26 62 38 97 75	84 16 07 44 99	83 11 46 32 24	20 14 85 88 45	10 93 72 88 71
23 42 40 64 74	82 97 77 77 81	07 45 32 14 08	32 98 94 07 72	93 85 79 10 75
52 36 28 19 95	50 92 26 11 97	00 56 76 31 38	80 22 02 53 53	86 60 42 04 53
37 85 94 35 12	83 39 50 08 30	42 34 07 96 88	54 42 06 87 98	35 85 29 48 39
70 29 17 12 13	40 33 20 38 26	13 89 51 03 74	17 76 37 13 04	07 74 21 19 30
56 62 18 37 35	96 83 50 87 75	97 12 25 93 47	70 33 24 03 54	97 77 46 44 80
99 49 57 22 77	88 42 95 45 72	16 64 36 16 00	04 43 18 66 79	94 77 24 21 90
16 08 15 04 72	33 27 14 34 09	45 59 34 68 49	12 72 07 34 45	99 27 72 95 14
31 16 93 32 43	50 27 89 87 19	20 15 37 00 49	52 85 66 60 44	38 68 88 11 80
68 34 30 13 70	55 74 30 77 40	44 22 78 84 26	04 33 46 09 52	68 07 97 06 57
74 57 25 65 76	59 29 97 68 60	71 91 38 67 54	13 58 18 24 76	15 54 55 95 52
27 42 37 86 53	48 55 90 65 72	96 57 69 36 10	96 46 92 42 45	97 60 49 04 91
00 39 68 29 61	66 37 32 20 30	77 84 57 03 29	10 45 65 04 26	11 04 96 67 24
29 94 98 94 24	68 49 69 10 82	53 75 91 93 30	34 25 20 57 27	40 48 73 51 92
16 90 82 66 59	83 62 64 11 12	67 19 00 71 74	60 47 21 29 68	02 02 37 03 31
11 27 94 75 06	06 09 19 74 66	02 94 37 34 02	76 70 90 30 86	38 45 94 30 38
35 24 10 16 20	33 32 51 26 38	79 78 45 04 91	16 92 53 56 16	02 75 50 95 98
38 23 16 86 38	42 38 97 01 50	87 75 66 81 41	40 01 74 91 62	48 51 84 08 32
31 96 25 91 47	96 44 33 49 13	34 86 82 53 91	00 52 43 48 85	27 55 26 89 62
66 67 40 67 14	64 05 71 95 86	11 05 65 09 68	76 83 20 37 90	57 16 00 11 66
14 90 84 45 11	75 73 88 05 90	52 27 41 14 86	22 98 12 22 08	07 52 74 95 80
68 05 51 18 00	33 96 02 75 19	07 60 62 93 55	59 33 82 43 90	49 37 38 44 59
20 46 78 73 90	97 51 40 14 02	04 02 33 31 08	39 54 16 49 36	47 95 93 13 30
64 19 58 97 79	15 06 15 93 20	01 90 10 75 06	40 78 78 89 62	02 67 74 17 33
05 26 93 70 60	22 35 85 15 13	92 03 51 59 77	59 56 78 06 83	52 91 05 70 74
07 97 10 88 23	09 98 42 99 64	61 71 62 99 15	06 51 29 16 93	58 05 77 09 51
68 71 86 85 85	54 87 66 47 54	73 32 08 11 12	44 95 92 63 16	29 56 24 29 48
26 99 61 65 53	58 37 78 80 70	42 10 50 67 42	32 17 55 85 74	94 44 67 16 94
14 65 52 68 75	87 59 36 22 41	26 78 63 06 55	13 08 27 01 50	15 29 39 39 43
17 53 77 58 71	71 41 61 50 72	12 41 94 96 26	44 95 27 36 99	02 96 74 30 83
90 26 59 21 19	23 52 23 33 12	96 93 02 18 39	07 02 18 36 07	25 99 32 70 23
41 23 52 55 99	31 04 49 69 96	10 47 48 45 88	13 41 43 89 20	97 17 14 49 17
60 20 50 81 69	31 99 73 68 68	35 81 33 03 76	24 30 12 48 60	18 99 10 72 34
91 25 38 05 90	94 58 28 41 36	45 37 59 03 09	90 35 57 29 12	82 62 54 65 60
34 50 57 74 37	98 80 33 00 91	09 77 93 19 82	74 94 80 04 04	45 07 31 66 49
85 22 04 39 43	73 81 53 94 79	33 62 46 86 28	08 31 54 46 31	53 94 13 38 47
09 79 13 77 48	73 82 97 22 21	05 03 27 24 83	72 89 44 05 60	35 80 39 94 88
88 75 80 18 14	22 95 75 42 49	39 32 82 22 49	02 48 07 70 37	16 04 61 67 87
90 96 23 70 00	39 00 03 06 90	55 85 78 38 36	94 37 30 69 32	90 89 00 76 33

*Abridged from Table XXXIII of R. A. Fisher and Frank Yates, *Statistical Tables for Biological, Agricultural and Medical Research* (1953 edition), published by Oliver & Boyd, Ltd., Edinburgh, and by permission of the authors and publishers.

Table 6   Random Numbers II

53 74 23 99 67	61 32 28 69 84	94 62 67 86 24	98 33 41 19 95	47 53 53 38 09
63 38 06 86 54	99 00 65 26 94	02 82 90 23 07	79 62 67 80 60	75 91 12 81 19
35 30 58 21 46	06 72 17 10 94	25 21 31 75 96	49 28 24 00 49	55 65 79 78 07
63 43 36 82 69	65 51 18 37 88	61 38 44 12 45	32 92 85 88 65	54 34 81 85 35
98 25 37 55 26	01 91 82 81 46	74 71 12 94 97	24 02 71 37 07	03 92 18 66 75
02 63 21 17 69	71 50 80 89 56	38 15 70 11 48	43 40 45 86 98	00 83 26 91 03
64 55 22 21 82	48 22 28 06 00	61 54 13 43 91	82 78 12 23 29	06 66 24 12 27
85 07 26 13 89	01 10 07 82 04	59 63 69 36 03	69 11 15 83 80	13 29 54 19 28
58 54 16 24 15	51 54 44 82 00	62 61 65 04 69	38 18 65 18 97	85 72 13 49 21
34 85 27 84 87	61 48 64 56 26	90 18 48 13 26	37 70 15 42 57	65 65 80 39 07
03 92 18 27 46	57 99 16 96 56	30 33 72 85 22	84 64 38 56 98	99 01 30 98 64
62 95 30 27 59	37 75 41 66 48	86 97 80 61 45	23 53 04 01 63	45 76 08 64 27
08 45 93 15 22	60 21 75 46 91	98 77 27 85 42	28 88 61 08 84	69 62 03 42 73
07 08 55 18 40	45 44 75 13 90	24 94 96 61 02	57 55 66 83 15	73 42 37 11 61
01 85 89 95 66	51 10 19 34 88	15 84 97 19 75	12 76 39 43 78	64 63 91 08 25
72 84 71 14 35	19 11 58 49 26	50 11 17 17 76	86 31 57 20 18	95 60 78 46 75
88 78 28 16 84	13 52 53 94 53	75 45 69 30 96	73 89 65 70 31	99 17 43 48 76
45 17 75 65 57	28 40 19 72 12	25 12 74 75 67	60 40 60 81 19	24 62 01 61 16
96 76 28 12 54	22 01 11 94 25	71 96 16 16 88	68 64 36 74 45	19 59 50 88 92
43 31 67 72 30	24 02 94 08 63	38 32 36 66 02	69 36 38 25 39	48 03 45 15 22
50 44 66 44 21	66 06 58 05 62	68 15 54 35 02	42 35 48 96 32	14 52 41 52 48
22 66 22 15 86	26 63 75 41 99	58 42 36 72 24	58 37 52 18 51	03 37 18 39 11
96 24 40 14 51	23 22 30 88 57	95 67 47 29 83	94 69 40 06 07	18 16 36 78 86
31 73 91 61 19	60 20 72 93 48	98 57 07 23 69	65 95 39 69 58	56 80 30 19 44
78 60 73 99 84	43 89 94 36 45	56 69 47 07 41	90 22 91 07 12	78 35 34 08 72
84 37 90 61 56	70 10 23 98 05	85 11 34 76 60	76 48 45 34 60	01 64 18 39 96
36 67 10 08 23	98 93 35 08 86	99 29 76 29 81	33 34 91 58 93	63 14 52 32 52
07 28 59 07 48	89 64 58 89 75	83 85 62 27 89	30 14 78 56 27	86 63 59 80 02
10 15 83 87 60	79 24 31 66 56	21 48 24 06 93	91 98 94 05 49	01 47 59 38 00
55 19 68 97 65	03 73 52 16 56	00 53 55 90 27	33 42 29 38 87	22 13 88 83 34
53 81 29 13 39	35 01 20 71 34	62 33 74 82 14	53 73 19 09 03	56 54 29 56 93
51 86 32 68 92	33 98 74 66 99	40 14 71 94 58	45 94 19 38 81	14 44 99 81 07
35 91 70 29 13	80 03 54 07 27	96 94 78 32 66	50 95 52 74 33	13 80 55 62 54
37 71 67 95 13	20 02 44 95 94	64 85 04 05 72	01 32 90 76 14	53 89 74 60 41
93 66 13 83 27	92 79 64 64 72	28 54 96 53 84	48 14 52 98 94	56 07 93 89 30
02 96 08 45 65	13 05 00 41 84	93 07 54 72 59	21 45 57 09 77	19 48 56 27 44
49 83 43 48 35	82 88 33 69 96	72 36 04 19 76	47 45 15 18 60	82 11 08 95 97
84 60 71 62 46	40 80 81 30 37	34 39 23 05 38	25 15 35 71 30	88 12 57 21 77
18 17 30 88 71	44 91 14 88 47	89 23 30 63 15	56 34 20 47 89	99 82 93 24 98
79 69 10 61 78	71 32 76 95 62	87 00 22 58 40	92 54 01 75 25	43 11 71 99 31
75 93 36 57 83	56 20 14 82 11	74 21 97 90 65	96 42 68 63 86	74 54 13 26 94
38 30 92 29 03	06 28 81 39 38	62 25 06 84 63	61 29 08 93 67	04 32 92 08 09
51 29 50 10 34	31 57 75 95 80	51 97 02 74 77	76 15 48 49 44	18 55 63 77 09
21 31 38 86 24	37 79 81 53 74	73 24 16 10 33	52 83 90 94 76	70 47 14 54 36
29 01 23 87 88	58 02 39 37 67	42 10 14 20 92	16 55 23 42 45	54 96 09 11 06
95 33 95 22 00	18 74 72 00 18	38 79 58 69 32	81 76 80 26 92	82 80 84 25 39
90 84 60 79 80	24 36 59 87 38	82 07 53 89 35	96 35 23 79 18	05 98 90 07 35
46 40 62 98 82	54 97 20 56 95	15 74 80 08 32	16 46 70 50 80	67 72 16 42 79
20 31 89 03 43	38 46 82 68 72	32 14 82 99 70	80 60 47 18 97	63 49 30 21 30
71 59 73 05 50	08 22 23 71 77	91 01 93 20 49	82 96 59 26 94	66 39 67 98 60

Table 7 Table of N^2 and $\sqrt{N}$ *

Number	Square	Square root	Number	Square	Square root
1	1	1.0000	41	16 81	6.4031
2	4	1.4142	42	17 64	6.4807
3	9	1.7321	43	18 49	6.5574
4	16	2.0000	44	19 36	6.6332
5	25	2.2361	45	20 25	6.7082
6	36	2.4495	46	21 16	6.7823
7	49	2.6458	47	22 09	6.8557
8	64	2.8284	48	23 04	6.9282
9	81	3.0000	49	24 01	7.0000
10	1 00	3.1623	50	25 00	7.0711
11	1 21	3.3166	51	26 01	7.1414
12	1 44	3.4641	52	27 04	7.2111
13	1 69	3.6056	53	28 09	7.2801
14	1 96	3.7417	54	29 16	7.3485
15	2 25	3.8730	55	30 25	7.4162
16	2 56	4.0000	56	31 36	7.4833
17	2 89	4.1231	57	32 49	7.5498
18	3 24	4.2426	58	33 64	7.6158
19	3 61	4.3589	59	34 81	7.6811
20	4 00	4.4721	60	36 00	7.7460
21	4 41	4.5826	61	37 21	7.8102
22	4 84	4.6904	62	38 44	7.8740
23	5 29	4.7958	63	39 69	7.9373
24	5 76	4.8990	64	40 96	8.0000
25	6 25	5.0000	65	42 25	8.0623
26	6 76	5.0990	66	43 56	8.1240
27	7 29	5.1962	67	44 89	8.1854
28	7 84	5.2915	68	46 24	8.2462
29	8 41	5.3852	69	47 61	8.3066
30	9 00	5.4472	70	49 00	8.3666
31	9 61	5.5678	71	50 41	8.4261
32	10 24	5.6569	72	51 84	8.4853
33	10 89	5.7446	73	53 29	8.5440
34	11 56	5.8310	74	54 76	8.6023
35	12 25	5.9161	75	56 25	8.6603
36	12 96	6.0000	76	57 76	8.7178
37	13 69	6.0828	77	59 29	8.7750
38	14 44	6.1644	78	60 84	8.8318
39	15 21	6.2450	79	62 41	8.8882
40	16 00	6.3246	80	64 00	8.9443

*From Herbert Sorenson, *Statistics for Students of Psychology and Education,* McGraw-Hill Book Co., New York, 1936, Table 72, pp. 347–359, with the permission of the author.

Table 7 (continued)

Number	Square	Square root	Number	Square	Square root
81	65 61	9.0000	121	1 46 41	11.0000
82	67 24	9.0554	122	1 48 84	11.0454
83	68 89	9.1104	123	1 51 29	11.0905
84	70 56	9.1652	124	1 53 76	11.1355
85	72 25	9.2195	125	1 56 25	11.1803
86	73 96	9.2736	126	1 58 76	11.2250
87	75 69	9.3274	127	1 61 29	11.2694
88	77 44	9.3808	128	1 63 84	11.3137
89	79 21	9.4340	129	1 66 41	11.3578
90	81 00	9.4868	130	1 69 00	11.4018
91	82 81	9.5394	131	1 71 61	11.4455
92	84 64	9.5917	132	1 74 24	11.4891
93	86 49	9.6437	133	1 76 89	11.5326
94	88 36	9.6954	134	1 79 56	11.5758
95	90 25	9.7468	135	1 82 25	11.6190
96	92 16	9.7980	136	1 84 96	11.6619
97	94 09	9.8489	137	1 87 69	11.7047
98	96 04	9.8995	138	1 90 44	11.7473
99	98 01	9.9499	139	1 93 21	11.7898
100	1 00 00	10.0000	140	1 96 00	11.8322
101	1 02 01	10.0499	141	1 98 81	11.8743
102	1 04 04	10.0995	142	2 01 64	11.9164
103	1 06 09	10.1489	143	2 04 49	11.9583
104	1 08 16	10.1980	144	2 07 36	12.0000
105	1 10 25	10.2470	145	2 10 25	12.0416
106	1 12 36	10.2956	146	2 13 16	12.0830
107	1 14 49	10.3441	147	2 16 09	12.1244
108	1 16 64	10.3923	148	2 19 04	12.1655
109	1 18 81	10.4403	149	2 22 01	12.2066
110	1 21 00	10.4881	150	2 25 00	12.2474
111	1 23 21	10.5357	151	2 28 01	12.2882
112	1 25 44	10.5830	152	2 31 04	12.3288
113	1 27 69	10.6301	153	2 34 09	12.3693
114	1 29 96	10.6771	154	2 37 16	12.4097
115	1 32 25	10.7238	155	2 40 25	12.4499
116	1 34 56	10.7703	156	2 43 36	12.4900
117	1 36 89	10.8167	157	2 46 49	12.5300
118	1 39 24	10.8628	158	2 49 64	12.5698
119	1 41 61	10.9087	159	2 52 81	12.6095
120	1 44 00	10.9545	160	2 56 00	12.6491

Table 7 (continued)

Number	Square	Square root	Number	Square	Square root
161	2 59 21	12.6886	201	4 04 01	14.1774
162	2 62 44	12.7279	202	4 08 04	14.2127
163	2 65 69	12.7671	203	4 12 09	14.2478
164	2 68 96	12.8062	204	4 16 16	14.2829
165	2 72 25	12.8452	205	4 20 25	14.3178
166	2 75 56	12.8841	206	4 24 36	14.3527
167	2 78 89	12.9228	207	4 28 49	14.3875
168	2 82 24	12.9615	208	4 32 64	14.4222
169	2 85 61	13.0000	209	4 36 81	14.4568
170	2 89 00	13.0384	210	4 41 00	14.4914
171	2 92 41	13.0767	211	4 45 21	14.5258
172	2 95 84	13.1149	212	4 49 44	14.5602
173	2 99 29	13.1529	213	4 53 69	14.5945
174	3 02 76	13.1909	214	4 57 96	14.6287
175	3 06 25	13.2288	215	4 62 25	14.6629
176	3 09 76	13.2665	216	4 66 56	14.6969
177	3 13 29	13.3041	217	4 70 89	14.7309
178	3 16 84	13.3417	218	4 75 24	14.7648
179	3 20 41	13.3791	219	4 79 61	14.7986
180	3 24 00	13.4164	220	4 84 00	14.8324
181	3 27 61	13.4536	221	4 88 41	14.8661
182	3 31 24	13.4907	222	4 92 84	14.8997
183	3 34 89	13.5277	223	4 97 29	14.9332
184	3 38 56	13.5647	224	5 01 76	14.9666
185	3 42 25	13.6015	225	5 06 25	15.0000
186	3 45 96	13.6382	226	5 10 76	15.0333
187	3 49 69	13.6748	227	5 15 29	15.0665
188	3 53 44	13.7113	228	5 19 84	15.0997
189	3 57 21	13.7477	229	5 24 41	15.1327
190	3 61 00	13.7840	230	5 29 00	15.1658
191	3 64 81	13.8203	231	5 33 61	15.1987
192	3 68 64	13.8564	232	5 38 24	15.2315
193	3 72 49	13.8924	233	5 42 89	15.2643
194	3 76 36	13.9284	234	5 47 56	15.2971
195	3 80 25	13.9642	235	5 52 25	15.3297
196	3 84 16	14.0000	236	5 56 96	15.3623
197	3 88 09	14.0357	237	5 61 69	15.3948
198	3 92 04	14.0712	238	5 66 44	15.4272
199	3 96 01	14.1067	239	5 71 21	15.4596
200	4 00 00	14.1421	240	5 76 00	15.4919

Table 7 (continued)

Number	Square	Square root	Number	Square	Square root
241	5 80 81	15.5242	281	7 89 61	16.7631
242	5 85 64	15.5563	282	7 95 24	16.7929
243	5 90 49	15.5885	283	8 00 89	16.8226
244	5 95 36	15.6205	284	8 06 56	16.8523
245	6 00 25	15.6525	285	8 12 25	16.8819
246	6 05 16	15.6844	286	8 17 96	16.9115
247	6 10 09	15.7162	287	8 23 69	16.9411
248	6 15 04	15.7480	288	8 29 44	16.9706
249	6 20 01	15.7797	289	8 35 21	17.0000
250	6 25 00	15.8114	290	8 41 00	17.0294
251	6 30 01	15.8430	291	8 46 81	17.0587
252	6 35 04	15.8745	292	8 52 64	17.0880
253	6 40 09	15.9060	293	8 58 49	17.1172
254	6 45 16	15.9374	294	8 64 36	17.1464
255	6 50 25	15.9687	295	8 70 25	17.1756
256	6 55 36	16.0000	296	8 76 16	17.2047
257	6 60 49	16.0312	297	8 82 09	17.2337
258	6 65 64	16.0624	298	8 88 04	17.2627
259	6 70 81	16.0935	299	8 94 01	17.2916
260	6 76 00	16.1245	300	9 00 00	17.3205
261	6 81 21	16.1555	301	9 06 01	17.3494
262	6 86 44	16.1864	302	9 12 04	17.3781
263	6 91 69	16.2173	303	9 18 09	17.4069
264	6 96 96	16.2481	304	9 24 16	17.4356
265	7 02 25	16.2788	305	9 30 25	17.4642
266	7 07 56	16.3095	306	9 36 36	17.4929
267	7 12 89	16.3401	307	9 42 49	17.5214
268	7 18 24	16.3707	308	9 48 64	17.5499
269	7 23 61	16.4012	309	9 54 81	17.5784
270	7 29 00	16.4317	310	9 61 00	17.6068
271	7 34 41	16.4621	311	9 67 21	17.6352
272	7 39 84	16.4924	312	9 73 44	17.6635
273	7 45 29	16.5227	313	9 79 69	17.6918
274	7 50 76	16.5529	314	9 85 96	17.7200
275	7 56 25	16.5831	315	9 92 25	17.7482
276	7 61 76	16.6132	316	9 98 56	17.7764
277	7 67 29	16.6433	317	10 04 89	17.8045
278	7 72 84	16.6733	318	10 11 24	17.8326
279	7 78 41	16.7033	319	10 17 61	17.8606
280	7 84 00	16.7332	320	10 24 00	17.8885

Table 7 (continued)

Number	Square	Square root	Number	Square	Square root
321	10 30 41	17.9165	361	13 03 21	19.0000
322	10 36 84	17.9444	362	13 10 44	19.0263
323	10 43 29	17.9722	363	13 17 69	19.0526
324	10 49 76	18.0000	364	13 24 96	19.0788
325	10 56 25	18.0278	365	13 32 25	19.1050
326	10 62 76	18.0555	366	13 39 56	19.1311
327	10 69 29	18.0831	367	13 46 89	19.1572
328	10 75 84	18.1108	368	13 54 24	19.1833
329	10 82 41	18.1384	369	13 61 61	19.2094
330	10 89 00	18.1659	370	13 69 00	19.2354
331	10 95 61	18.1934	371	13 76 41	19.2614
332	11 02 24	18.2209	372	13 83 84	19.2873
333	11 08 89	18.2483	373	13 91 29	19.3132
334	11 15 56	18.2757	374	13 98 76	19.3391
335	11 22 25	18.3030	375	14 06 25	19.3649
336	11 28 96	18.3303	376	14 13 76	19.3907
337	11 35 69	18.3576	377	14 21 29	19.4165
338	11 42 44	18.3848	378	14 28 84	19.4422
339	11 49 21	18.4120	379	14 36 41	19.4679
340	11 56 00	18.4391	380	14 44 00	19.4936
341	11 62 81	18.4662	381	14 51 61	19.5192
342	11 69 64	18.4932	382	14 59 24	19.5448
343	11 76 49	18.5203	383	14 66 89	19.5704
344	11 83 36	18.5472	384	14 74 56	19.5959
345	11 90 25	18.5742	385	14 82 25	19.6214
346	11 97 16	18.6011	386	14 89 96	19.6469
347	12 04 09	18.6279	387	14 97 69	19.6723
348	12 11 04	18.6548	388	15 05 44	19.6977
349	12 18 01	18.6815	389	15 13 21	19.7231
350	12 25 00	18.7083	390	15 21 00	19.7484
351	12 32 01	18.7350	391	15 28 81	19.7737
352	12 39 04	18.7617	392	15 36 64	19.7990
353	12 46 09	18.7883	393	15 44 49	19.8242
354	12 53 16	18.8149	394	15 52 36	19.8494
355	12 60 25	18.8414	395	15 60 25	19.8746
356	12 67 36	18.8680	396	15 68 16	19.8997
357	12 74 49	18.8944	397	15 76 09	19.9249
358	12 81 64	18.9209	398	15 84 04	19.9499
359	12 88 81	18.9473	399	15 92 01	19.9750
360	12 96 00	18.9737	400	16 00 00	20.0000

Table 7 (continued)

Number	Square	Square root	Number	Square	Square root
401	16 08 01	20.0250	441	19 44 81	21.0000
402	16 16 04	20.0499	442	19 53 64	21.0238
403	16 24 09	20.0749	443	19 62 49	21.0476
404	16 32 16	20.0998	444	19 71 36	21.0713
405	16 40 25	20.1246	445	19 80 25	21.0950
406	16 48 36	20.1494	446	19 89 16	21.1187
407	16 56 49	20.1742	447	19 98 09	21.1424
408	16 64 64	20.1990	448	20 07 04	21.1660
409	16 72 81	20.2237	449	20 16 01	21.1896
410	16 81 00	20.2485	450	20 25 00	21.2132
411	16 89 21	20.2731	451	20 34 01	21.2368
412	16 97 44	20.2978	452	20 43 04	21.2603
413	17 05 69	20.3224	453	20 52 09	21.2838
414	17 13 96	20.3470	454	20 61 16	21.3073
415	17 22 25	20.3715	455	20 70 25	21.3307
416	17 30 56	20.3961	456	20 79 36	21.3542
417	17 38 89	20.4206	457	20 88 49	21.3776
418	17 47 24	20.4450	458	20 97 64	21.4009
419	17 55 61	20.4695	459	21 06 81	21.4243
420	17 64 00	20.4939	460	21 16 00	21.4476
421	17 72 41	20 5183	461	21 25 21	21.4709
422	17 80 84	20.5426	462	21 34 44	21.4942
423	17 89 29	20.5670	463	21 43 69	21.5174
424	17 97 76	20.5913	464	21 52 96	21.5407
425	18 06 25	20.6155	465	21 62 25	21.5639
426	18 14 76	20.6398	466	21 71 56	21.5870
427	18 23 29	20.6640	467	21 80 89	21.6102
428	18 31 84	20.6882	468	21 90 24	21.6333
429	18 40 41	20.7123	469	21 99 61	21.6564
430	18 49 00	20.7364	470	22 09 00	21.6795
431	18 57 61	20.7605	471	22 18 41	21.7025
432	18 66 24	20.7846	472	22 27 84	21.7256
433	18 74 89	20.8087	473	22 37 29	21.7486
434	18 83 56	20.8327	474	22 46 76	21.7715
435	18 92 25	20.8567	475	22 56 25	21.7945
436	19 00 96	20.8806	476	22 65 76	21.8174
437	19 09 69	20.9045	477	22 75 29	21.8403
438	19 18 44	20.9284	478	22 84 84	21.8632
439	19 27 21	20.9523	479	22 94 41	21.8861
440	19 36 00	20.9762	480	23 04 00	21.9089

Table 7 (continued)

Number	Square	Square root	Number	Square	Square root
481	23 13 61	21.9317	521	27 14 41	22.8254
482	23 23 24	21.9545	522	27 24 84	22.8473
483	23 32 89	21.9773	523	27 35 29	22.8692
484	23 42 56	22.0000	524	27 45 76	22.8910
485	23 52 25	22.0227	525	27 56 25	22.9129
486	23 61 96	22.0454	526	27 66 76	22.9347
487	23 71 69	22.0681	527	27 77 29	22.9565
488	23 81 44	22.0907	528	27 87 84	22.9783
489	23 91 21	22.1133	529	27 98 41	23.0000
490	24 01 00	22.1359	530	28 09 00	23.0217
491	24 10 81	22.1585	531	28 19 61	23.0434
492	24 20 64	22.1811	532	28 30 24	23.0651
493	24 30 49	22.2036	533	28 40 89	23.0868
494	24 40 36	22.2261	534	28 51 56	23.1084
495	24 50 25	22.2486	535	28 62 25	23.1301
496	24 60 16	22.2711	536	28 72 96	23.1517
497	24 70 09	22.2935	537	28 83 69	23.1733
498	24 80 04	22.3159	538	28 94 44	23.1948
499	24 90 01	22.3383	539	29 05 21	23.2164
500	25 00 00	22.3607	540	29 16 00	23.2379
501	25 10 01	22.3830	541	29 26 81	23.2594
502	25 20 04	22.4054	542	29 37 64	23.2809
503	25 30 09	22.4277	543	29 48 49	23.3024
504	25 40 16	22.4499	544	29 59 36	23.3238
505	25 50 25	22.4722	545	29 70 25	23.3452
506	25 60 36	22.4944	546	29 81 16	23.3666
507	25 70 49	22.5167	547	29 92 09	23.3880
508	25 80 64	22.5389	548	30 03 04	23.4094
509	25 90 81	22.5610	549	30 14 01	23.4307
510	26 01 00	22.5832	550	30 25 00	23.4521
511	26 11 21	22.6053	551	30 36 01	23.4734
512	26 21 44	22.6274	552	30 47 04	23.4947
513	26 31 69	22.6495	553	30 58 09	23.5160
514	26 41 96	22.6716	554	30 69 16	23.5372
515	26 52 25	22.6936	555	30 80 25	23.5584
516	26 62 56	22.7156	556	30 91 36	23.5797
517	26 72 89	22.7376	557	31 02 49	23.6008
518	26 83 24	22.7596	558	31 13 64	23.6220
519	26 93 61	22.7816	559	31 24 81	23.6432
520	27 04 00	22.8035	560	31 36 00	23.6643

Table 7 (continued)

Number	Square	Square root	Number	Square	Square root
561	31 47 21	23.6854	601	36 12 01	24.5153
562	31 58 44	23.7065	602	36 24 04	24.5357
563	31 69 69	23.7276	603	36 36 09	24.5561
564	31 80 96	23.7487	604	36 48 16	24.5764
565	31 92 25	23.7697	605	36 60 25	24.5967
566	32 03 56	23.7908	606	36 72 36	24.6171
567	32 14 89	23.8118	607	36 84 49	24.6374
568	32 26 24	23.8328	608	36 96 64	24.6577
569	32 37 61	23.8537	609	37 08 81	24.6779
570	32 49 00	23.8747	610	37 21 00	24.6982
571	32 60 41	23.8956	611	37 33 21	24.7184
572	32 71 84	23.9165	612	37 45 44	24.7385
573	32 83 29	23.9374	613	37 57 69	24.7588
574	32 94 76	23.9583	614	37 69 96	24.7790
575	33 06 25	23.9792	615	37 82 25	24.7992
576	33 17 76	24.0000	616	37 94 56	24.8193
577	33 29 29	24.0208	617	38 06 89	24.8395
578	33 40 84	24.0416	618	38 19 24	24.8596
579	33 52 41	24.0624	619	38 31 61	24.8797
580	33 64 00	24.0832	620	38 44 00	24.8998
581	33 75 61	24.1039	621	38 56 41	24.9199
582	33 87 24	24.1247	622	38 68 84	24.9399
583	33 98 89	24.1454	623	38 81 29	24.9600
584	34 10 56	24.1661	624	38 93 76	24.9800
585	34 22 25	24.1868	625	39 06 25	25.0000
586	34 33 96	24.2074	626	39 18 76	25.0200
587	34 45 69	24.2281	627	39 31 29	25.0400
588	34 57 44	24.2487	628	39 43 84	25.0599
589	34 69 21	24.2693	629	39 56 41	25.0799
590	34 81 00	24.2899	630	39 69 00	25.0998
591	34 92 81	24.3105	631	39 81 61	25.1197
592	35 04 64	24.3311	632	39 94 24	25.1396
593	35 16 49	24.3516	633	40 06 89	25.1595
594	35 28 36	24.3721	634	40 19 56	25.1794
595	35 40 25	24.3926	635	40 32 25	25.1992
596	35 52 16	24.4131	636	40 44 96	25.2190
597	35 64 09	24.4336	637	40 57 69	25.2389
598	35 76 04	24.4540	638	40 70 44	25.2587
599	35 88 01	24.4745	639	40 83 21	25.2784
600	36 00 00	24.4949	640	40 96 00	25.2982

Table 7 (continued)

Number	Square	Square root	Number	Square	Square root
641	41 08 81	25.3180	681	46 37 61	26.0960
642	41 21 64	25.3377	682	46 51 24	26.1151
643	41 34 49	25.3574	683	46 64 89	26.1343
644	41 47 36	25.3772	684	46 78 56	26.1534
645	41 60 25	25.3969	685	46 92 25	26.1725
646	41 73 16	25.4165	686	47 05 96	26.1916
647	41 86 09	25.4362	687	47 19 69	26.2107
648	41 99 04	25.4558	688	47 33 44	26.2298
649	42 12 01	25.4755	689	47 47 21	26.2488
650	42 25 00	25.4951	690	47 61 00	26.2679
651	42 38 01	25.5147	691	47 74 81	26.2869
652	42 51 04	25.5343	692	47 88 64	26.3059
653	42 64 09	25.5539	693	48 02 49	26.3249
654	42 77 16	25.5734	694	48 16 36	26.3439
655	42 90 25	25.5930	695	48 30 25	26.3629
656	43 03 36	25.6125	696	48 44 16	26.3818
657	43 16 49	25.6320	697	48 58 09	26.4008
658	43 29 64	25.6515	698	48 72 04	26.4197
659	43 42 81	25.6710	699	48 86 01	26.4386
660	43 56 00	25.6905	700	49 00 00	26.4575
661	43 69 21	25.7099	701	49 14 01	26.4764
662	43 82 44	25.7294	702	49 28 04	26.4953
663	43 95 69	25.7488	703	49 42 09	26.5141
664	44 08 96	25.7682	704	49 56 16	26.5330
665	44 22 25	25.7876	705	49 70 25	26.5518
666	44 35 56	25.8070	706	49 84 36	26.5707
667	44 48 89	25.8263	707	49 98 49	26.5895
668	44 62 24	25.8457	708	50 12 64	26.6083
669	44 75 61	25.8650	709	50 26 81	26.6271
670	44 89 00	25.8844	710	50 41 00	26.6458
671	45 02 41	25.9037	711	50 55 21	26.6646
672	45 15 84	25.9230	712	50 69 44	26.6833
673	45 29 29	25.9422	713	50 83 69	26.7021
674	45 42 76	25.9615	714	50 97 96	26.7208
675	45 56 25	25.9808	715	51 12 25	26.7395
676	45 69 76	26.0000	716	51 26 56	26.7582
677	45 83 29	26.0192	717	51 40 89	26.7769
678	45 96 84	26.0384	718	51 55 24	26.7955
679	46 10 41	26.0576	719	51 69 61	26.8142
680	46 24 00	26.0768	720	51 84 00	26.8328

Table 7 (continued)

Number	Square	Square root	Number	Square	Square root
721	51 98 41	26.8514	761	57 91 21	27.5862
722	52 12 84	26.8701	762	58 06 44	27.6043
723	52 27 29	26.8887	763	58 21 69	27.6225
724	52 41 76	26.9072	764	58 36 96	27.6405
725	52 56 25	26.9258	765	58 52 25	27.6586
726	52 70 76	26.9444	766	58 67 56	27.6767
727	52 85 29	26.9629	767	58 82 89	27.6948
728	52 99 84	26.9815	768	58 98 24	27.7128
729	53 14 41	27.0000	769	59 13 61	27.7308
730	53 29 00	27.0185	770	59 29 00	27.7489
731	53 43 61	27.0370	771	59 44 41	27.7669
732	53 58 24	27.0555	772	59 59 84	27.7849
733	53 72 89	27.0740	773	59 75 29	27.8029
734	53 87 56	27.0924	774	59 90 76	27.8209
735	54 02 25	27.1109	775	60 06 25	27.8388
736	54 16 96	27.1293	776	60 21 76	27.8568
737	54 31 69	27.1477	777	60 37 29	27.8747
738	54 46 44	27.1662	778	60 52 84	27.8927
739	54 61 27	27.1846	779	60 68 41	27.9106
740	54 76 00	27.2029	780	60 84 00	27.9285
741	54 90 81	27.2213	781	60 99 61	27.9464
742	55 05 64	27.2397	782	61 15 24	27.9643
743	55 20 49	27.2580	783	61 30 89	27.9821
744	55 35 36	27.2764	784	61 46 56	28.0000
745	55 50 25	27.2947	785	61 62 25	28.0179
746	55 65 16	27.3130	786	61 77 96	28.0357
747	55 80 09	27.3313	787	61 93 69	28.0535
748	55 95 04	27.3496	788	62 09 44	28.0713
749	56 10 01	27.3679	789	62 25 21	28.0891
750	56 25 00	27.3861	790	62 41 00	28.1069
751	56 40 01	27.4044	791	62 56 81	28.1247
752	56 55 04	27.4226	792	62 72 64	28.1425
753	56 70 09	27.4408	793	62 88 49	28.1603
754	56 85 16	27.4591	794	63 04 36	28.1780
755	57 00 25	27.4773	795	63 20 25	28.1957
756	57 15 36	27.4955	796	63 36 16	28.2135
757	57 30 49	27.5136	797	63 52 09	28.2312
758	57 45 64	27.5318	798	63 68 04	28.2489
759	57 60 81	27.5500	799	63 84 01	28.2666
760	57 76 00	27.5681	800	64 00 00	28.2843

Table 7 (continued)

Number	Square	Square root	Number	Square	Square root
801	64 16 01	28.3019	841	70 72 81	29.0000
802	64 32 04	28.3196	842	70 89 64	29.0172
803	64 48 09	28.3373	843	71 06 49	29.0345
804	64 64 16	28.3549	844	71 23 36	29.0517
805	64 80 25	28.3725	845	71 40 25	29.0689
806	64 96 36	28.3901	846	71 57 16	29.0861
807	65 12 49	28.4077	847	71 74 09	29.1033
808	65 28 64	28.4253	848	71 91 04	29.1204
809	65 44 81	28.4429	849	72 08 01	29.1376
810	65 61 00	28.4605	850	72 25 00	29.1548
811	65 77 21	28.4781	851	72 42 01	29.1719
812	65 93 44	28.4956	852	72 59 04	29.1890
813	66 09 69	28.5132	853	72 76 09	29.2062
814	66 25 96	28.5307	854	72 93 16	29.2233
815	66 42 25	28.5482	855	73 10 25	29.2404
816	66 58 56	28.5657	856	73 27 36	29.2575
817	66 74 89	28.5832	857	73 44 49	29.2746
818	66 91 24	28.6007	858	73 61 64	29.2916
819	67 07 61	28.6082	859	73 78 81	29.3087
820	67 24 00	28.6356	860	73 96 00	29.3258
821	67 40 41	28.6531	861	74 13 21	29.3428
822	67 56 84	28.6705	862	74 30 44	29.3598
823	67 73 29	28.6880	863	74 47 69	29.3769
824	67 89 76	28.7054	864	74 64 96	29.3939
825	68 06 25	28.7228	865	74 82 25	29.4109
826	68 22 76	28.7402	866	74 99 56	29.4279
827	68 39 29	28.7576	867	75 16 89	29.4449
828	68 55 84	28.7750	868	75 34 24	29.4618
829	68 72 41	28.7924	869	75 51 61	29.4788
830	68 89 00	28.8097	870	75 69 00	29.4958
831	69 05 61	28.8271	871	75 86 41	29.5127
832	69 22 24	28.8444	872	76 03 84	29.5296
833	69 38 89	28.8617	873	76 21 29	29.5466
834	69 55 56	28.8791	874	76 38 76	29.5635
835	69 72 25	28.8964	875	76 56 25	29.5804
836	69 88 96	28.9137	876	76 73 76	29.5973
837	70 05 69	28.9310	877	76 91 29	29.6142
838	70 22 44	28.9482	878	77 08 84	29.6311
839	70 39 21	28.9655	879	77 26 41	29.6749
840	70 56 00	28.9828	880	77 44 00	29.6648

Table 7 (continued)

Number	Square	Square root	Number	Square	Square root
881	77 61 61	29.6816	921	84 82 41	30.3480
882	77 79 24	29.6985	922	85 00 84	30.3645
883	77 96 89	29.7153	923	85 19 29	30.3809
884	78 14 56	29.7321	924	85 37 76	30.3974
885	78 32 25	29.7489	925	85 56 25	30.4138
886	78 49 96	29.7658	926	85 74 76	30.4302
887	78 67 69	29.7825	927	85 93 29	30.4467
888	78 85 44	29.7993	928	86 11 84	30.4631
889	79 03 21	29.8161	929	86 30 41	30.4795
890	79 21 00	29.8329	930	86 49 00	30.4959
891	79 38 81	29.8496	931	86 67 61	30.5123
892	79 56 64	29.8664	932	86 86 24	30.5287
893	79 74 49	29.8831	933	87 04 89	30.5450
894	79 92 36	29.8998	934	87 23 56	30.5614
895	80 10 25	29.9166	935	87 42 25	30.5778
896	80 28 16	29.9333	936	87 60 96	30.5941
897	80 46 09	29.9500	937	87 79 69	30.6105
898	80 64 04	29.9666	938	87 98 44	30.6268
899	80 82 01	29.9833	939	88 17 21	30.6431
900	81 00 00	30.0000	940	88 36 00	30.6594
901	81 18 01	30.0167	941	88 54 81	30.6757
902	81 36 04	30.0333	942	88 73 64	30.6920
903	81 54 09	30.0500	943	88 92 49	30.7083
904	81 72 16	30.0666	944	89 11 36	30.7246
905	81 90 25	30.0832	945	89 30 25	30.7409
906	82 08 36	30 0998	946	89 49 16	30.7571
907	82 26 49	30.1164	947	89 68 09	30.7734
908	82 44 64	30.1330	948	89 87 04	30.7896
909	82 62 81	30.1496	949	90 06 01	30.8058
910	82 81 00	30.1662	950	90 25 00	30.8221
911	82 99 21	30.1828	951	90 44 01	30.8383
912	83 17 44	30.1993	952	90 63 04	30.8545
913	83 35 69	30.2159	953	90 82 09	30.8707
914	83 53 96	30.2324	954	91 01 16	30.8869
915	83 72 25	30.2490	955	91 20 25	30.9031
916	83 90 56	30.2655	956	91 39 36	30.9192
917	84 08 89	30.2820	957	91 58 49	30.9354
918	84 27 24	30.2985	958	91 77 64	30.9516
919	84 45 61	30.3150	959	91 96 81	30.9677
920	84 64 00	30.3315	960	92 16 00	30.9839

Table 7 (continued)

Number	Square	Square root	Number	Square	Square root
961	92 35 21	31.0000	981	96 23 61	31.3209
962	92 54 44	31.0161	982	96 43 24	31.3369
963	92 73 69	31.0322	983	96 62 89	31.3528
964	92 92 96	31.0483	984	96 82 56	31.3688
965	93 12 25	31.0644	985	97 02 25	31.3847
966	93 31 56	31.0805	986	97 21 96	31.4006
967	93 50 89	31.0966	987	97 41 69	31.4166
968	93 70 24	31.1127	988	97 61 44	31.4325
969	93 89 61	31.1288	989	97 81 21	31.4484
970	94 09 00	31.1448	990	98 01 00	31.4643
971	94 28 41	31.1609	991	98 20 81	31.4802
972	94 47 84	31.1769	992	98 40 64	31.4960
973	94 67 29	31.1929	993	98 60 49	31.5119
974	94 86 76	31.2090	994	98 80 36	31.5278
975	95 06 25	31.2250	995	99 00 25	31.5436
976	95 25 76	31.2410	996	99 20 16	31.5595
977	95 45 29	31.2570	997	99 40 09	31.5753
978	95 64 84	31.2730	998	99 60 04	31.5911
979	95 84 41	31.2890	999	99 80 01	31.6070
980	96 04 00	31.3050	1000	100 00 00	31.6228

Index